TURBULENT DRAG REDUCTION BY SURFACTANT ADDITIVES

TURBULENT DRAG REDUCTION BY SURFACTANT ADDITIVES

Feng-Chen Li
Harbin Institute of Technology, China

Bo Yu
China University of Petroleum (Beijing), China

Jin-Jia Wei
Xi'an Jiaotong University, China

Yasuo Kawaguchi
Tokyo University of Science, Japan

John Wiley & Sons Singapore Pte. Ltd.

高等教育出版社
HIGHER EDUCATION PRESS

This edition first published 2012

Published by John Wiley & Sons Singapore Pte. Ltd., 1 Fusionopolis Walk, #07-01 Solaris South Tower, Singapore 138628, under exclusive license by Higher Education Press in all media and all languages throughout the world excluding Mainland China and excluding Simplified and Traditional Chinese languages.

For details of our global editorial offices, for customer services and for information about how to apply for permission to reuse the copyright material in this book please see our website at www.wiley.com.

Library of Congress Cataloging-in-Publication Data

Turbulent drag reduction by surfactant additives / Feng-Chen Li ... [et al.].
p. cm.
Includes bibliographical references and index.
ISBN 978-1-118-18107-2 (cloth)
1. Drag (Aerodynamics) 2. Turbulence. 3. Frictional resistance (Hydrodynamics) 4. Surface active agents. I. Li, Feng-Chen, 1971-
TL574.D7T87 2011
629.132′34–dc23

2011034428

Print ISBN: 9781118181072

Set in 11/13 pt Times by Thomson Digital, Noida, India

Contents

Preface

The subject of this book is the presentation of detailed information on turbulence characteristics, theories, special techniques, and application issues for drag-reducing flows by surfactant additives, mainly based on state-of-the-art research results by the authors through experimental studies, numerical simulations, and theoretical analyses.

The phenomenon of turbulent drag reduction by additives has attracted the interest of researchers in the fields of chemical engineering, turbulence, fluid dynamics, rheology, petroleum, municipal and environmental engineering, and so on, for more than half a century. Turbulent drag reducing flows are especially complex due to the twofold effects of the turbulence itself and the drag reduction phenomenon induced by viscoelacticity. This complexity, combined with the great potential for drag reduction in industrial applications, has driven researchers to continuously approach the problem in an interdisciplinary fashion. Researchers are working to clarify the mechanism of drag reduction by additives, developing particular research approaches for this unique phenomenon, and exploring the characteristics of drag-reduced turbulent flows. Throughout this process, they have established increasingly exhaustive theoretical descriptions for the rheological properties of the fluid and drag reduction phenomenon of the flow, promoting the applications of this phenomenon in practical systems. This book provides important information on turbulent drag reduction by surfactant additives, particularly information covering introductions to experimental studies using laser techniques (laser Doppler velocimetry and particle image velocimetry), direct numerical simulations with special treatment of constitutive equations for viscoelastic fluid and numerical simulation algorithms, Brownian dynamic simulations of the rheological properties of surfactant solution with consideration of microstructures in the fluid, field tests and other issues associated with practical applications, and elucidations and summarizations of state-of-the-art results obtained from those studies. The contents of this book are the central concerns of the interdisciplinary community related to turbulent drag reduction by additives.

The authors do not claim that they have addressed in this book all the relevant issues of turbulent drag reduction by additives. Particularly unbounded flows with drag-reducing effects, drag-reducing polymer solution flows, the chemistry of drag-reducing additives, and so on were not mentioned or elaborated in detail in this book.

F.-C. Li composed Sections 1.1–1.3, 2.1, 2.2, 2.4, Chapter 3, Section 6.1 and the index. He also coordinated all procedures involved in the publication of this book. He would like to acknowledge the support from National Natural Science Foundation of China (NSFC; Grant No. 10872060 and 51076036) and Fundamental Research Funds for the Central Universities (Grant No. HIT.BRET1.2010008). For Chapter 4, he acknowledges the support from NSFC (Grant No. 50506017, 50876114 and 51134006). J.-J. Wei composed Sections 1.4, 2.3, Chapter 5 and Sections 6.2–6.4. He acknowledges the support from NSFC (Grant No. 51076124 and 50821064). Y. Kawaguchi contributed to modifying the organizations of the book and correcting the contents; he is also one of the key contributors of most of the studies involved in this book.

F.-C. Li
B. Yu
J.-J. Wei
Y. Kawaguchi

1

Introduction

The problems associated with energy sources comprise one of the most important issues accompanying the development of economy all over the world. Energy saving itself has been treated as "the fifth biggest energy source," following coal, petroleum, natural gas, and electricity, by scientists in China [1]. Much more importantly, the utilization of energy saving, that is, "this fifth energy source," does not generate any emission of harmful gases such as CO_2, NO_x, and so on, but the reduction of equivalent amounts of emitted harmful gases can be obtained when gaining the energy source of "energy saving." In developing countries such as China, both the energy consumption for productions and the energy consumption for economic output are much higher than in developed countries. Therefore, the potential of energy saving is huge if energy-saving approaches can be executed in many aspects. Turbulent drag reduction (DR) by additives is one such energy-saving approach used in the long-distance transportation of liquid or in the circulation systems of liquid. It has bright application prospectives.

1.1 Background

The turbulent DR technique is of great significance for improving energy utilization efficiency, protecting the ecological environment, and so on. In recent years, the international and academic community has been attaching more and more importance to fundamental and applicable studies on turbulent DR. Every year, specific academic conferences associated with turbulent DR are held or symposiums are opened for topics related to turbulent DR in the international conferences (or congresses) on fluid dynamics or fluid engineering. Turbulent DR has been a hot research topic in the field of fluid mechanics and fluid engineering. As an important branch of the field of turbulent DR, "turbulent DR by additives" herein refers to the liquid transportation technique that adding a minute amount (generally at a level of

Turbulent Drag Reduction by Surfactant Additives, First Edition. Feng-Chen Li, Bo Yu, Jin-Jia Wei and Yasuo Kawaguchi.

parts per million, or ppm) of additives may reduce the frictional drag greatly. Compared with other turbulent DR techniques, the salient features of this approach are its most obvious turbulent drag-reducing effect, its price (the lowest), and ease of operation.

As early as 1931, Forrest and Grierson [2] found that flow resistance could be reduced at the same flow rate when pulp fibers floated in the water turbulent flow in a pipe. But this phenomenon did not receive enough attention. The first one who observed the turbulent DR phenomenon by polymer additives was the American scholar Mysels [3,4]. Mysels and his assistants found that after dissolving aluminum disoap into the gasoline flow in a pipe, the flow drag could be decreased at the same flow rate. However, due to the Second World War, this discovery wasn't published until 1949. In the first International Rheological Congress held in 1948 (its proceedings were published in 1949), the English scholar Toms reported the turbulent DR phenomenon of dilute polymer solution flow [5]. Hence, the turbulent DR phenomenon is often named the Toms effect. Since then, a large amount of investigations have been carried out all over the world on the mechanisms of turbulent DR and the real applications of the turbulent DR technique. The turbulent DR phenomenon has been well recognized, and significant progress for this technique has been made in real applications.

A typical example of a commercial application of the Toms effect was reported in 1982. Polymer additives were utilized to obtain the effects of reducing drag and increasing flow transportation efficiency in US Alaskan petroleum pipelines [6]. A relatively dense polymer solution was injected into the petroleum transportation piping system at four different pumping stations, which increased 25% of the flow rate of petroleum. For the Alaskan petroleum transportation piping system with an inner diameter of 48 in, the transportation capacity has been enlarged by 100 000 barrels. Presently, the turbulent drag-reducing effect of polymer additives has been widely used for long-distance petroleum transportation systems in order to increase their transportation capacity. However, exerted with some factors such as strong shear stress and high temperature, the flexible long-chain structure of polymer molecules, which plays the main role in the turbulent drag-reducing effect, can be destroyed. The destroyed structures cannot self-repair, resulting in the permanent loss of their drag-reducing effect. Therefore, polymer drag reducer is usually inapplicable for a liquid circulation system with a pump.

From about the 1970s, it was found that some kinds of surfactant additives also have the effect of reducing the frictional drag for turbulent flow. In later investigations, it was observed that the key factor for turbulent drag-reducing effect is the microstructures formed in the surfactant solution, that is, the so-called shear-induced structures (SIS). This is because after surfactant is dissolved into water, with the aid of a stabilizer (e.g., providing counterions for the cationic surfactant), the rod-like microstructures can be formed from the small molecules, which can further form the network microstructures under proper shear stress. Like the flexible long-chain microstructures in the polymer solution, the network microstructures formed in the surfactant solution can generate the so-called viscoelasticity rheological property, making the solution flow display non-Newtonian fluid characteristics. This kind of fluid is thus named "viscoelastic fluid." Due to the viscoelasticity generated in the solution flow, which then interacts with turbulence, the turbulent drag-reducing effect is brought forth. Since the microstructures in solution are essentially formed with small molecular units of surfactant, when exerted with strong shear stress (e.g., when passing through the centrifugal pump in the liquid-recirculating system), the network microstructures in the solution can also be disassociated and destroyed. But after the disappearance of the strong shear, the microstructures can be automatically reformed within a short time in the order of seconds. This indicates that the microstructures in surfactant solution have self-reparability, and so does the turbulent drag-reducing effect. Based on this phenomenon, in a real liquid recirculation system with a pump, surfactant drag reducer is more appropriate to use for the purposes of reducing the flow friction drag in the long term and saving the pumping power. However, surfactant drag reducer also has an effective temperature range. Out of this temperature range, it loses its drag-reducing effect only temporarily. Once the fluid temperature returns back to its effective range, the drag-reducing ability can be automatically recovered.

The largest potential application area in industry and daily life for the turbulent drag-reducing effect of surfactant additives is a district heating and cooling recirculation system. There have been successful application examples for this case. For example, in July 2002, surfactant drag reducer was utilized in a central water-cooled air-conditioning system with 60 m^3 capacity in the Tochigi factory of the Sony Corporation, Japan. During the first year, the average DR reached 57%. The saved electricity used for driving the pump was 0.61 million kWh. Calculated with the local price per unit of electricity as 12 yen/kWh, the saved money for one year was 7.32 million yen (about 0.55 million RMB) [7]. The expenses for the first injection were, however, only 0.65 million yen (about 50 000 RMB). Presently, this water-cooled air-conditioning system in Sony Corporation's Tochigi factory is still operating normally. Although periodically, due to leakage of the system, an additional injection of additive is needed to keep the system's concentration of DR, and considering the effective lifetime of the additive itself and other cost factors, the cost savings from reducing the pumping power are much larger than the consumption cost of additives. The authors of this book performed, for the first time in China, a field test of drag-reducing effect by surfactant additives in a real district heating system in December 2007 [8]. The expected energy-saving effect has been obtained. At the same time, some practical problems to be solved have also been found. This application test has established a foundation for spreading out the real applications of the turbulent drag reduction technique by additives in a district heating system. Details of this application test will be described in Chapter 6.

This book mainly summarizes and generalizes turbulent drag reduction theories and techniques systematically, covering several aspects including the characteristics of surfactant drag reducers, mechanisms of turbulent DR, characteristics of the convective heat transfer of turbulent drag-reducing flows, practical application studies, and so on, combining this with research conducted by the authors for many years. It is aiming at establishing a relatively complete system for the field of turbulent DR by surfactant additives.

1.2 Surfactant Solution

Whether or not the surfactant solution displays drag-reducing effect depends on whether or not the shear-induced rod-like and then network microstructures can be formed. Such microstructures are essentially micelles composed of small surfactant molecules. The characteristics of microstructures in a solution of surfactant drag reducers have already been depicted in detail in several review papers [9–11]. The surfactant molecule has its hydrophobic "tail" and hydrophilic "head" spontaneously. In general, the hydrophobic tail is a long chain of alkyl bases, while the hydrophilic head is ionizable, being polar or polarizable. When surfactant molecules are in an aqueous environment, the hydrophilic heads prefer to stay in the water phase while the hydrophobic tails try to leave the water phase. There are only two choices for preventing the hydrophobic tails from contacting the water phase. One is that the hydrophobic tails orient to the nonpolar phase, for example the gas phase, nonpolar solid phase, or liquid phase with hydrophobicity. The other is that the surfactant molecule groups realign and reassemble to form micelles, and so the nonpolar parts (hydrophobic tails) of molecules are concentrated in the center while the polar parts (hydrophilic heads) of molecules are concentrated at the surface and in contact with the water phase. Figure 1.1 schematically shows the formed microstructures.

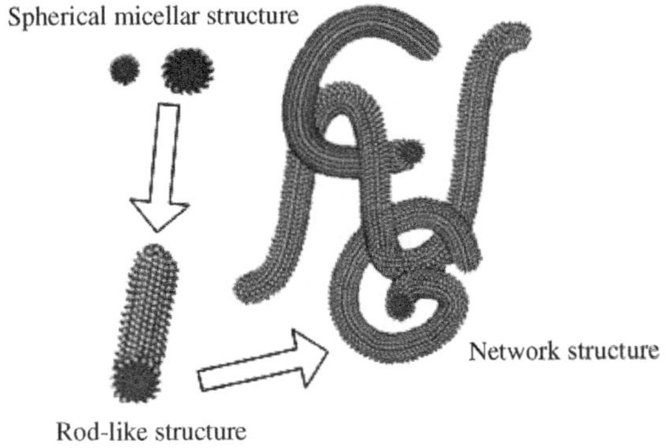

Figure 1.1 Schematics of microstructures formed in the surfactant solution. Surfactant molecules assemble to form spherical and rod-like micelles, and then form network structures under shear in the flow

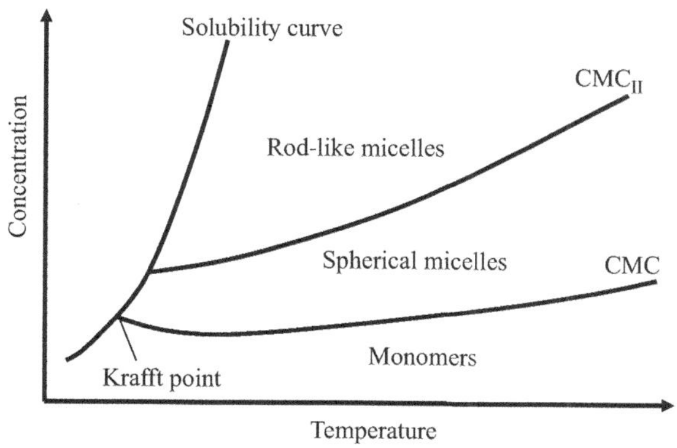

Figure 1.2 Schematic phase diagram for surfactant solution [9]

The ability to form micelle structures is the typical feature that the surfactant solution displays after it exceeds the critical micelle concentration (CMC). Above the CMC, the surfactant monomer molecules and the formed micelle structures are always in a thermodynamic equilibrium state. Figure 1.2 schematically shows a simplified phase diagram [9]. Based on Figure 1.2, Zakin *et al.* made the following summarization in reference [9]. The surfactant in its solution will be partially in the crystal or gel state when the temperature is lower than the so-called Krafft point. When the temperature is higher than the Krafft point, surfactant molecules will aggregate and form spherical micelles at concentrations over the CMC; with further increase of the solution concentration to be over the CMC_{II}, surfactants become rod-like micelles. Figure 1.2 also shows that the CMC of aqueous solution of surfactant has nearly no change in temperature, whereas CMC_{II} enlarges with increases in temperature. Per the hydrophilic or polar head of the surfactant molecule, it can be categorized into two groups: ionic and nonionic surfactants, respectively. The former can be further categorized into two types: anionic and cationic surfactants. The latter can be further categorized into three groups: zwitterionic, semipolar, and single-bond surfactants [12].

At the early stage, anionic and nonionic surfactants are considered to be turbulent drag-reducing candidates. However, in experimental studies on mechanisms of turbulent drag-reducing flows up to the present, overwhelmingly researchers have been using cationic surfactant, with the addition of other additives, as turbulent drag reducers. Since a cationic surfactant molecule usually involves tetra-ammonium and its anaerobic degradation is very slow, some environmental supervising organizations have begun to appreciate the negative side of using cationic surfactant in district heating and cooling systems as a turbulent drag reducer. Therefore, in recent decades, researchers have made efforts to exploit some representative surfactant drag reducers and found that the zwitterionic and amphoteric surfactants are more appropriate for use in district heating and cooling systems for energy-saving purposes. In reference [10], different kinds of surfactants suitable for turbulent drag reduction have been reviewed and summarized.

1.2.1 Anionic Surfactant

In earlier experimental studies on turbulent drag-reducing flows, several anionic surfactants were used as drag reducers. For example, Pilpel investigated the visco-elastic rheological properties of sodium oleate, sodium chloride, and C_6–C_{12} alcohols and found that the salt and the alcohols had the ability to help the micelles formed by surfactant molecules to change from spherical structures to long and cylindrical ones [13]. Although Pilpel did not measure the turbulent DR of the solution flow of his investigated surfactants, the existence of viscoelasticity itself had proved that this kind of fluid can be used as a turbulent drag reducer. In 1967, Savins reported the earliest experimental results about turbulent DR using anionic surfactant [14]. He used an aqueous solution of 10% potassium chloride dissolved with 0.2% sodium oleate as working fluid and obtained an 82% turbulent DR rate. Tsujii *et al.* reported that partially neutralized alkenylsuccinates solution also displayed strong viscoelasticity and so can also be used as a turbulent drag reducer [15]. But the optimum viscoelastic behaviors of this surfactant solution were at a pH of about 6; hence, it was basically precluded from use in steel tubes for turbulent DR due to its strong corrosion rate at that acidic state.

1.2.2 Cationic Surfactant

When used as a turbulent drag reducer, the cationic surfactant usually needs to be aided by additional additives for providing counterions. The paper published in 1967 by White was one of the earliest reports on turbulent DR flow of aqueous solution of cationic surfactant [16]. The drag-reducing fluid used in this study was 508 ppm cetyltrimethylammonium bromide and 1-naphthol. At that time, White already observed an important phenomenon: Turbulent DR stops at a critical Reynolds number at a certain concentration of surfactant solution and a certain pipe diameter. This phenomenon was clarified afterward: Because of the microstructures in the surfactant solution responsible for generating viscoelasticity, the turbulent drag-reducing effect will be destroyed at a critical Reynolds number (critical shear rate), which makes the solution flow lose its viscoelasticity and drag-reducing ability. This phenomenon has recurred in a later large amount of experimental studies on turbulent drag-reducing surfactant solution flows, including those done by the present authors [17–20]. This is of particular importance in determining the solution concentration of drag reducer and other related flow parameters (pipe diameter, flow speed, etc.) when turbulent drag-reducing effect is applied to real systems. An important advance in the turbulent DR of cationic surfactant was to use aromatic compounds instead of 1-naphthol as the additions to tetra-ammonium surfactant solutions. For example, Gravsholt reported in his paper published in 1976 that aqueous solution of cetyltrimethylammonium salicylate showed viscoelastic rheological properties at a concentration as low as 67 ppm [21], indicating that this kind of fluid has turbulent drag-reducing ability. In later experiments on turbulent drag-reducing flows of cationic surfactant solution, aromatic compounds were then mostly utilized as additional additives to provide counterions; for example, in a series of experimental studies on turbulent DR by additives performed by the present authors, sodium salicylate (NaSal) was chosen to supply counterions for cetyltrimethylammonium chloride (CTAC) surfactant, and both were used to make aqueous solution with turbulent drag-reducing effect [17–20].

1.2.3 Nonionic Surfactant

In 1972, Zakin and Chiang for the first time reported the turbulent DR characteristics of a nonionic surfactant solution flow [22]. They dissolved the ethoxylated mixture of C_{12} and C_{14} alcohols into water and obtained an aqueous solution with 1% concentration. With the further addition of sodium sulfate, they found that, at a constant temperature of 30°C, the flow of this solution showed turbulent DR phenomena after the added salt concentration was larger than 0.1 M, and the turbulent drag-reducing effect was best when the salt concentration reached 0.175 M. This kind of nonionic surfactant belongs to the class of hydroxyethyl fatty alcohol ethoxylates. Another group of nonionic surfactant with turbulent drag-reducing effect is hydroxyethyl fatty acid amide ethoxylates. In the compound of fatty acid mono-ethanolamide hydroxyethyl, when the degree of ethoxylation reaches a certain level, its aqueous solution flow can display turbulent DR, and the effective temperature of its turbulent drag-reducing effect can even extend to below 0°C. Also, a mixture of the above-mentioned two types of nonionic surfactants can possess excellent turbulent drag-reducing effects [10].

1.2.4 Amphoteric Surfactant

"Amphoteric surfactant" refers to those surfactants with both acidity and alkalinity. Reactions between alkane epoxide and sodium sarcosinate may generate amphoteric surfactant. When the aqueous solution of such a product is at a state of $pH < 9$, it is found to be viscoelastic, implying that its flow may display a turbulent drag-reducing effect; for example, Harwigsson and Hellsten obtained turbulent DR in their experimental study on aqueous solution flow of amphoteric surfactant [23].

1.2.5 Zwitterionic Surfactant

Since the 2-hydroxyalkyl sarcosinates display viscoelasticity and turbulent drag-reducing ability only in the zwitterionic structure, researchers were then reminded to investigate and develop zwitterionic surfactant drag reducers. Saul *et al.* studied a mixture of hexadecyl dimethyl ammoniopropane sulfonate and sodium dodecyl sulfate and found that their aqueous solution showed obvious viscoelasticity when the molar ratio of these two chemicals was 1.7:1 [24]. Tsuji *et al.* [25] and Iwasaki *et al.* [26] also investigated viscoelastic properties of the mixture of different kinds of zwitterionic surfactants and anionic surfactants, respectively. Harwigsson and Hellsten [23] and Myska and Chara [27] measured the turbulent DR characteristics of an aqueous solution flow of N-alkylbetaine and sodium dodecyl benzene sulfonate at a molar ratio of 4:1 in their experimental studies. In the later studies, it was found that both alkyl sulfates and alkyl polyglycol ether sulfates, together with N-alkylbetaine, can be used as anionic surfactant drag reducers [10].

1.3 Mechanism and Theory of Drag Reduction by Surfactant Additives

Although the application study, or industrial application of chemical drag reducers (herein particularly high-polymer drag reducers and surfactant drag reducers) has been considerably explored and some influencing factors on DR have been understood, the mechanism of turbulent DR caused by the addition of drag reducers has actually not been so clear up to now due to the nonthorough understanding of turbulence itself. Some hypotheses and models have been proposed for the DR mechanism. But there is no single theory that can fully explain all the experimental phenomena appearing in turbulent drag-reducing flows by additives, so further studies are needed on the mechanism of turbulent DR. Explanations about the turbulent DR mechanism that provide the viewpoint of the physics of turbulence apply to both polymer and surfactant drag reducers. Therefore, the statements below will not be classified based on categorizations of chemical drag reducers, though some turbulent DR mechanisms have been proposed for polymers. When qualitatively explaining the mechanisms of turbulent DR from the viewpoint of microstructures in drag reducer solution, however, it is different for polymer and surfactant drag reducers. In this book, only surfactant drag reducers will be mentioned in this respect.

1.3.1 Explanations of the Turbulent DR Mechanism from the Viewpoint of Microstructures

The microstructures in drag-reducing surfactant solutions might be worm-like micelles only, network structures formed by worm-like micelles with branches or vesicular structures. Each microstructure with a different form may impart different rheological properties into the solution and result in different extents of turbulent drag-reducing effect. Zakin *et al.* [11] classified surfactant drag reducers into three categories according to the above-mentioned three possible microstructure forms at the static state, and correlated the turbulent drag-reducing effect with rheological properties.

At the static state, worm-like micellar structures can be formed in most of the surfactant drag reducer solutions. Exerted with shear stress, the worm-like micellar structures are apt to align with the flow direction, resulting in the occurrence of turbulent drag-reducing effect and a larger critical shear stress or critical Reynolds number (the turbulent DR rate increases with the increase of the flow Reynolds number at first and reaches the maximum level at the critical Reynolds number; after that, the DR rate decreases with the further increase of the flow Reynolds number until it reaches zero). Surfactant drag reducer solutions with inner worm-like micellar structures usually display obvious rheological properties, such as relatively large zero-shear viscosity, shear thinning properties (the shear viscosity decreases with the increase of shear rate), large viscoelasticity, a rapid rebounding phenomenon after the cease of rotation driving, a large ratio of extensional viscosity to shear viscosity (normally larger than 100), and so on.

In some turbulent drag-reducing fluid systems, worm-like micellar structures with branches, that is, worm-like structures with three branches joined together, can be formed. When the energy necessary for forming the semispherical head of a micellar structure becomes high enough to form a saddle-shaped branch joint, the branch structure can thus be generated. Comparing this with the former case, the number of ends of the worm-like micellar microstructures in the solution decreases. It has been observed from experiments that the joints of branches can freely move along the axial direction of the worm-like micellar structure. Hence, when exerted by shear, the shear stress can be immediately released, and so the shear viscosity of solution decreases [28]. The turbulent drag-reducing effect is also obvious for the surfactant solution flow with branched microstructures. But its maximum DR rate is smaller compared with that of nonbranched microstructures, while its critical Reynolds number is larger. Moreover, the complicated behavior of the free movement of the branch joints along the axial direction of micellar structures also induces much more complex rheological properties for such surfactant solutions.

There are also some kinds of surfactant solutions in which vesicular or crystal structures can be formed at the static state. In turbulent flows, when the exerted shear rate exceeds the critical value, these structures can change to worm-like micellar structures and make the surfactant solution flow display the turbulent DR phenomenon. This transition process of surfactant solution, from a state without drag-reducing effect to one with drag-reducing effect, is analogous to the inception process of DR in a turbulent flow of surfactant solution. The difference is that the inception of DR cannot be observed when the critical shear rate for the change of microstructures in surfactant solution is smaller than the critical wall shear for the laminar-to-turbulent transition. For the normal rheology measurements, if the applied shear rate by rheometer is usually not large enough to reach the critical shear rate for the change of microstructures in surfactant solution, the measured solution may display Newtonian fluid properties, that is, a relatively small shear viscosity, no generation of SIS, the first normal stress difference is 0, a relatively small ratio of extensional viscosity to shear viscosity, and so on [29].

1.3.2 Explanations of the Turbulent DR Mechanism from the Viewpoint of the Physics of Turbulence

Several typical theories for turbulent DR published up to the present are summarized below.

1.3.2.1 Pseudo-plasticity

Early on, Toms [5] proposed that polymer solutions have pseudo-plasticity. The larger the shear rate is, the smaller the apparent viscosity of a polymer solution becomes. Hence, when a solution flows in a pipe, its apparent viscosity decreases with proximity to the wall due to the local large shear rate, and so the flow resistance is decreased. From then on, through a large amount of experimental and theoretical studies, it has been shown that the mechanisms of turbulent DR by polymer additives are much more complicated, and this theory is currently denied.

1.3.2.2 Effective Slip

Virk proposed the elastic sublayer model [30] based on velocity measurement results for pipe flows. This model states that an elastic sublayer is formed in between the laminar substrate layer and turbulent bulk region with the addition of polymer. The velocity gradient in this layer is large, and so the velocity profile in the turbulent bulk region is elevated to some distance (i.e., the increment of velocity level in the turbulent bulk region) as compared with that of Newtonian fluid flows. This distance or increment of velocity level was then named "effective slip" (as schematically shown in Figure 1.3). Since the velocity in the turbulent bulk region is increased, the flow rate is thus increased at the same flow conditions (the same driving power herein), indicating the occurrence of DR. Here, the elastic sublayer is the region where polymer molecules react with the flow, that is, the region where DR occurs.

According to Virk's hypothesis, the thickness of the elastic sublayer is increased with the increase of drag reducer solution concentration. When the elastic sublayer is extended to the pipe axis, the DR rate reaches its maximum value. This hypothesis was successful in explaining the phenomenon of maximum DR and the diameter effect. Based on Virk's hypothesis, the concentrations corresponding to the maximum DR and maximum heat transfer reduction (HTR), respectively, should be the same; however, the experimental results showed that they were different.

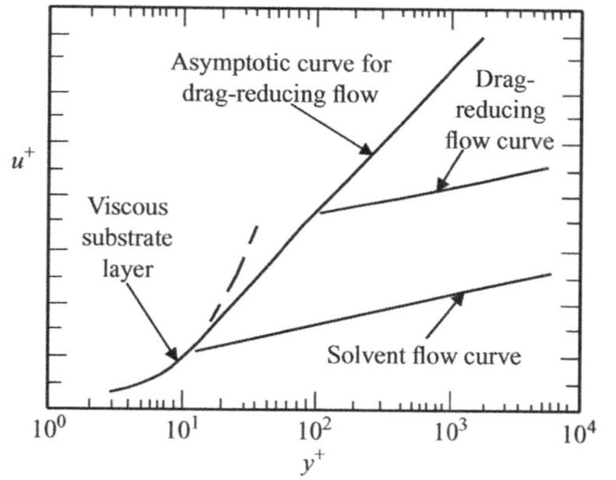

Figure 1.3 Model of elastic sublayer [30]

1.3.2.3 Suppression of Turbulent Fluctuations

Polymer and surfactant drag reducers have a drag-reducing effect only when their solution flows are at a turbulent state. Based on this fact, the mechanism for the occurrence of DR has been proposed as follows: The drag reducer suppresses the generation of turbulent vortices, resulting in the decrease of turbulent fluctuation intensity and turbulent kinetic energy, and finally inducing the occurrence of a turbulent DR phenomenon. In connection with the correctness of this proposal, there have been two different statements. In all the reported studies on turbulent drag-reducing flow by additives, it has been concluded that the turbulent intensity of the velocity component in the wall–normal direction is depressed, and some researchers also reported that the turbulent intensity in the streamwise direction is enhanced. The latter statement was based on comparisons between nondimensionalized streamwise velocity fluctuation intensities in Newtonian fluid turbulent flow and turbulent drag-reducing flow [31–34]. Note here that the nondimensionalization mentioned here is based on the friction velocity at its individual flow condition. Li *et al.* analyzed this argument in their paper [17] and thought that it was not so appropriate to use the friction velocity at the individual flow condition as the nondimensionalization parameter, and posited that one should use the bulk velocity (which also reflects the flow Reynolds number) at the individual flow condition. When Li *et al.* re-examined all the reported results using bulk velocity as the nondimensionalization parameter, they found that turbulence intensities in all directions are decreased in turbulent drag-reducing flows. Therefore, the viewpoint of "suppression of turbulent fluctuations" can be thought to be correct in qualitatively explaining the turbulent DR phenomenon. However, it is necessary to point out that the effect of turbulent drag reducers not only depresses the intensities of turbulent fluctuations but also modifies the turbulence structures [19,20,35–37].

1.3.2.4 Decoupling of Turbulent Fluctuations

It has been indicated from many studies that the effect of drag reducer on turbulent flows also appears as the decreased correlation between the axial and radial fluctuations. This effect is named "decoupling." The decoupling of turbulent fluctuations can decrease the Reynolds stress. According to the quantitative relationship between Reynolds shear stress and the turbulent contribution to frictional drag coefficient deduced by Fukagata *et al.* (i.e., the FIK equation) [38], a decrease of Reynolds shear stress directly results in a decrease of the friction factor of turbulent flow, and so turbulent DR. Actually, a decrease of Reynolds stress is caused by twofold effects, that is, the decoupling of turbulent fluctuations and turbulence suppression [17,33,39–41]. This postulation is also correct qualitatively.

1.3.2.5 Viscoelasticity

All polymer and surfactant solutions with turbulent drag-reducing effects display viscoelastic rheological properties. With the development of viscoelastic fluid mechanics, some researchers proposed that the drag-reducing effect of polymer and surfactant solutions is the result of the interaction between viscoelasticity and turbulent vortices. The microstructures (polymer molecule chains or network structures in surfactant solution) in the drag reducer solution at a high-shear-rate region can absorb the turbulent kinetic energy of small vortices within the energy-containing range and store it. When the microstructures are diffused or convected to a low-shear-rate region, they will be relaxed to a random threadlike entanglement and the stored energy will be released to the low-wave-number vortices (large-scaled vortices) in the form of elastic stress waves, which greatly decreases the dissipation of turbulent kinetic energy and induces turbulent DR.

The viscoelastic theory for the mechanism of turbulent DR by additives was proposed by DeGennes [42]. The viscoelasticity postulation not only explains the turbulent DR phenomenon in many polymer and surfactant solution flows with viscoelasticity, but also estimates the DR rate quantitatively. It is also a powerful tool for studying the mechanism of turbulent DR from the viewpoint of the physics of turbulence and developing new quantitative analysis theories for turbulent drag-reducing flows. However, this postulation was challenged by the "anisotropic stresses" hypothesis proposed by Toonder [43].

1.3.2.6 Anisotropic Stresses

The anisotropic stresses hypothesis posited that the main reason for turbulent DR by polymer additives is the anisotropy of stresses induced by the re-orientation of polymer molecules due to stretch in the turbulent boundary layer. In order to verify this opinion,

Toonder [43] measured the turbulence statistics and spectra in the near-wall region of turbulent drag-reducing flow by means of laser Doppler velocimetry, through which two models were established. One is the viscous anisotropic stress (VA) model, and the other is the viscoelastic anisotropic stress (VEA) model. Direct numerical simulations were also performed by Toonder to obtain the influence of polymer additives on the near-wall stress tensor. These indicated that the DNS results using the VA model were in good agreement with the experimental data, whereas when using the VEA model with the introduction of elasticity of the polymer molecule, contrarily, the DNS results differed from the experiments quite a bit. Those investigation results re-inforced the anisotropic stresses hypothesis by Toonder, and challenged the conventional visco-elasticity hypothesis.

Bewersdoff *et al.* [44] utilized the small-angle neutron-scattering technique to observe the deformation characteristics of micelles formed in a heavy water solution of surfactant. It was found that the micelle was elliptically spherical in still solution and the longer axis is random in its direction; when the solution flowed in a turbulent flow state, the orientation of longer axes of elliptical spheres became directional. This conclusion is in accordance with the hypothesis of anisotropic stresses. Presently, the conclusions have not yet been finalized in this research area, calling for further studies to clarify whether the viscoelasticity hypothesis or the anisotropic stresses hypothesis is correct.

1.3.2.7 Stabilization of Turbulent Vortices

In turbulent flows, the principal portion of flow resistance stems from turbulent vortices with random sizes and random motions. Although turbulent vortices are randomly generated, they are always decomposed into smaller and smaller vortices (by the cascading of turbulent kinetic energy) until the smallest vortex (a Kolmogorov-scaled vortex) corresponds to the flow Reynolds number, and finally the kinetic energy of the vortex is dissipated by fluid viscosity.

This hypothesis points out that the existence of drag reducer makes the turbulent kinetic energy cascading process change a lot. When the Reynolds stress in the turbulent fluid motions is on a similar order as the elastic stress in drag reducer solution, the cascading process will be truncated and energy dissipation cannot reach the final state, resulting in a decrease of flow resistance [45–47].

1.3.2.8 Random Renewal of Fluids at the Wall Surface

The boundary layer in wall-bounded turbulent flows can be divided into three layers: the viscous substrate layer, buffer layer, and turbulent core region. The viscous substrate is a thin layer closest to the wall surface in which intermittent turbulent fluctuations exist. If the momentum transport boundary layer of wall-bounded turbulent flow is assumed to be composed of a series of momentum transport patches, then these patches are randomly renewed by the fluid elements coming from the turbulent core region, and decompose into new fluid elements to form turbulent eddies. The turbulent energy cascading process thus occurs, and the momentum transport and energy diffusion can be accomplished.

On the basis of turbulent random theory and the wall-surface renewal theory by Dankwests [48], this model posits that, after adding drag reducer into the Newtonian turbulent flow, the drag reducer molecules can form a liquid film on the surface of the pipe wall; the fluid elements in the proximity of the wall are thus difficult to renew, hindering the turbulent eddies being formed and then resulting in the decrease of energy dissipation and so turbulent DR.

In recent years, Li *et al.* [20,49] carried out a series of experimental studies by means of particle image velocimetry. They indirectly measured the occurrence frequency and strength of the near-wall coherent structures in turbulent channel flows (Newtonian fluid turbulent flows and turbulent drag-reducing flows of surfactant solution). The near-wall turbulent coherent structures were then treated as energy quanta and directly contributed to the turbulent portion of frictional drag. Based on this conjecture, the researchers concluded that the turbulent contribution to the friction factor is linearly proportional to the occurrence frequency and strength of the turbulent bursting events (coherent structures). As compared with Newtonian fluid turbulent flows, the occurrence frequency and strength of the near-wall coherent structures are greatly decreased, which directly induces significant decrease of the frictional drag coefficient, that is, the occurrence of turbulent DR.

1.3.2.9 Others

From the viewpoint of molecule deformation, Keyes and Abernathy [50] analyzed the evolution process of vortex tubes in a shear flow. Two effects can be brought about from the deformation motion of large molecules: inhibition of the formation of vortex tubes and inhibition of the growth of the formed vortex tubes.The combined effect of these two aspects is to reduce the bursting event frequency in the turbulent flow of polymer dilute solution, inducing the turbulent DR.

1.4 Application Techniques of Drag Reduction by Surfactant Additives

Although high drag-reducing performance for surfactant drag-reducing flow has been obtained in experimental facilities, some problems must be solved before large-scale industrial applications will be possible. These problems include the scale-up method caused by the pipe diameter effect, heat transfer reduction of surfactant solution, separation and recovery of surfactant and counterion, stability and toxic effects of surfactant, long-term effectiveness and corrosion of the drag reduction system, and so on.

1.4.1 Heat Transfer Reduction of Surfactant Drag-reducing Flow

Surfactant can reduce the frictional drag of the flow by suppressing the turbulence, and thus lead to a large reduction of the heat transfer coefficient. Under some conditions, the heat transfer reduction is much larger than the drag reduction, deteriorating the heat transfer performance of heat exchangers. Although the addition of surfactant can reduce the pump energy consumption, it makes the heat exchanger perform heat transfer very badly, leading to low thermal efficiency of the system, and thus limiting the application of drag reduction technology to the heat transfer field. Therefore, it is necessary to develop an effective method and device with low pressure loss to solve the heat transfer reduction problem.

1.4.2 Diameter Effect of Surfactant Drag-reducing Flow

Even the same surfactant solution may have different drag-reducing effects in pipelines with different diameters, indicating that the friction factor of the drag-reducing flow is related not only to the Reynolds number but also to the pipe diameter. This indicates that the drag reduction data obtained in small-sized pipes cannot be used directly for the application of surfactant drag reduction to real pipe systems. Therefore, it is necessary to develop a universal scale-up method.

1.4.3 Toxic Effect of Cationic Surfactant Solution

To avoid environmental pollution in the large-scale application of surfactant drag reduction, the surfactant solution should be processed by separating the surfactant and counterion from the solution before it is discharged to the environment. Therefore, it is necessary to develop a separation and recovery technology for surfactants. The toxicity of nonionic surfactants is relatively low compared with that of cationic surfactants, and it can be biodegraded rapidly.

1.4.4 Chemical Stability of Surfactant Solution

It has been observed from experiments that drag reduction ability will be weakened after a certain running period of the surfactant drag-reducing flow, and then the drag reduction ability will disappear completely. It was found that surfactant solutions of Harbon G and Ethoquad T/13-50/Nasal have drag reduction effects for a very long period in experiments (longer than one month), but lose their drag reduction ability after only a three-day run in the actual process. Some investigation showed that the reason for these solutions losing their drag reduction ability is that the surfactant molecules are absorbed by the iron rust particles and $Cu(OH)_2$ particles in the pipe flow. The drag reduction can be recovered by adding some chemical additives that can decrease the corrosion of metal walls (i.e., corrosion inhibitor).

1.4.5 Corrosion of Surfactant Solution

Surfactant solutions are more or less corrosive, and can corrode the carbon steel pipe and pump in a flow system. Moreover, many pipes are pressure bearing in flow systems, and the combined action of a corrosive environment and stress may aggravate the corrosion. If the stress is a tensile one, the corrosion will become more severe, which greatly threatens the life and safety of the equipment and the entire system.

In summary, the above problems must be solved before attempting a large-scale industrial application of surfactant drag reduction, and in order to make it an effective drag-reducing technology.

References

1. Wang, G.H., Tian, Y.L., and Yuan, J.T. (2005) *Energy and Sustainable Development*, Chemical Industry Press, Beijing (in Chinese).
2. Forrest, F. and Grierson, G.A. (1931) Friction losses in cast iron pipe carrying *paper*stock. *Paper Trade Journal*, **92**, 39–41.
3. Mukerjee, P. and Hofmann, A.F. (2000) In memoriam Karl Joseph Mysels (1914–1998). *Journal of Colloid and Interface Science*, **224**, 1–3.
4. Mysels, K.J. (1949) Flow of thickened fluid, December 27, US Patent 2, 492, 173.
5. Toms, B.A. (1949) Some observations on the flow of linear polymer solutions through straight tubes at large Reynolds number. Proceedings of the 1st International Rheological Congress, II, Part 2, North-Holland Publishing Co., Amsterdam, pp. 135–142.
6. Burger, E.D., Munk, W.R. and Wahl, H.A. (1982) Flow increase in trans-Alaska pipeline through use of polymeric drag-reducing additive. *Journal of Petroleum Technology*, **34**, 377–386.
7. Kawaguchi, Y. (2003) *Drag-Reducing Surfactant Additives and Its Applications*, Invited Lecture in Shanghai Jiao Tong University.
8. Jiao, L.F., Li, F.C., Su, W.T. *et al.* (2008) Experimental study on surfactant drag-reduction applying to some district heating system. *Energy Conservation Technology*, **26**, 195–201 (in Chinese).
9. Zakin, J.L., Lu, B. and Bewersdorff, H.W. (1998) Surfactant drag reduction. *Reviews in Chemical Engineering*, **14**, 253–318.
10. Hellsten, M. (2002) Drag-reducing surfactants. *Journal of Surfactants and Detergents*, **4**, 65–70.
11. Zakin, J.L., Zhang, Y. and Ge, W. (2007) Drag reduction by surfactant giant micelles, in *Giant Micelles: Properties and Applications*, Surfactant Science Series, vol. **140** (eds R. Zana and E.W. Kaler), Taylor and Francis, New York.
12. Laughlin, R.G. (1991) Aqueous phase science of cationic surfactant salts, in *Cationic Surfactants: Physical Chemistry*, Surfactant Science Series, vol. **37** (eds D.N. Rubingh and P.M. Holland), Marcel Dekker Inc., New York.

13. Pilpel, N. (1956) Viscoelasticity in aqueous soap solutions. *Journal of Physical Chemistry*, **60**, 779–782.
14. Savins, J. (1967) A stress-controlled drag-reduction phenomenon. *Rheologica Acta*, **6**, 323–330.
15. Tsujii, K., Saito, N. and Takeuchi, T. (1984) Viscoelastic and some colloid chemical properties of partially neutralized alkenylsuccinates in dilute aqueous solutions. *Journal of Colloid and Interface Science*, **99**, 553.
16. White, A. (1967) Flow characteristics of complex soap systems. *Nature*, **214**, 585.
17. Li, F-C., Kawaguchi, Y. and Hishida, K. (2004) Investigation on the characteristics of turbulence transport for momentum and heat in a drag-reducing surfactant solution flow. *Physics of Fluids*, **16**, 3281–3295.
18. Li, F-C., Kawaguchi, Y., Segawa, T. and Hishida, K. (2005) Reynolds-number dependence of turbulence structures in a drag-reducing surfactant solution channel flow investigated by PIV. *Physics of Fluids*, **17**, 075104.
19. Li, F-C., Kawaguchi, Y., Hishida, K. and Oshima, M. (2006) Investigation of turbulence structures in a drag-reduced turbulent channel flow with surfactant additive using stereoscopic particle image velocimetry. *Experiments in Fluids*, **40**, 218–230.
20. Li, F-C., Yu, B., Wei, J-J. *et al.* (2008) Experimental study of drag-reduction mechanics for a dilute surfactant solution flow. *International Journal of Heat and Mass Transfer*, **51**, 835–843.
21. Gravshlot, S. (1976) Viscoelasticity in highly dilute aqueous solutions of pure cationic detergents. *Journal of Colloid and Interface Science*, **57**, 575.
22. Zakin, J.L. and Chiang, J.L. (1972) Non-ionic surfactants as drag reducing additives. *Nature Physical Science*, **239**, 26.
23. Harwigsson, I. and Hellsten, M. (1996) Environmental acceptable drag-reducing surfactants for district heating and cooling. *Journal of American Oil Chemical Society*, **73**, 921.
24. Saul, D., Tiddy, G.J.T., Wheeler, B.A. *et al.* (1974) Phase structure and rheological properties of a mix zwitterionic/anionic surfactant system. *Journal of Chemical Society of Faraday Transactions I*, **70**, 163.
25. Tsujii, K., Okahashi, K. and Takeuchi, T. (1982) Addition-compound formation between anionic and zwitterionic surfactants in water. *Journal of Physical Chemistry*, **86**, 1437.
26. Iwasaki, T., Ogawa, W., Esumi, K. and Meguro, K. (1991) Interactions between betaine-type zwitterionic and anionic surfactants in mixed micelles. *Langmuir*, **7**, 30.
27. Myska, J. and Chara, Z. (2001) The effect of a zwitterionic and cationic surfactant in turbulent flow. *Experiments in Fluids*, **30**, 229.
28. Ali, A.A. and Makhloufi, R. (1999) Effect of organic salts on micellar growth and structure studied by rheology. *Colloid and Polymer Science*, **277**, 270–275.
29. Qi, Y. and Zakin, J.L. (2002) Chemical and rheological characterization of drag-reducing cationic surfactant systems. *Industrial and Engineering Chemistry Research*, **41**, 6326–6336.
30. Virk, P.S. (1975) Drag reduction fundamentals. *AIChE Journal*, **21**, 625–656.
31. Pinho, F.T. and Whitelaw, J.H. (1990) Flow of non-Newtonian fluids in a pipe. *Journal of Non-Newtonian Fluid Mechanics*, **34**, 129–144.
32. Walker, D.T. and Tiederman, W.G. (1990) Turbulent structures in a channel flow with polymer injection at the wall. *Journal of Fluid Mechanics*, **218**, 377–403.
33. Wei, T. and Willmarth, W.W. (1992) Modifying turbulent structure with drag-reducing polymer additives in turbulent channel flows. *Journal of Fluid Mechanics*, **245**, 619–641.
34. Warholic, M.D., Schmidt, G.M. and Hanratty, T.J. (1999) The influence of a drag-reducing surfactant on a turbulent velocity field. *Journal of Fluid Mechanics*, **388**, 1–20.
35. Hetsroni, G., Zakin, J.L. and Mosyak, A. (1997) Low-speed streaks in drag reduced turbulent flow. *Physics of Fluids*, **9**, 2397–2404.
36. Kawaguchi, Y., Segawa, T., Feng, Z.P. and Li, P.W. (2002) Experimental study on drag-reducing channel flow with surfactant additives – spatial structure of turbulent investigation by PIV system. *International Journal of Heat and Fluid Flow*, **23**, 700–709.

37. Kim, K. and Sirviente, A.I. (2005) Turbulence structure of polymer turbulent channel flow with and without macromolecular polymer structures. *Experiments in Fluids*, **38**, 739–749.
38. Fukagata, K., Iwamoto, K. and Kasagi, N. (2002) Contribution of Reynolds stress distribution to the skin friction in wall-bounded flows. *Physics of Fluids*, **14**, L73–L76.
39. Chen, R.F., Jiang, Z.D. and Wu, J. (1998) Experimental study on turbulence structures in turbulent drag-reducing polymer solution flow. *Journal of Dalian University of Technology*, **35**, 525–528 (in Chinese).
40. Li, F-C., Wang, D-Z., Kawaguchi, Y. and Hishida, K. (2004) Simultaneous measurements of velocity and temperature fluctuations in thermal boundary layer in a drag-reducing surfactant solution flow. *Experiments in Fluids*, **36**, 131–140.
41. Li, F-C., Kawaguchi, Y. and Hishida, K. (2005) Structural analysis of turbulent transport in a heated drag-reducing channel flow with surfactant additives. *International Journal of Heat and Mass Transfer*, **48**, 965–973.
42. DeGennes, P.G. (1990) *Introduction to Polymer Dynamics*, Cambridge University Press.
43. Den Toonder, J.M.J., Hulsen, M.A., Kuiken, G.D.C. and Nieuwstadt, F.T.M. (1997) Drag reduction by polymer additives in a turbulent pipe flow: numerical and laboratory experiments. *Journal of Fluid Mechanics*, **337**, 193–231.
44. Bewersdoff, H.W., Frings, B., Lindner, P. *et al.* (1986) The conformation of drag reducing micelles from small-angle-neutron-scattering experiments. *Rheologica Acta*, **25**, 642–646.
45. Rabin, Y. and Zielinska, B.J. (1989) Scale-dependent enhancement and damping of vorticity distrubuances by polymers in elongational flow. *Physical Review Letters*, **63**, 512–515.
46. van Dam, P.H.J., Wegdam, G.H. and van der Elsken, B. (1994) The structure of turbulence in a dilute polymer solution. *Journal of Non-Newtonian Fluid Mechanics*, **53**, 215–225.
47. Hoyt, J.W. and Sellin, R.H.J. (1991) Polymer ‘threads’ and drag reduction. *Rheologica Acta*, **30**, 307–315.
48. Danckwerts, P.V. (1995) Continuous flow systems: distribution of residence times. *Chemical Engineering Science*, **50**, 3857–3866.
49. Li, F-C., Kawaguchi, Y., Hishida, K. and Oshima, M. (2006) On turbulent contribution to frictional drag in wall-bounded turbulent flow. *Chinese Physics Letters*, **23**, 1226–1229.
50. Keyes, D.E. and Abernathy, F.H. (1987) A model for the dynamics of polymers in a laminar shear flows. *Journal of Fluid Mechanics*, **185**, 503–522.

2

Drag Reduction and Heat Transfer Reduction Characteristics of Drag-Reducing Surfactant Solution Flow

2.1 Fundamental Concepts of Turbulent Drag Reduction

In this section, some fundamentals of fluid mechanics associated with turbulent drag-reducing flows are introduced. When fluid flows in a pipe or channel, frictional drag will be generated at the passage wall surface due to the fluid's viscosity. In order to keep a constant flow rate, driving power needs to be added from outside, such as the pumping power in a liquid circulation system. Through analyzing the force balance within the fluid flowing in the passage, the relationship between the wall shear stress exerted by the fluid flowing in a certain interval and the pressure difference in this certain distance can thus be established. The wall shear stress is defined as

$$\tau_W = \frac{D}{4L}\Delta P \tag{2.1}$$

where D is the hydraulic diameter of the flow passage, L is the specified distance, and ΔP is the pressure difference in length of L. The above equation applies for both pipe flow and channel flow. In some turbulent channel flows, when one side of the rectangular cross-section of the channel (W) is much larger than the other side (H), a

Turbulent Drag Reduction by Surfactant Additives, First Edition. Feng-Chen Li, Bo Yu, Jin-Jia Wei and Yasuo Kawaguchi.

two-dimensional (2D) assumption is often made. Hence, Equation 2.1 can also be simplified as

$$\tau_W = \frac{\Delta P \cdot W \cdot H}{2(W+H) \cdot L} \approx \frac{H}{2L} \Delta P \tag{2.2}$$

Here, H is often named as the height of the 2D channel. After measuring the wall shear stress of the flow, the frictional resistance coefficient of the wall flow can thus be calculated. The Fanning friction factor is often used and defined as

$$f = \frac{\tau_W}{\frac{1}{2}\rho V^2} \tag{2.3}$$

where V is the mean velocity in the channel, and ρ is the fluid density. In studies of turbulent drag-reducing flows, after obtaining the friction factor at different Reynolds numbers (Re) for turbulent drag-reducing flows either by experimental measurement or by numerical simulation, the turbulent drag reduction (DR) rate can be obtained by comparing the friction factors with those for the Newtonian fluid flow (usually the solvent flow) at the same Re. The Fanning friction factor is estimated with the following theoretical equation (this equation can also be used to testify the reliability of the measurement system in experimental studies):

$$f = 0.0791\,\mathrm{Re}^{-1/4}, \text{ turbulent flow in a pipe} \tag{2.4a}$$

$$f = 0.073\,\mathrm{Re}^{-1/4}, \text{ turbulent flow in a channel (Dean's equation [1])} \tag{2.4b}$$

The Re here is defined by the properties of the Newtonian fluid of solvent:

$$\mathrm{Re} = \frac{\rho V D}{\mu} \tag{2.5}$$

where μ is the dynamic viscosity of solvent. Li *et al.* [2–8] used cetyltrimethylammonium chloride (CTAC)/sodium salicylate (NaSal) as a drag reducer in their series of experimental studies on turbulent DR, and found that the aqueous solution of CTAC/NaSal still had very good turbulent drag-reducing effects even at concentrations as low as 25 wppm (parts per million by weight) [4]. For dilute drag reducer solution at very low concentrations, the viscosity as measured with a rheometer usually has no obvious difference compared with that of its solvent (usually water). Under this situation, it is naturally reasonable to use solvent properties to define Re. The other cases involve a drag reducer solution with a relatively large concentration, and the shear viscosity is much higher than that of the solvent. Thus, the viscosity variation cannot be neglected anymore when performing a physical analysis of the flow characteristics. But when calculating the turbulent DR rate, since it is used to make comparisons with the flow characteristics of the Newtonian fluid of the solvent, the solvent-Re has been used to do data analysis in most studies of DR.

The turbulent drag-reducing effect of drag reducer solution flows, as compared with the Newtonian fluid solvent flow, displays the pressure difference in the same interval along the streamwise direction and at the same flow rate (the same solvent-Re) is much lower for the former, or the flow rate is much larger for the former at the same pressure difference and in the same streamwise interval. The DR rate is usually defined as in the studies of turbulent drag-reducing flows:

$$DR = \frac{f_0 - f}{f_0} \times 100, (\%) \tag{2.6}$$

Here, f_0 represents the friction factor of the Newtonian fluid solvent flow and f represents the friction factor of the solution flow at the same flow rate. For turbulent drag-reducing solution flows, DR rates at different Reynolds numbers have maximum asymptotic values. Virk [9] and Zakin *et al.* [10] proposed the asymptote of the friction factor for polymer drag reducer and surfactant drag reducer solution flows, respectively, as follows:

$$1/\sqrt{f} = 19.0\lg\left(\mathrm{Re}\sqrt{f}\right) - 32.4, \text{ Virk asymptote} \tag{2.7}$$

$$f = 0.315\ \mathrm{Re}^{-0.55}, \text{ Zakin } et\ al. \text{ asymptote} \tag{2.8}$$

In Chapter 1, it was mentioned that the largest potential application area for surfactant drag reducers is to use them to reduce the flow resistance and save pumping power for district heating and cooling systems. For any district heating and cooling system, it has to include heat exchangers. Hence, issues on convective heat transfer also have to be considered in studies of turbulent drag-reducing flow. The occurrence of turbulent DR is eventually because the turbulence structures in the flow are changed and the turbulence intensity is greatly suppressed at the same Re. In consequence, the convective heat transfer rate for turbulent drag-reducing flows is also greatly reduced. In order to quantitatively describe the reduction of the convective heat transfer ability for turbulent drag-reducing flows, the following heat transfer reduction (HTR) rate is defined as follows:

$$HTR = \frac{Nu_0 - Nu}{Nu_0} \times 100, (\%) \tag{2.9}$$

Here, Nu_0 and Nu represent the Nusselt numbers for the Newtonian fluid solvent and drag reducer solution flows, respectively. The definition of Nu is as follows:

$$Nu = \frac{hD}{\lambda} \tag{2.10}$$

where h is the convective heat transfer coefficient of the flow and λ is the thermal conduction of the fluid (usually that for solvent is used). For convective heat transfer in Newtonian fluid flows, the following correlations by Gnielinski [11] can be used to estimate Nu:

$$Nu = 0.012\left(\text{Re}^{0.87} - 280\right)\text{Pr}^{0.4}\left[1 + (D/L_H)^{2/3}\right] \tag{2.11}$$

where L_H is the length of heating section of the flow passage. After obtaining Nu, the Colburn factor, j_H, can be calculated:

$$j_H = Nu \cdot \text{Re} \cdot \text{Pr}^{-1/3} \tag{2.12}$$

where Pr is the Prandtl number based on solvent properties. For turbulent drag-reducing flow by additives, similar to the existence of an asymptote of friction factor, there also exists an asymptote of heat transfer coefficient, that is, a reduction of the heat transfer coefficient cannot be lower than the asymptote. Cho and Hartnett obtained the following correlation for an asymptote of j_H through experimental studies on turbulent drag-reducing heated pipe flows with additives [12]:

$$j_H = 0.03\,\text{Re}^{-0.45} \tag{2.13}$$

2.2 Characteristics of Drag Reduction by Surfactant Additives and Its Influencing Factors

The normal characteristics of turbulent drag-reducing flow of surfactant solutions will usually experience several processes as shown in Figure 2.1. From the qualitative viewpoint, the variation process of flow characteristics with the Reynolds number for different kinds of surfactant drag reducers is similar, that is, the DR onset point (Re_1), the maximum DR point (Re_C), and the DR disappearance point (Re_2). In this section, taking only the cationic surfactant drag reducer as an example, the flow characteristics of turbulent drag-reducing flows and their influencing factors are illustrated based on a series of experimental studies on turbulent DR of surfactant solution flows [2,4,13,14].

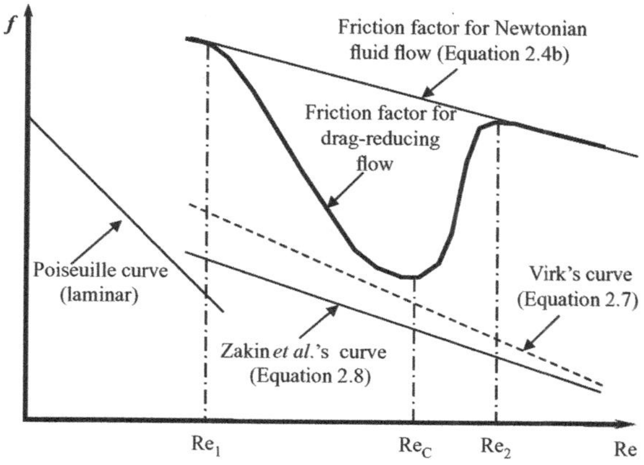

Figure 2.1 Schematic showing the flow characteristics of turbulent drag-reducing flows with surfactant additives

2.2.1 Characteristics of Drag Reduction by Surfactant Additives

2.2.1.1 Introduction to the Experimental Facility

The experimental set-up for the exampled studies on turbulent drag-reducing surfactant solution flows is shown in Figure 2.2. This is a closed-circuit water channel flow experimental facility, composed of a water tank, a circulating pump, a 2D channel, piping lines, and the related valves. The flow channel is made of transparent acrylic resin. It can be used for experimental studies using laser Doppler velocimetry (LDV) or particle image velocimetry (PIV) at the same time. This part will be introduced in Chapter 3. The dimensions of the channel are as follows: length, 10 m (smoothly inner-tenon-connected with 5 units of 2 m length each); width, 0.5 m; and height, 0.04 m. A heating plate is also set up in the test section, which is used for the measurement of convective heat transfer characteristics and the turbulent temperature field in a turbulent drag-reducing flow (convective heat transfer characteristics will be elucidated in Section 2.4, and the characteristics of turbulent temperature fields will be illustrated in Chapter 3). An electromagnetic flow meter (with an accuracy of $\pm 0.01\ m^3/min$) is installed in between the outlet of the circulating pump and the inlet of the 2D channel in order to accurately measure the volumetric flow rate. The wall shear stress is estimated from the pressure difference between two fixed points in the streamwise direction (by means of Equation 2.2). The precise differential pressure gauge that was used has an uncertainty of ± 0.1 Pa. The volume of the water tank is $2\ m^3$. The fluid temperature in the water tank is controlled by a temperature-controlling system (composed of an electric heater, cooling coils, and a propeller agitator) within ± 0.1 K. A honeycomb rectifier with a length of 0.15 m is set at the entrance of the water channel for removing large eddies. The test section is the last unit of the channel. Hence, the starting point of the test section is located at 8 m ($= 200\ H$) downstream of

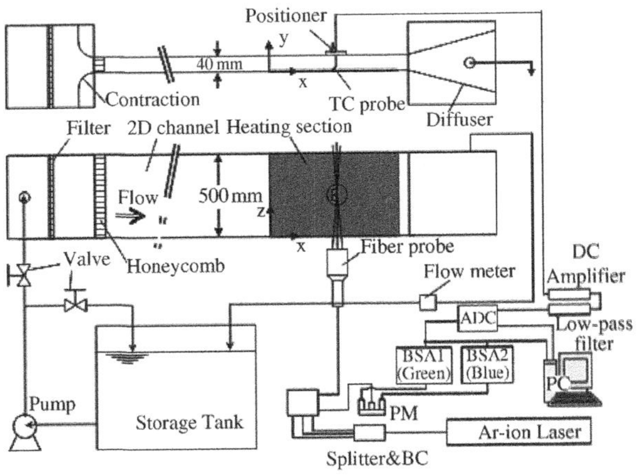

Figure 2.2 Schematic showing the experimental set-up for studying the characteristics of turbulent DR and HTR in drag-reducing surfactant solution flows

the channel entrance. This location has been in the hydrodynamic fully developed region for turbulent flow. The LDV and fine-wire thermocouple probe schematically shown in Figure 2.2 are for the simultaneous measurement of velocity and temperature fluctuations, so that the turbulent momentum and heat transports can be obtained. The relative contents of this part will be elucidated in detail in Chapter 3.

2.2.1.2 Introduction to Working Fluids for Tested Turbulent Drag-reducing Flows

The tested turbulent drag reducer is a cationic surfactant, CTAC (with the chemical formula of $C_{16}H_{33}N(CH_3)_3Cl$ and a molecular weight of 320), aided with NaSal (molecular weight of 160) for providing counterions. In this series of experiments, the ratio of CTAC to NaSal is CTAC:NaSal = 1:1 by weight, and so 1:2 by mole. From the viewpoint of chemistry, the optimum should be when the molar ratio of CTAC to NaSal is 1:1. But Lu *et al.* [15] showed that the optimum molar ratio of surfactant to counterions provider for turbulent drag-reducing effect does not have a 1:1 proportion, although the solution of both chemicals demonstrates obvious turbulent drag-reducing ability when the molar ratio is 1:1. As described in references [13,14,16], they also used the CTAC/NaSal system with a molar ratio of 1:2 as a drag reducer in their experimental studies on turbulent drag-reducing surfactant solution flows and obtained an excellent drag-reducing effect. Therefore, the series of experimental studies to be introduced here also chose this ratio.

As already described in Chapter 1, the reason why the aqueous solution of a surfactant drag reducer can generate a turbulent drag-reducing effect is eventually because the surfactant molecules assemble together to form rod-like micellar structures and then network structures (SIS structure) under shear. After adding NaSal into an aqueous solution of CTAC, the negatively charged counterions generated during the dissolution of NaSal molecules have hydrophobicity. They can penetrate into the center of micelles and neutralize the cations at the CTAC molecule head. Thus, the micellar structures formed in the aqueous solution of the CTAC/NaSal system become more stable. Even at very low concentrations, a large enough amount of stable micellar structures can still be formed which can further generate SIS structures and induce the turbulent drag-reducing effect [17].

Before performing experiments on turbulent DR, the rheological properties of drag-reducing fluids were measured at first [18]. The shear viscosity, extensional viscosity, and storage and loss moduli of an aqueous solution of CTAC/NaSal at concentrations of 30 ppm, 50 ppm, 75 ppm, 100 ppm, and 200 ppm have been measured. The shear viscosity and viscous moduli are measured by an ARES rheometer (Rheometrics Scientific F.E. Ltd., Japan), and the extensional viscosity by an RFX rheometer (Rheometric Inc., United States). The measurement results show that the measured rheological properties for very dilute solutions (the two cases of 30 ppm and 50 ppm) have no obvious differences from those of water. This conflicts with the aforementioned mechanism of turbulent DR (the formation of SIS structures in solution, which imparts viscoelastic rheological properties to the solution flow). Thus, it is natural to think that this phenomenon is most probably due to the fact that the difference between the properties of very dilute CTAC/NaSal solution and water is too small to be precisely distinguished by the normal rheometer. Li *et al.* [19] designed an indirect determination method for weak viscoelasticity of a very dilute solution of surfactant drag reducer by means of a free-surface swirling flow in a cylindrical container. In this method, the bottom of the open cylindrical container is designed to be a rotatable driving disc. A free-surface swirling flow of liquid put into the container can thus be driven by the rotating bottom disc. An obvious phenomenon in this swirling flow facility is that, even for the very dilute aqueous solution of CTAC/NaSal (such as at only 40 ppm concentration), the depth of the dip formed at the center region of the free surface was decreased, and the shape of free surface (or curvature of the free surface) was also different compared with water flow at the same rotation speed of the driving bottom disc (this rotation speed must be within a certain range without exceeding the critical rotating speed at which the shear stress may destroy the SIS structure in solution). Through analysis of the differences appearing in the depth of dip at the free surface and the overall force balance compared with water flow, the weak viscoelasticity can be distinguished for the very dilute solution of CTAC/NaSal. By this indirect method, the first normal stress difference was obtained for the 40 ppm CTAC/NaSal solution, which testifies to the existence of viscoelasticity in very dilute surfactant drag reducer solutions. Therefore, in the experimental studies to be described below, for the cases using very dilute CTAC/NaSal solutions, such as 25 ppm, 30 ppm, 40 ppm, and 50 ppm, the working fluids should be viscoelastic even though the normal rheometer cannot resolve it, since an obvious turbulent drag-reducing effect has been obtained for these very dilute solution flows.

2.2.1.3 Overall Characteristics of Turbulent Drag-reducing Surfactant Solution Flows

Figure 2.3 depicts the variation trend of the frictional drag coefficient and DR rate with the Reynolds number of a 25 ppm CTAC/NaSal solution at 30°C [4]. This figure shows the typical characteristics of turbulent drag-reducing surfactant solution flows.

The turbulent drag-reducing effect of surfactant drag reducer solution flow and polymer solution flow always starts to appear when the flow enters into the turbulent state, that is, from the Reynolds number of Re_1 shown in Figure 2.3. From Re_1, the friction factor of the solution flow begins to deviate from that of the Newtonian fluid solvent flow, and continuously decreases with Re before reaching to Re_C. Since in the experimental study reported in reference [4], measurement was not performed for low Reynolds number flows, the position of the onset Re for turbulent DR, Re_1, is just schematically shown in Figure 2.3 with a dash-and-dot line. Before Re_1, the flow is at laminar and laminar-to-turbulent transition states, at which the friction factor for surfactant solution flow is similar to or even higher than that of water flow. Kawaguchi *et al.* [20] has measured the friction factor for CTAC/NaSal solution flows in detail even at low Reynolds numbers, and they obtained conclusions similar to that mentioned above. Re_C is a critical Reynolds number. It can be thought that, at this flow condition, the network microstructures (shear induced structure, or SIS) are at the optimum state for turbulent DR and the friction factor and DR rate reach the minimum and maximum values, respectively, at that solution concentration and fluid temperature.

With a further increase of Re from Re_C, the frictional factor of solution flow begins to increase and the DR rate decreases. After Re reaches to Re_2, the friction factor profiles for solution and solvent flows collapse onto each other, indicating that surfactant solution loses its drag-reducing ability. From Re_C to Re_2, the microstructures in

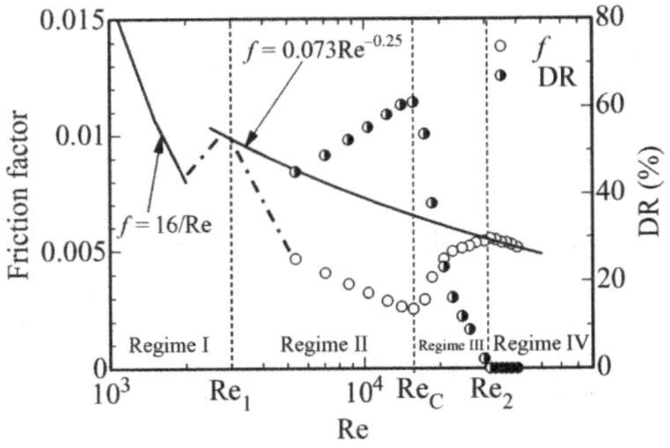

Figure 2.3 Characteristics of turbulent drag-reducing surfactant solution flows [4] (Reprinted with permission from Li, F.-C., Kawaguchi, Y., Segawa, T. and Hishida, K. "Reynolds-number dependence of turbulence structures in a drag-reducing surfactant solution channel flow investigated by PIV," *Physics of Fluids* **17**, 075104, 2005. © 2005 American Institute of Physics.)

surfactant solution flow that generate viscoelasticity and then turbulent drag-reducing effect are at a state of being continuously destroyed. The higher the Reynolds number is, the larger the shear stress becomes, and the more serious the destruction to the microstructures is. Over Re_2, the microstructures in solution have been so destroyed as to not interact with turbulent vortical structures and modify turbulence structures and contributions of turbulent kinetic energy, making the solution flow lose its drag-reducing effect.

According to the variation trend of friction factor for surfactant solution flow with Re, the drag-reducing surfactant solution flow can be classified into four different regions, Regimes I–IV, as shown in Figure 2.3. In different flow regimes, particularly in Regime II and Regime III in which the flow friction factor for surfactant solution flow differs from that of the Newtonian fluid flow, the turbulent structures are indeed changed. Reference [4] has experimentally investigated this issue in detail. The relative contents of this part will be described in the next chapter in detail.

2.2.2 Influencing Factors of Drag Reduction by Surfactant Additives

2.2.2.1 Influence of Solution Concentration

Figure 2.4 shows the turbulent DR characteristics of aqueous solution flow of CTAC/NaSal with different concentrations at 30°C [13] (the 2D channel used in the

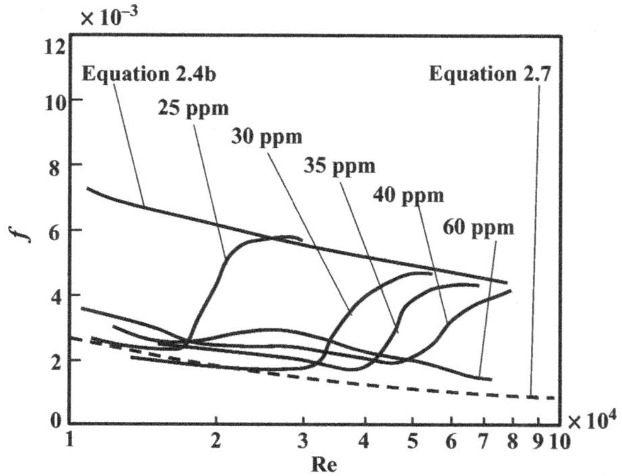

Figure 2.4 Influence of concentration on the characteristics of turbulent drag-reducing surfactant solution flows [13] (Reprinted with permission from Li, P.W., Kawaguchi, Y., Daisaka, H., Yabe, A., Hishida, K. and Maeda, M., "Heat transfer enhancement to the drag-reducing flow of surfactant solution in two-dimensional channel with mesh-screen inserts at the inlet," *Journal of Heat Transfer* **123**, 779–789, 2001. © 2001 ASME International.)

experimental study of reference [13] was 6 m in length). The influence of concentration of surfactant solution on its turbulent drag-reducing flow characteristics is obvious. The denser the solution is, the larger the concentration and strength of the network SIS microstructures formed by worm-like micelles in the solution become, and, in consequence, the higher the critical Reynolds number, Re_C, and the Reynolds number at which the turbulent drag-reducing effect disappears, Re_2, for turbulent drag-reducing flow are. As shown in Figure 2.4, for turbulent drag-reducing flow of a 25 ppm CTAC/NaSal solution, its Re_C is about 1.5×10^4 and its Re_2 about 2.5×10^4. Note that, compared with Figure 2.2, for the same 25 ppm CTAC/NaSal aqueous solution with the same ratio of CTAC to NaSal, the Re_C and Re_2 are different for the two different experiments. The reason might be as follows. The 2D channels used in these two experiments have different lengths. The piping length between the outlet of the pump and the inlet of the 2D channel is also different. This causes a big difference in the adsorption effect of surfactant molecules onto a solid wall surface for the two experiments, and finally results in a different effective Reynolds number corresponding to the drag-reducing effect. Issues on the influence of the adsorption effect will be discussed in a later chapter about the application of surfactant drag reducers. The Re_C and Re_2 for a 40 ppm solution are 5.0×10^4 and 8.0×10^4, respectively, whereas for the two cases of 60 ppm and 90 ppm solutions, the critical Re phenomenon does not appear within the reachable range of a Reynolds number for that experimental facility. Obviously, for the 90 ppm CTAC/NaSal solution case, it can be ascertained that its effective Re can range to an order between 10^5 and 10^6. Note that, for the main supplying pipe flow in a real district heating system, the Reynolds number is usually at an order of 10^6. Solution concentration has no obvious influence on the onset Reynolds number, Re_1 [20].

2.2.2.2 Influence of Fluid Temperature

There is an effective temperature range of turbulent drag-reducing effect for surfactant drag reducers. This temperature range and the upper and lower limitations of this range are also related to the surfactant solution concentration, the ion polarity of the surfactant, the molecular structures of the same type of surfactant, and so on.

Figure 2.5 shows the friction factor profile as a function of the Reynolds number for a 30 ppm CTAC/NaSal aqueous solution at different temperatures. It can be seen that the upper limit of the effective temperature range for the aqueous solution of CTAC/NaSal at 30 ppm is about 39°C (the upper limit of the effective temperature range of turbulent DR is slightly increased with an increase of the solution concentration). When the temperature is close to the upper limitation (when the temperature is 35°C, as shown in the figure), the applicable range of the Reynolds number for a turbulent drag-reducing effect also becomes narrow and the DR rate is decreased at the same Re. On the other hand, when the temperature is close to the upper limit of the effective temperature

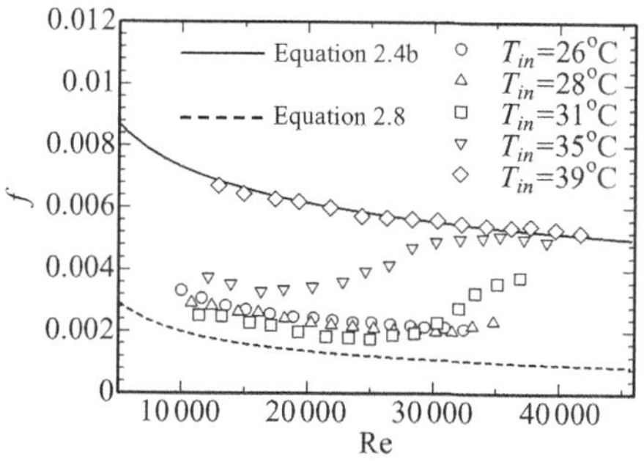

Figure 2.5 Influence of fluid temperature on the characteristics of turbulent drag-reducing surfactant solution flows [2] (Reprinted with permission from Li, F.-C., Kawaguchi, Y., and Hishida, K.,"Investigation on the characteristics of turbulence transport for momentum and heat in a drag-reducing surfactant solution flow," *Physics of Fluids* **16**, 3281–3295, 2004. © 2004 American Institute of Physics.)

range, the effective lifetime of surfactant drag reducers is also shrunk [21]. The lower limit of effective temperature range for this kind of surfactant drag reducer is about 10°C [13].

2.2.2.3 Influence of Added Amounts of Counterions

For cationic surfactant drag reducers, it is necessary to add some material for providing counterions to its dilute aqueous solution in order to bring about the turbulent drag-reducing effect. Without the aid of counterions, the small surfactant molecules cannot effectively form the rod-like and then network micelles in its dilute solution, so there is no way to generate SIS structures and viscoelasticity.

Lu *et al.* [15] utilized experimental approaches, including rheological property measurement, the measurement of turbulent DR characteristics, and Cryo-TEM observation, to investigate the viscoelasticity, turbulent DR range, and SIS structure features of surfactant solutions (5 mM Arquad 16–50/NaSal and 5 mM Ethoquad O/12/NaSal system); note that the surfactant with the commercial name of Arquad 16–50 is just surfactant CTAC) with a different added ratio of counterions (provided by NaSal). The molar ratio of NaSal to surfactant changes from 0.6 to 2.5 in their experimental study. The experimental results show that within the measured range of the molar ratio, both of the measured surfactant solutions have visco-elasticity and turbulent DR characteristics. Furthermore, the viscoelasticity and DR increase with the added amount of counterions. This is because the increase of the added amount of counterions will more effectively neutralize the positive charges at the head of surfactant molecules and stabilize the network micellar structures. Although from the theoretical viewpoint, when the molar ratio of additive-providing counterions to surfactant drag reducer is 1, it can satisfy the request of neutralization of positive charges at surfactant molecules' heads, experimental results show that an exceeded amount of counterions can augment the viscoelasticity of solution and increase its turbulent drag-reducing effect.

Usui *et al.* [22] also investigated the influence of an added amount of counterions (provided by NaSal) on the relative characteristics of a surfactant drag reducer (oleylbishydroxyethyl methyl ammonium chloride) through the measurement of turbulent drag-reducing characteristics and rheological properties. The measured surfactant solution concentration is 500 ppm, and the molar ratio (ξ) of added NaSal to surfactant varies from 0 to 100. The experimental results show that when $\xi = 0 \sim 0.5$, no DR occurs; after $\xi > 0.7$, the surfactant solution flow begins to display the turbulent drag-reducing phenomenon and DR increases with ξ. The optimum value of ξ is 1.5, at which point the solution displays strong viscoelasticity. When ξ is significantly increased ($\xi > 10$), the measured first normal stress difference (which characterizes viscoelasticity) decreases, but the solution flow still has an obvious turbulent drag-reducing effect.

2.2.2.4 Influence of Metallic Ions and Metallic Compounds

Consideration of the influence of metallic ions on the turbulent drag-reducing effect of surfactant drag reducer solutions is taken because, in the real application of drag-reducing additives to something like a district heating system, it is inevitable that one encounters metallic ions such as calcium and magnesium from tap water; iron, copper, and zinc from the corrosion of metallic piping; potassium from surfactant or the additives providing counterions and so on; as well as the metallic compounds of these ions. And the influence of metallic ions and metallic compounds on the turbulent drag-reducing effect of surfactant additives, particularly on the effective lifetime of drag-reducing ability, is usually negative.

Zakin *et al.* [23,24] pointed out that anionic surfactant drag reducers are very sensitive to the calcium and magnesium ions coming from tap water and seawater. These metallic ions can cause precipitation in anionic surfactant solutions. As for the influence of metallic ions and compounds on cationic surfactant, Hu and Matthys [25] have carried out specific experimental studies. The surfactant they studied is TTAA, tallowalkyl-*N*(C2H4OH)3Ac, with the commercial name of Ethoquad T/13–50. Counterions are provided by NaSal. The metallic ions are provided by their chloride solutions (NaCl, $CaCl_2$, $MgCl_2$, $ZnCl_2$, and $CuCl_2$), and metallic compounds include $Fe(OH)_3$, $Cu(OH)_2$, $Zn(OH)_2$, $Cu(OH)_2CO_3$, and so on. It shows that the principal parameter characterizing the viscoelasticity of drag-reducing surfactant solution (i.e., the first normal stress difference) is not sensitive to $Zn(OH)_2$. But with an increase of the added concentration of $Fe(OH)_3$, $Cu(OH)_2$, and $Cu(OH)_2CO_3$, the first normal stress difference of an aqueous solution of TTAA/NaSal with a concentration of 5 mM/12.5 mM greatly decreases, indicating that the viscoelasticity and turbulent drag-reducing ability are weakened greatly. Among the metallic ions, Mg^{2+} and Cu^{2+} have no obvious influences on the steady viscosity and the first normal stress difference, but Na^{+}, Zn^{2+}, and Ca^{2+} make the two above-mentioned parameters decrease.

2.3 The Diameter Effect of Surfactant Drag-reducing Flow and Scale-up Methods

2.3.1 The Diameter Effect and Its Influence

2.3.1.1 The Diameter Effect

As is well known, for Newtonian turbulent flow in a smooth circular pipe, the friction coefficient is only the function of the Reynolds number, Re:

$$1/\sqrt{\lambda} = 0.873 \ln\left[\mathrm{Re}\sqrt{\lambda}\right] - 0.8 \tag{2.14}$$

The above equation is the Karman–Prandtl friction coefficient formula, where λ is the engineering friction coefficient for flow in pipes, defined as $\lambda = 2(\Delta P/L)D/\rho u^2 = 4f$ (here ΔP is the pressure drop in pipe length L).

However, for drag-reducing fluids in the non-asymptotic region this is not the case, even for the same drag reducer; the effects of friction reduction are different for different pipe diameters. This is the so-called diameter effect for drag-reducing fluids [26]. From Figure 2.6, we can see clearly the "diameter effect" for both polymer and surfactant drag-reducing flows. This effect indicates that the friction factor, even under the same concentration (C_m) and temperature (T), is dependent on not only the Reynolds number, Re, but also on the pipe diameter, D. The friction factor f can be expressed as

$$f = f(\mathrm{Re}, D).$$

Savins [27] was the first to notice this phenomenon in 1964: for the drag-reducing flows at the same Reynolds number, the flow in smaller pipes shows a larger drag reduction effect than that in larger pipes. Savins considered that this could be interpreted as due to the action of drag reducer molecules on the boundary layer flow, and the boundary layer takes a larger portion of the total flow in the smaller pipes, resulting in a greater drag reduction.

2.3.1.2 Influence of the Diameter Effect

Diameter effect is a vexing issue from a practical point of view. It is not very difficult to measure the drag reduction (DR) level for a given fluid in tubes with diameters from 10 to 60 mm in the laboratory, but it is not easy to do this for much larger pipes. In a hydronic heating system of a building, the pipe diameters vary from 40 mm to 350 mm,

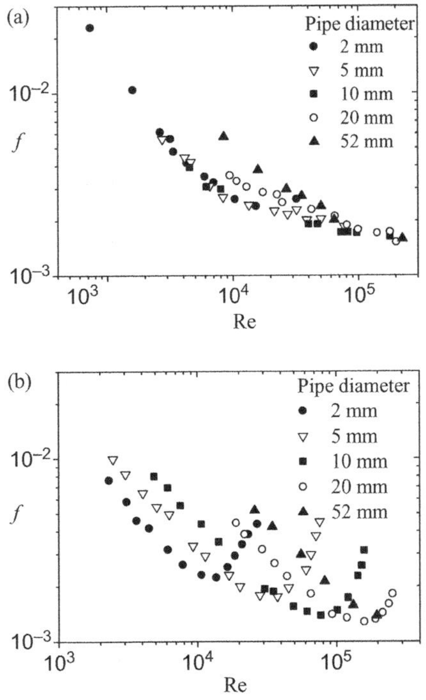

Figure 2.6 Diameter effect on drag-reducing flows. (a) Pipe diameter effect on a polymer drag-reducing flow; and (b) pipe diameter effect on a surfactant drag-reducing flow

so it is better to predict DR in such a building system by using a scale-up method based on the measured data from drag-reducing flow in small pipes.

2.3.2 Scale-up Methods

To solve the problem of the diameter effect existing in drag-reducing flows, some scale-up models and mathematical derivation methods have been developed in order to reduce or eliminate the diameter effect of the fluids. These methods are mainly for polymer and surfactant drag-reducing flows.

2.3.2.1 The DR-u_p^* Model

In 1969, Whitsitt *et al.* [28] proposed a method that correlates DR with the solution friction velocity (u_p^*), based on the assumption that the wall shear stress is the mechanism controlling DR, where the friction velocity is defined as $u_p^* = \sqrt{\tau_w/\rho}$, with τ_w and u_ρ^* the wall shear stress and fluid density, respectively, then plotted the DR level as a function of the solution friction velocity (u_p^*). This is the so-called DR-u_p^*correlation model.

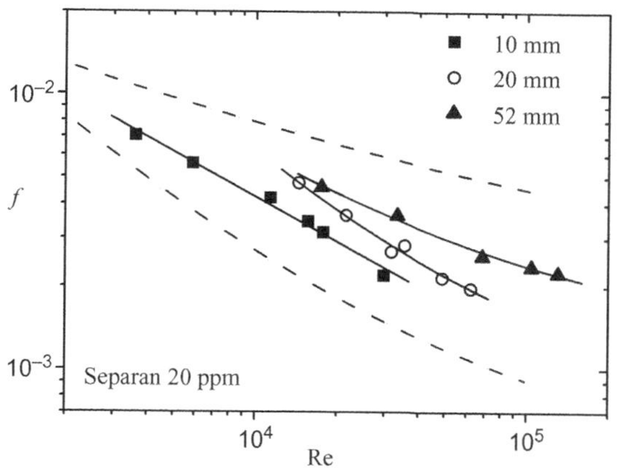

Figure 2.7 Drag reduction of a polymer solution in different-sized pipes

Gasljevic *et al.* [29] studied the drag-reducing flow of 20 ppm Separan AP-273 polymer aqueous solution in pipes with different diameters of 2, 10, 22, and 52 mm. The relationship of friction factors with the Reynolds number is shown in Figure 2.7, and the diameter effect can be seen obviously.

Figure 2.8 shows the fitting results of the measured data of polymer drag-reducing flow in different pipes by use of the DR-u_p^* correlation model. It can be seen that the deviations of the data are within 15% for the solution friction velocity less than 0.1 m/s.

Furthermore, Gasljevic *et al.* [29] studied experimentally the drag reduction of 2300 ppm Ethoquad T-13 surfactant aqueous solution in pipes with different diameters of 2, 10, 20, and 52 mm; the measured data are shown in Figure 2.9. They fitted the experimental data by use of the *DR*-u_p^* correlation model, as shown in Figure 2.10. This was the first to use the *DR*-u_p^* correlation model for surfactant drag-reducing flow.

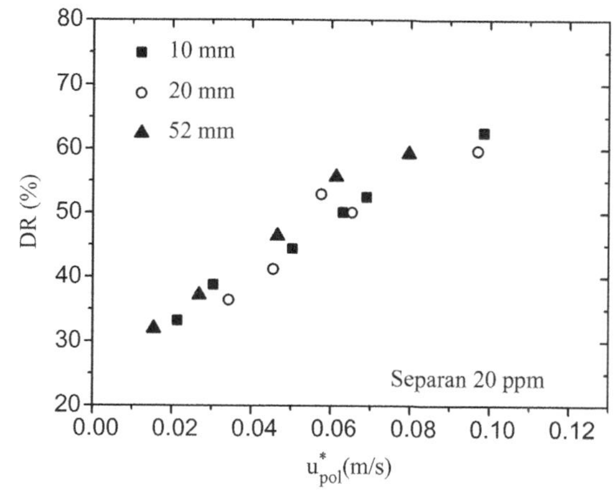

Figure 2.8 Relationship between the DR and friction velocity u_p^* of a solution flow

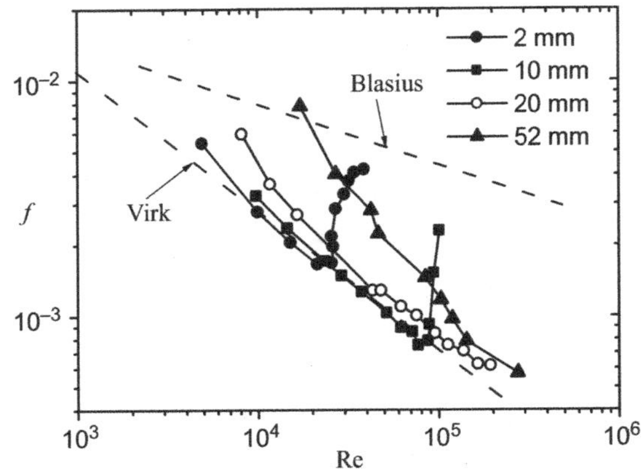

Figure 2.9 Drag reduction of a surfactant solution in different-sized pipes

From Figure 2.10, we can clearly see that except for the 2 mm pipe, the deviations of the measured data are within 15% for the other, larger pipes.

From the fitting results for both polymer and surfactant drag-reducing flows (Figures 2.8 and 2.10), we can see that the DR-u_p^* correlation model can reduce the scatter of measured data obviously. This method has the drawback that both variables (DR and u_{p}^*) contain the unknown parameter (τ_{w}), and an iterative procedure is therefore necessary to predict the DR of the solution.

2.3.2.2 The DR-u_w^* Correlation Model

To avoid the drawback of the DR-u_p^* correlation model, a number of other researchers – for example, Astarita *et al.* [30], Lee *et al.* [31], and Savins and Seyer [32]–then

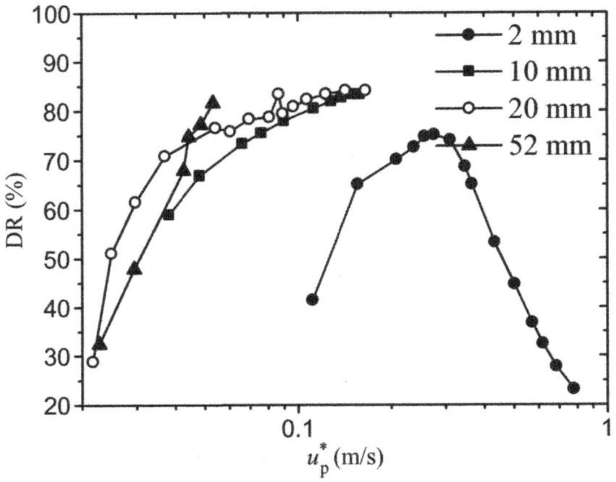

Figure 2.10 DR versus u_p^*

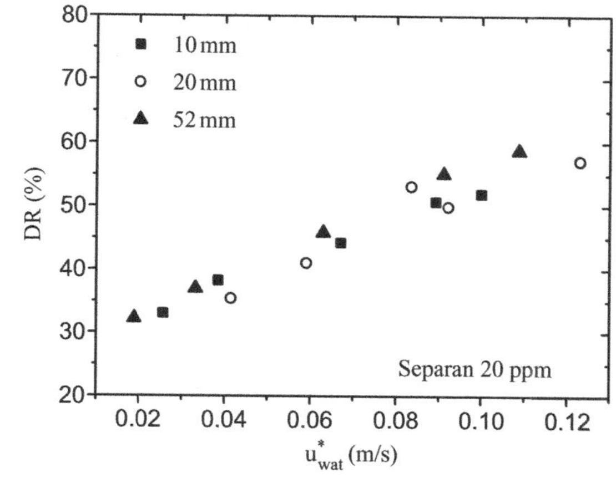

Figure 2.11 *DR* versus u_w^*

simplified Whitsitt's model by using the solvent friction velocity (u_w^*) instead of the solution friction velocity (u_p^*) and proposed the DR-u_w^* correlation model. The solvent friction velocity u_w^* is calculated from the wall shear stress τ_w for the same Reynolds number of the drag-reducing solution.

Figure 2.11 shows the fitting results of the measured data of 20 ppm Separan AP-273 polymer aqueous solution in pipes with different diameters of 2, 10, 22, and 52 mm, by using the DR-u_w^* correlation model. Compared with Figure 2.8, we can see that the deviation of the DR-u_w^* correlation model is smaller than that of the DR-u_p^* correlation model.

The DR-u_w^* correlation model was also used to correlate the measured DR data of 2300 ppm surfactant Ethoquad T-13 solution in pipes with different diameters of 2, 10, 20, and 52 mm, and the fitting results are shown in Figure 2.12. We can see that the

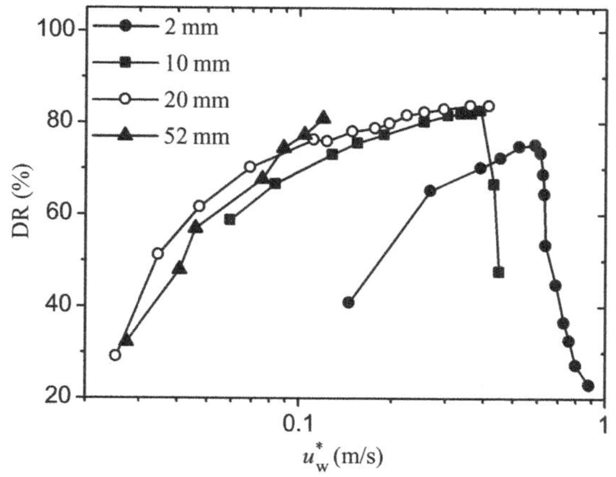

Figure 2.12 *DR*-u_w^* correlation model

maximum deviation is less than 10%, and the model shows a smaller deviation than that of the DR-u_p^* correlation model shown in Figure 2.10.

From the fitting results for both polymer and surfactant drag-reducing flows by using the above two correlation models, we can see that the *DR*-u_w^* correlation model is better than the *DR*-u_p^* model for correlating the measured DR data of different diameter pipes.

After Whitsitt *et al.* proposed the *DR*-u_p^* correlation model, Taylor and Sabersky in 1974 [33], Granville in 1977 [34] and 1984 [35], Matthys and Sabersky in 1982 [36], and Selline and Ollis in 1983 [37] used a map, iteration, or proposed empirical formula, respectively, to study the pipe diameter effect, but there was no distinct progress [38].

2.3.2.3 The DR-V Correlation Model

After that, Schmitt *et al.* [39] proposed two different empirical correlations for surfactants depending upon the shear stress level: DR versus $\tau_w$for high stresses, and τ_w versus *V* (bulk velocity) for low stresses. In 1995, Gasljevic *et al.* [29] proposed a new correlation model on the diameter effect of drag-reducing flow by correlating drag reduction level (DR) with bulk velocity. The fitting results of the new model for different diameter pipes are shown in Figure 2.13. We can see that the maximum deviation is less than 7% when the bulk velocity is not larger than the critical velocity of 7 m/s.

In addition, Gasljevic *et al.* [40] used the DR-V correlation model to correlate the measured data of a 2000 ppm Ethoquad surfactant solution in pipes of different diameters, and the results are shown in Figure 2.14. We can see that all measured data collapse to one curve when the bulk velocity is less than 6 m/s where the drag reduction is not decreased obviously. We can also see that the data of small-diameter (2 mm) pipes are well correlated, which is impossible for other correlation models. The most important is that the deviation of the data is random in this region, which indicates that

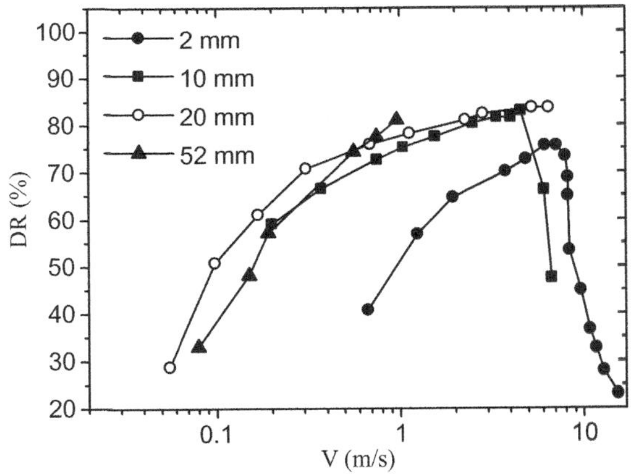

Figure 2.13 *DR-V* correlation model

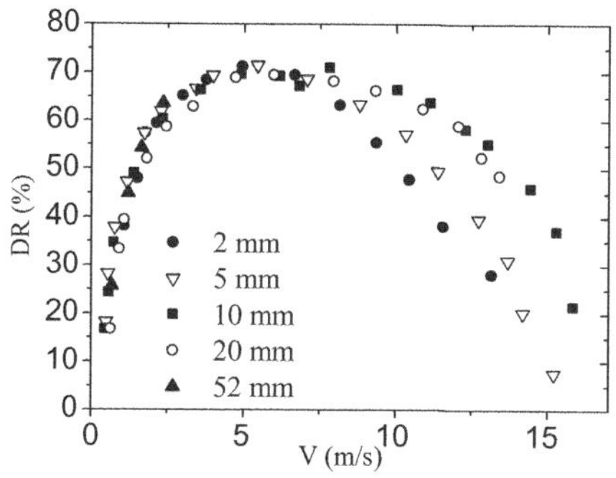

Figure 2.14 *DR* versus *V*

the correlated data are not relevant to the pipe diameter and are affected by the uncertainty in the measurement. When *V* is larger than 6 m/s, surfactant drag reduction decreases, and the deviation of data is large. However, the model is much more convenient to use than the previous models.

It is generally considered that the decrease of surfactant drag reduction starts at the critical shear stress, and the critical shear stress is independent of pipe size. From the DR-u_w^* correlation model (Figure 2.15), we can see that the drag reduction decreases for all pipes when u_w^* is larger than 0.3 m/s. A comparison of Figures 2.14 and 2.15 indicates that it is better to use shear stress (frictional velocity) than bulk velocity for the region where drag reduction decreases with fluid velocity under a given flow state; however, it is better to use the *DR*-V model than the DR-u_w^* model for the region where drag reduction increases with fluid velocity.

From the above, we can see that the *DR*-V model can correlate the pipe diameter effect of the above two surfactant solutions much better than the previous model. Now let us see its correlation effect on polymer drag-reducing solutions.

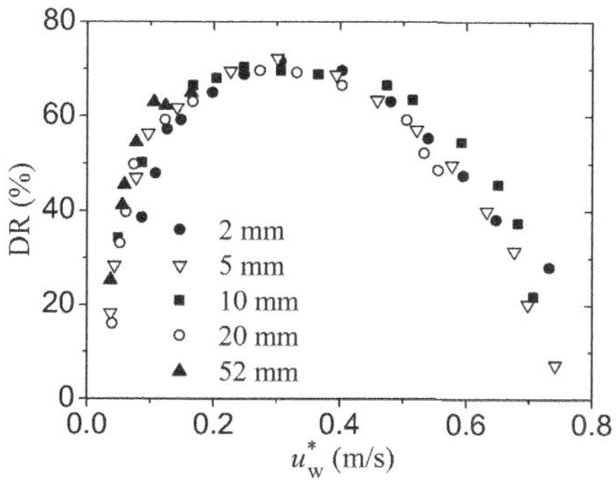

Figure 2.15 *DR* versus u_w^*

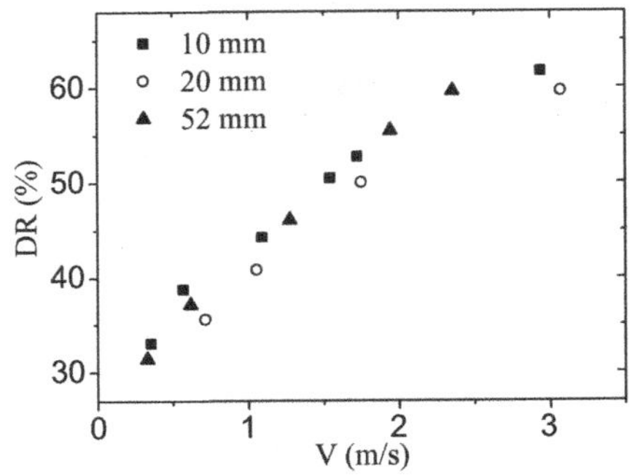

Figure 2.16 *DR* versus *V*

Gasljevic *et al.* [40] employed the DR-V model to correlate the measured data of 20 ppm Separan AP-273 polymer aqueous solution in the pipes with diameters of 10, 20, and 52 mm, and the results are shown in Figure 2.16. Compared with the previous two correlation models, the measured data correlated by the DR-V model are independent of pipe diameters in the low fluid velocity region ($V < 3$ m/s), and the deviation of the data is less than 5%, which can be considered as measurement uncertainty.

Gasljevic *et al.* [40] also used the DR-V model to correlate the measured data of 500 ppm guar gum polymer drag-reducing solution in large pipes (with the diameter from 52.5 to 208 mm) reported by DeLoof *et al.* [41], and the results are shown in Figure 2.17. We can see that the model can correlate the measured data in large pipes without systematic errors, and all deviations are caused by measurement uncertainty.

However, for some drag-reducing solutions, the deviation is very large by using the DR-V correlation model. Gasljevic *et al.* [40] used the DR-V model to correlate the experimental data of the nonionic surfactant solution SPE95285 (Figure 2.18). It can be clearly seen that it is not successful to use the DR-V correlation model, and the deviation of the data is very large: the DR of the 20 mm diameter pipe is lower than that

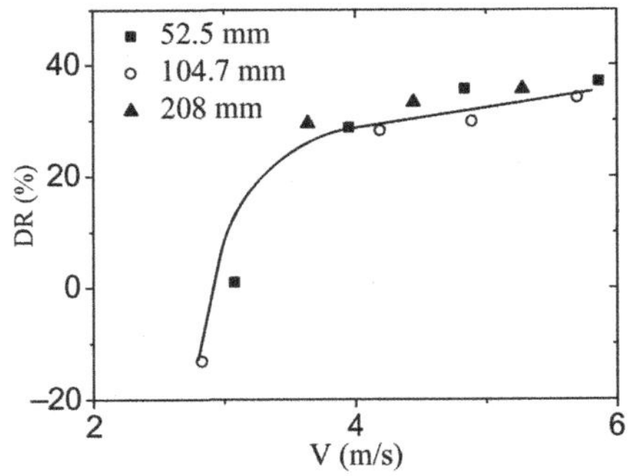

Figure 2.17 *DR* versus *V* for large-diameter pipes

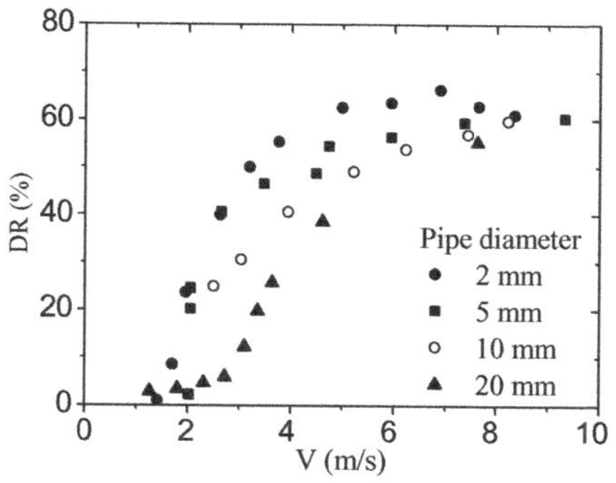

Figure 2.18 *DR* versus *V*

of the 2 mm diameter pipe by 30% for the same bulk velocity, indicating that the solution has a very strong pipe diameter effect.

The correlation effects of the DR-V model for polymer and most surfactant drag-reducing solutions are better than those of the DR-u^* model, indicating that the bulk velocity V is more important than the shear stress τ_w (frictional velocity u^*) in pipe flows. DR can be considered as a function of bulk velocity V for different diameter pipes. The cause of "pipe diameter effect" can be analyzed from two respects: (1) the flow velocity distribution in the pipe (local flow parameters); and (2) the relationship between different global flow parameters such as fluid bulk velocity, frictional velocity, and so on.

2.3.2.4 The τ_w-V Correlation Model

Two empirical theories proposed by Schmitt *et al.* [39] are related to the wall shear stress level of surfactant solution: the *DR*-τ_w correlation model is used at high wall shear stresses, and the τ_w-V correlation model is used at low shear stresses. These models are similar to the *DR*-u_p^* model proposed by Whitsitt *et al.* [28] and the *DR*-u_w^* model proposed by others. The τ_w-V model is good for surfactant solutions of C_{16}TaSal, but it has large-deviation predictions for some other drag-reducing solutions.

Gasljevic *et al.* [26] used the τ_w-V correlation model to correlate the experimental data of 2000 ppm Ethoquad surfactant solution in pipes of different diameters (Figure 2.19). We can see that the correlation effect is not so good. At a given bulk velocity, the wall shear stress of a 52 mm diameter pipe is twice that of a 2 mm diameter pipe, showing a very large deviation.

They also used the τ_w-V model to correlate the measured data of nonionic surfactant solutions of SPE95285 (Figure 2.20) [40]. It can be clearly seen that the correlation effect of the τ_w-V model is much better than that of the DR-V (Figure 2.18). Except for the data at low and high velocities, the data collapse to one curve of $\tau_w = 4.38 \times V^{1.25}$.

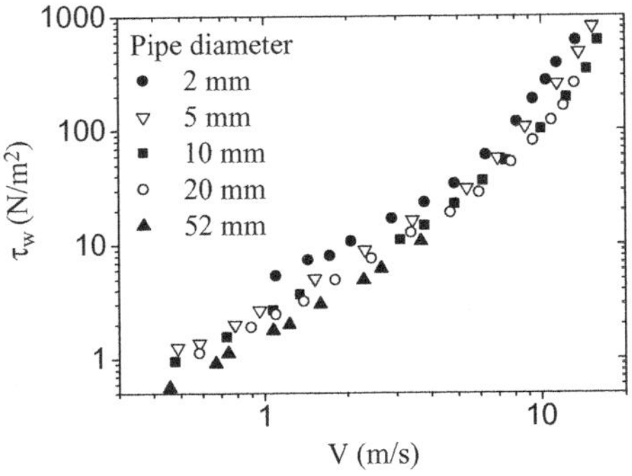

Figure 2.19 τ_w versus V

2.3.2.5 Other Scale-up Methods

Based on the relationship of rheology with the frictional velocity of drag-reducing flow, Wilson [42] proposed the following scale-up method:

$$V_2^+ = V_1^+ + 2.5\ln\left[\frac{D_2}{D_1}\right] \tag{2.15}$$

$$D_2^+ = D_1^+ \cdot \left[\frac{D_2}{D_1}\right] \tag{2.16}$$

where the superscript "+" indicates a dimensionless variable normalized by frictional velocity, V is the bulk velocity, D is the diameter, and subscripts 1 and 2 represent different pipe diameters.

Indartono *et al.* [43] conducted an experimental study for 900 ppm Ethoquad O/12 surfactant drag-reducing flow in different pipes with diameters of 13, 25, and 40 mm,

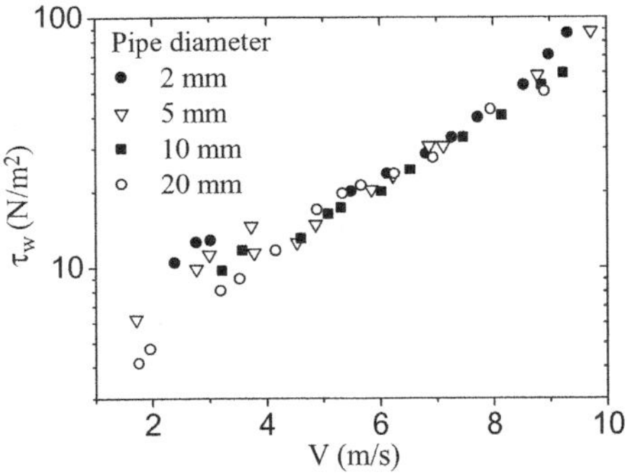

Figure 2.20 τ_w versus V

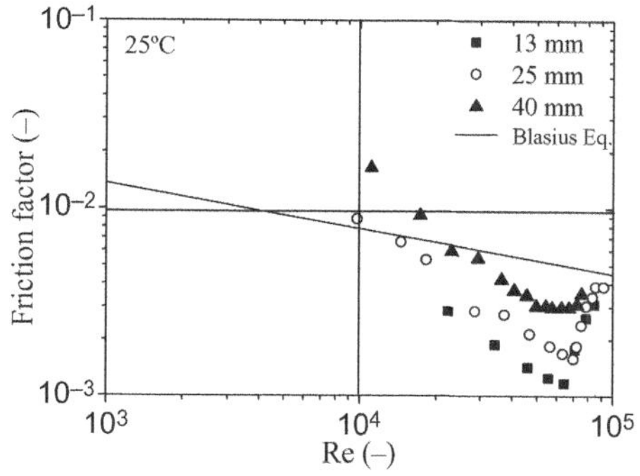

Figure 2.21 Friction factors of 900 ppm Ethoquad O/12 surfactant solution in pipes with different diameters

and the measured results are shown in Figure 2.21. The friction factors show an obvious pipe diameter effect. They correlated the data by using the scale-up method proposed by Wilson, and the results are shown in Figure 2.22. It can be seen that the correlation effect of 25 mm pipe data is better than that of 13 mm pipe data. The correlations are good in the region where drag reduction increases with bulk velocity, but there is a large deviation in the region where drag reduction decreases with bulk velocity.

Hoyt [38] proposed a new scale-up method for polymer solution. The velocity distribution of drag-reducing flow can be obtained by adding a parameter ΔB [44] to the velocity distribution of Newtonian turbulent flow in the core region.

$$u^+ = 2.5 \ln y^+ + 5.0 + \Delta B \tag{2.17}$$

where u^+ is the nondimensional velocity, and y^+ is the corresponding nondimensional distance from the wall. ΔB is the additional term accounting for drag

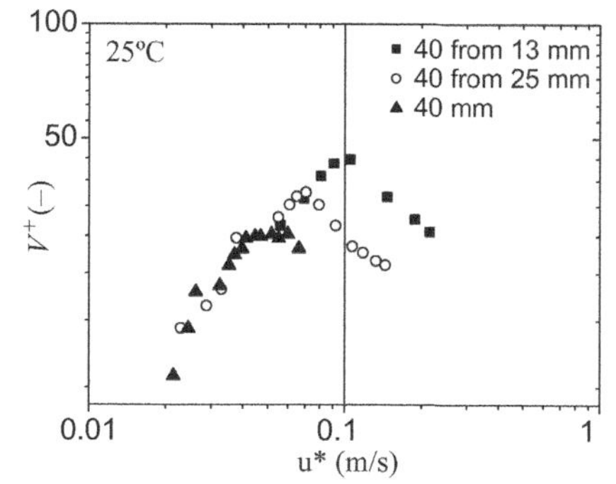

Figure 2.22 Scale-up results

reduction in the flow, which is a highly nonlinear function of the drag-reducing substance, its concentration, the pipe roughness, and the shear stress. ΔB is obtained by measuring the pressure drop in a given length of pipe, together with the average velocity. The values of ΔB obtained from the experiments are the same for the same fluid in pipes of different diameters, and the basic assumption in most scale-up techniques is that the values of ΔB are the same for equal values of shear stress in both small and large pipes.

Integrating the velocity distribution relationship, including ΔB, and introducing logs to base 10, one can get:

$$1/\sqrt{\lambda} = 2\log\left[\mathrm{Re}\sqrt{\lambda}/2.51\right] + \Delta B/8 \tag{2.18}$$

where the Reynolds number is defined as $\mathrm{Re} = uD/v$, D is the pipe diameter, u is the average fluid velocity, and v is the fluid viscosity.

Solving for ΔB, and according to the assumption that the values of ΔB are the same for small and large pipes, one can get:

$$1/\sqrt{\lambda_{large}} = 1/\sqrt{\lambda_{small}} + 2\log(D_{large}/D_{small}) \tag{2.19}$$

The final scale-up formula is:

$$\mathrm{Re}_{large} = \left(\sqrt{\lambda_{small}}\Big/\sqrt{\lambda_{large}}\right)(D_{large}/D_{small})\mathrm{Re}_{small} \tag{2.20}$$

Hoyt [38] testified to the above formula by using a series of experimental data, and further studied the scale-up method for the transport of crude oil with drag-reducing polymers. Due to different measurement temperatures, the dynamic viscosities are different. By retaining the individual viscosities in the preceding analysis, these difficulties can be avoided. Thus,

$$1/\sqrt{\lambda_{large}} = 1/\sqrt{\lambda_{small}} + 2\log(D_{large}/D_{small})\left[v_{small}/v_{large}\right] \tag{2.21}$$

and

$$\mathrm{Re}_{large} = \left(\sqrt{\lambda_{small}}\Big/\sqrt{\lambda_{large}}\right)(D_{large}/D_{small})\left[v_{small}/v_{large}\right]\mathrm{Re}_{small} \tag{2.22}$$

Figure 2.23 shows the results of scaling the smaller pipes of 2.664 cm and 5.250 cm to the 119.4 cm pipe size.

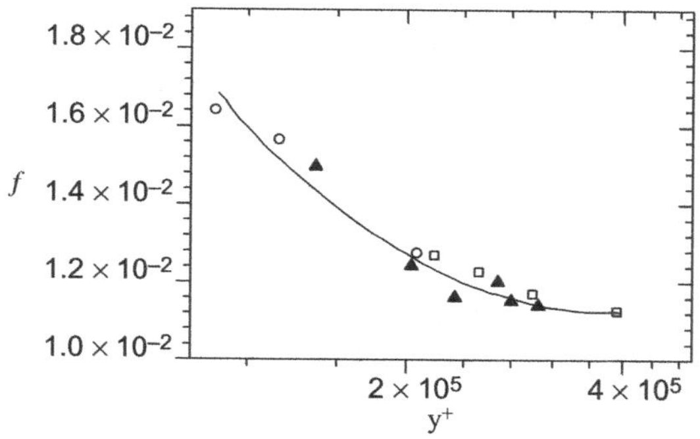

Figure 2.23 Scale-up results of a polymer solution ▲ indicate drag reduction data of 119.4 cm pipe diameter; □ represents the results of scaling the 2.664 cm pipe values to the 119.4 cm pipe size; and ○ represents the results of scaling the 5.250 cm pipe values to the 119.4 cm pipe size

Hoyt's scale-up method was only tested by polymer solution, and was not tested by surfactant solution. We used the scale-up method to correlate our experimental DR data of surfactant solution in pipes of different diameters, and the correlation results are shown in Figure 2.24. Figure 2.24a shows clearly the pipe diameter effect for the same surfactant solution. After scaling the 10 mm pipe size to the 40 mm pipe size by using Equations 2.24 and 2.25, as shown in Figure 2.24b, we can see that the diameter effect is greatly reduced, indicating that Hoyt's scaling method can also be applicable to surfactant solutions.

2.3.3 Evaluation of Different Scale-up Methods

From the analyses of different scale-up models and methods above, we can see that the DR-V model can correlate most polymer drag-reducing solutions (Separan AP-273 solution, guar glue, etc.) and surfactant solutions (Ethoquad and etc.) much better than the DR-u_p^* model and the DR-u_w^* model, whereas the τ_w-V correlation model has a large deviation for predicting these solutions. However, for some surfactant solutions such as SPE95285 and C16TaSal, the prediction of the DR-V model has a large deviation, while the τ_w-V model can predict these solutions very well.

Gasljevic *et al.* classified the drag-reducing flow into three types:

1. Fluids that comply with the three-layer velocity profile model proposed by Virk *et al.* [45] (a viscous sublayer, elastic layer, and turbulent core region; Figure 2.25), and the temperature profile is similar to the three-layer velocity profile: The elastic layer increases with increasing bulk velocity, and the slope of the temperature profile keeps constant in the elastic layer. These fluids are called "3L-type fluids."

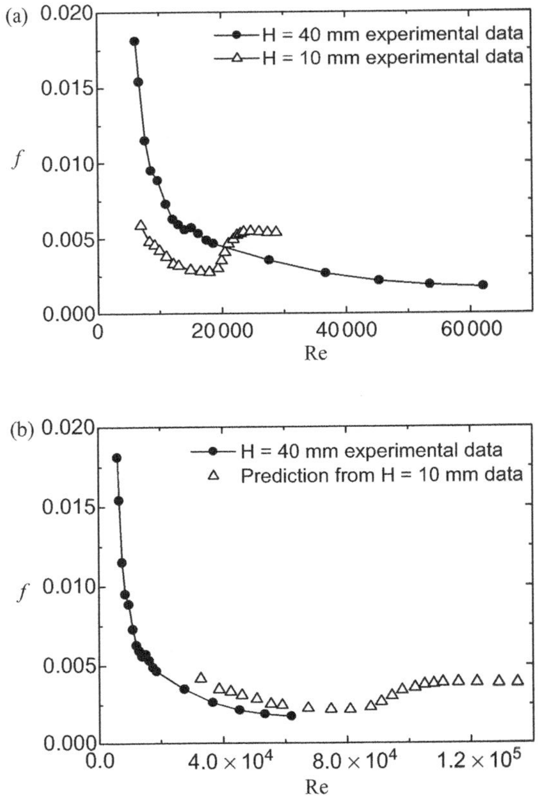

Figure 2.24 Scale-up results of surfactant solution. (a) Experimental data; and (b) scale-up results by using Equations 2.19 and 2.20

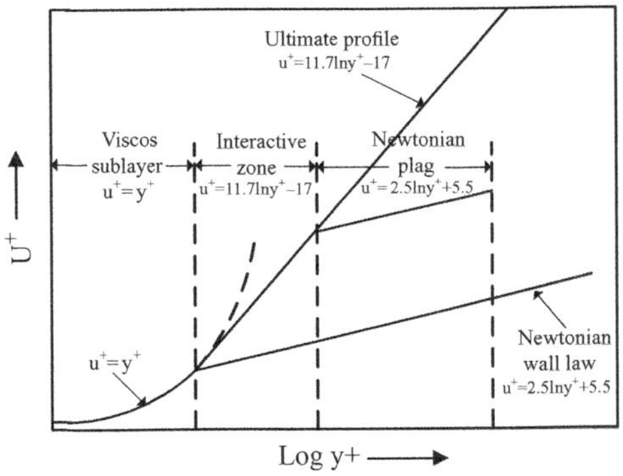

Figure 2.25 Three-layer velocity profile model of a drag-reducing fluid

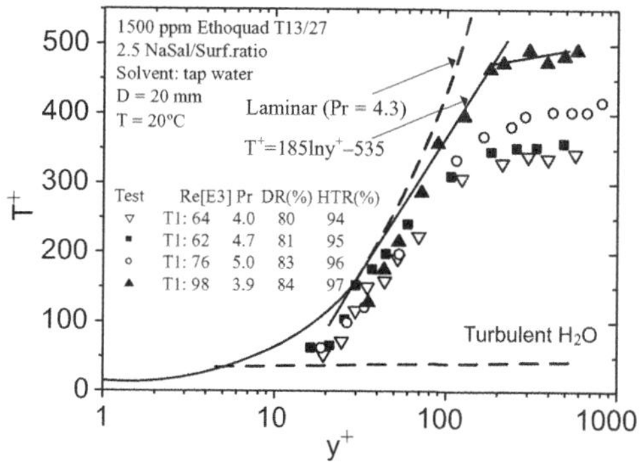

Figure 2.26 Dimensionless temperature curve of an Ethoquad surfactant solution

The study showed that if the drag-reducing solutions comply with the three-layer velocity profile model, the *DR-V* model can correlate well with the experimental data. For the Ethoquad cationic surfactant solution mentioned above, it can be seen from Figure 2.26 that the dimensionless temperature curve T^+ (the superscript "+" denotes a dimensionless variable normalized by frictional temperature, defined by $T_\tau = q_w/\rho c_p u_\tau$, where q_w is the heat flux of the heated wall, and c_p is the specific capacity of the fluid) is very similar to the three-layer velocity distribution, indicating that the drag-reducing fluid belongs to 3L-type fluids. The correlation effect of the DR-V model can be seen from Figure 2.14. The correlation results of the model are very good in the region where drag reduction increases with fluid velocity, and the prediction deviations are much smaller than those of the DR-u^* model.

2. Fluids that have no obvious elastic and turbulent core zones, and the entire cross-section is affected by the drag reduction effect. With increasing fluid velocity, the heat transfer reduction increases as the slope of the temperature curve becomes steep, and it does not increase with the increasing elastic layer. The extensions of both temperature and velocity curves go toward the center of the pipe. Different from 3L-type fluids, in the plot of dimensionless temperature T^+ (or dimensionless velocity V^+) versus dimensionless distance y^+, the gradient of temperature (or velocity) changes in a fan-type region with changing fluid velocity, and does not stay unchanged with an increasing viscous sublayer like 3L-type fluids; these kinds of fluids are called F-type fluids. For the SPE95285 nonionic surfactant solution mentioned above, the slope of the dimensionless temperature changes in a fan region with fluid velocity, as shown in Figure 2.27, indicating that the solution is a typical F-type fluid.

As mentioned above, the surfactant solution cannot be correlated well by the DR-V model (as shown in Figure 2.18), while the τ_w-V model can correlate very well (as shown in Figure 2.20).

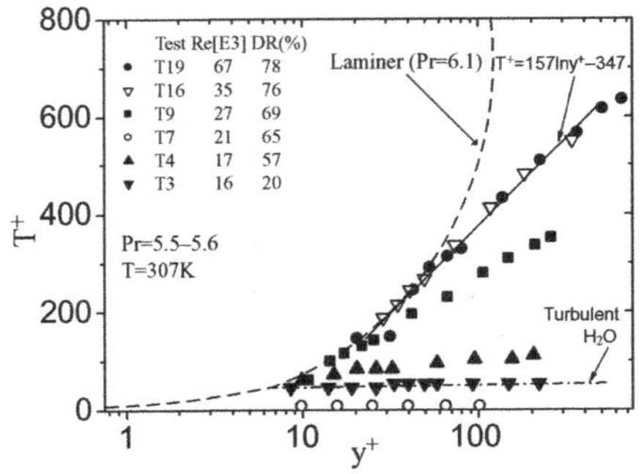

Figure 2.27 Dimensionless temperature curve of a SPE95285 surfactant solution at different Reynolds numbers

Bewersdorff and Ohlendorf [46] plotted velocity distribution curves of the C_{16}TaSal surfactant solution used by Schmitt *et al.* [39], as shown in Figure 2.28. It can be clearly seen that the velocity distribution curves show a fan region similar to that of nonionic surfactant solution. Schmitt *et al.* used the τ_w-V model to scale up the experimental data, and the predictions agree well with the experimental data. Therefore, for F-type surfactant fluids, the τ_w-V model is much better than the DR-V model for correlating the experimental data in the scale-up procedure.

3. Other-type fluids: It was found that the characteristics of some high-concentration degraded polymer solutions differ from those of other fluids, and the existing model cannot correlate their experimental data ideally. An effective correlation model needs to be developed for this kind of solution in the future study.

In the actual application, if we know which type the drag-reducing solution belongs to, we can select the corresponding scale-up method to correlate the experimental data.

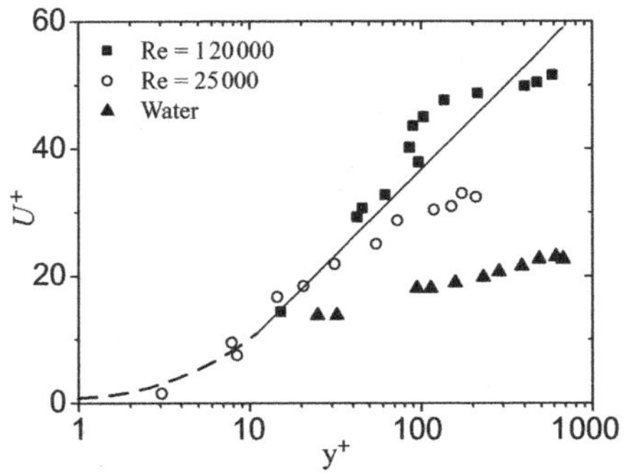

Figure 2.28 Dimensionless velocity distribution curves of a C_{16}TaSal solution at different Reynolds numbers

However, sometimes the characteristics of the studied solution are not reported yet in the published literatures, so we need to guess based on some evidences or carry out some simple experiments such as measurements of velocity or temperature distributions to determine the fluid type.

The scale-up method proposed by Hoyt is to introduce an additional term ΔB in the dimensionless velocity distribution formula, and to introduce the idea of "negative roughness" to simplify the method. The method assumed that the values of ΔB are equal for different pipe diameters at a given frictional velocity. Therefore, the method used the similarity of roughness effect on Newtonian fluid velocity distribution and the ΔB effect on drag-reducing fluid velocity distribution to simplify the scale-up procedure greatly, and the deviations of the predictions from the experimental data are very small, as shown in Figure 2.23, indicating that this method is an effective one.

Usui *et al.* [47] proposed a scale-up method for middle- to high-concentration Ethoquad O/12 surfactant solution. They studied the pipe diameter effect on surfactant drag reduction in pipes with diameters of 11∼150 mm. The scale-up method used a complicated six-step iterative procedure, but the prediction for small-diameter pipes agreed with the experimental data very well. The prediction for large pies deviated from the experimental data, and the deviation increases with increasing pipe diameter. The method was used only for surfactant solution, and its suitability to polymer solution needs to be further investigated.

2.4 Heat Transfer Characteristics of Drag-reducing Surfactant Solution Flow and Its Enhancement Methods

2.4.1 Convective Heat Transfer Characteristics of Drag-reducing Surfactant Solution Flow

The occurrence of turbulent DR is due to the significant suppression of momentum transports within the turbulent flow field and between the flow field and solid boundary surface. The convective heat transfer rate of the flow is always closely related to the momentum transport rate. Therefore, after the occurrence of turbulent DR, the heat transfer efficiency of the flow is bound to be decreased. In fact, for the turbulent drag-reducing surfactant solution flows, the HTR rate is even slightly higher than the DR rate. The reason for this will be discussed in later chapters of this volume that cover the structural and statistical characteristics of turbulent velocity and temperature fields.

In the experimental facility schematically shown in Figure 2.2, a heating plate is simultaneously installed in the test section, as shown in Figure 2.29 for its structure, which is for experimental study on the characteristics of convective heat transfer in turbulent drag-reducing flows. The heating section has a length of 0.9 m and is located

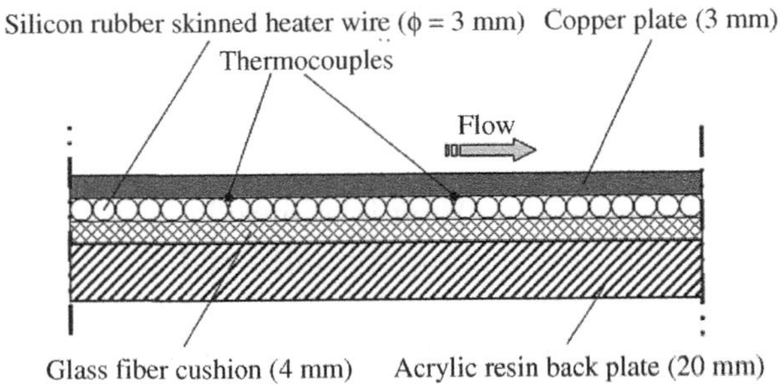

Figure 2.29 A schematic showing the heating plate for experiments on convective heat transfer in turbulent drag-reducing flows [3] (With kind permission from Springer Science +Business Media: *Experiments in Fluids,* "Simultaneous measurements of velocity and temperature fluctuations in thermal boundary layer in a drag-reducing surfactant solution flow," **36**, © 2004, 131–140, Li, F.-C., Wang, D.-Z., Kawaguchi, Y. and Hishida, K.)

8.2 m (measured from its front edge) downstream of the channel entrance. The heating plate is installed on only one side of the test section. The isoflux boundary condition is used for the heated wall, and the opposite wall has an adiabatic boundary condition. The measured flow field has been in its fully developed region, while the temperature field is still in the developing region. In the following description of experimental results, the convective heat transfer coefficient is chosen at a local point in the central plane in the spanwise direction and 20H downstream of the front edge of the heating section. Previous experimental studies [13,14] show that, at this location, the flow is hydrodynamically fully developed and thermally developing, but the local Nusselt number remains nearly constant (Nu changes significantly only in the entrance region of the heating section). Therefore, it is thought to be reasonable to take these local data for analysis of the convective heat transfer characteristics in turbulent drag-reducing flows. For a detailed description of experimental set-ups and measurement procedures, see references [2,3].

Figure 2.30 plots the convective heat transfer rate (Colburn factor) as a function of the Reynolds number for a 30 ppm CTAC/NaSal aqueous solution at different temperatures (Figure 2.30a) and a comparison of HTR and DR (Figure 2.30b). It can be seen that the variation trend of the Colburn factor with the Reynolds number and temperature is the same as that of the Fanning friction factor (also refer to Figure 2.5). Furthermore, within the effective Reynolds number range for turbulent DR, HTR is always slightly higher than DR.

For the Newtonian fluid turbulent flow with heat transfer, the so-called Colburn analogy satisfies (i.e., $j_H = f/2$). For an illustration of accordance with or departure from the Colburn analogy for drag-reducing surfactant solution flows, Figure 2.31 plots j_H/f versus Re. It is seen that the Colburn analogy is no longer valid for the turbulent drag-reducing surfactant solution flows, and the discrepancy is larger at a higher DR level.

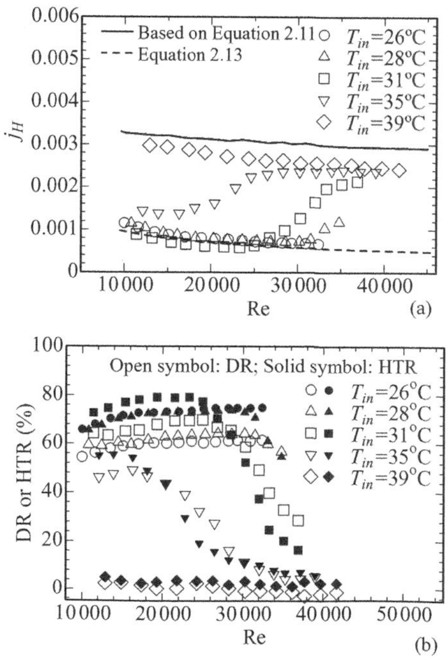

Figure 2.30 Characteristics of convective heat transfer in turbulent drag-reducing surfactant solution flows [2] (Reprinted with permission from Li, F.-C., Kawaguchi, Y., and Hishida, K.,"Investigation on the characteristics of turbulence transport for momentum and heat in a drag-reducing surfactant solution flow," *Physics of Fluids* **16**, 3281–3295, 2004. © 2004 American Institute of Physics.)

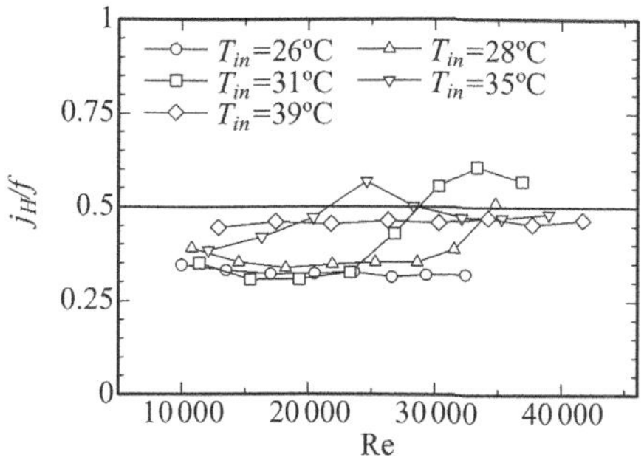

Figure 2.31 Ratio of the Colburn factor to the friction coefficient for turbulent drag-reducing surfactant solution flows [2] (Reprinted with permission from Li, F.-C., Kawaguchi, Y., and Hishida, K.,"Investigation on the characteristics of turbulence transport for momentum and heat in a drag-reducing surfactant solution flow," *Physics of Fluids* **16**, 3281–3295, 2004. © 2004 American Institute of Physics.)

2.4.2 *Heat Transfer Enhancement Methods for Drag-reducing Surfactant Solution Flows*

The application of the turbulent drag-reducing effect of surfactant additives can decrease the flow resistance and pumping power for a liquid circulation system and obtain a significant energy-saving outcome. However, when drag reducer is used in, for example, a district heating and cooling system, due to the existence of heat exchangers, the great decrease of the flow convective heat transfer rate may bring about negative influences on this kind of real application system. Therefore, it is a hot research topic to develop novel heat transfer enhancement techniques for turbulent drag-reducing surfactant solution flows.

Note that, when surfactant drag reducers are used in long-distance liquid transportation systems, the great reduction of heat transfer can, however, be used to decrease heat loss (or loss of cooling capacity). For example, when the long-distantly transported liquid is hot water from a heat resource (such as a heat and power plant) far away, if surfactant drag reducers are applied in such a system, it can reach an ideal situation for not only obtaining DR and increasing transportation capacity, or DR and energy-saving effects (a long straight pipe is the optimum flow passage for a turbulent drag-reducing effect by additives), but also obtaining a great decrease of heat loss using the HTR feature of drag-reducing flows.

As introduced in previous sections, the qualitative mechanism of occurrence of the drag-reducing effect for surfactant solution flow is due to the formation of network microstructures (SIS structures) in the solution. Seen from the characteristic curve shown in Figure 2.3, when the flow state is in Regime IV, the surfactant solution loses its turbulent drag-reducing ability due to the destruction of SIS structures caused by large shear stress. This phenomenon provides implications for developing heat transfer enhancement approaches for surfactant drag reducer solution flows. As schematically shown in Figure 2.32, a variety of devices can be installed in the entrance part of a heat exchanger in order to destroy the SIS structures in the drag reducer solution. If it can be realized, turbulent DR and HTR will not appear anymore when drag-reducing fluid flows through such heat exchangers. Furthermore, since surfactant drag reducers have the ability to self-repair, SIS structures in the solution can be automatically recovered after flowing through a heat exchanger at a not-so-far downstream distance (the automatic recover time for SIS structures is in the order of seconds), and DR and HTR phenomena appear in the solution flow again.

On the other hand, some researchers have even attempted to enhance convective heat transfer for turbulent drag-reducing surfactant solution flow through the temperature control method, that is, by locally heating the solution flow in a heat exchanger to make the temperature exceed the effective temperature range for DR, so as to get the result of eliminating DR and HTR phenomena locally. But this kind of method is very complicated, and it is also very harsh to the effective temperature range for DR. It is therefore not successful. In this book, descriptions about the heat transfer enhancement

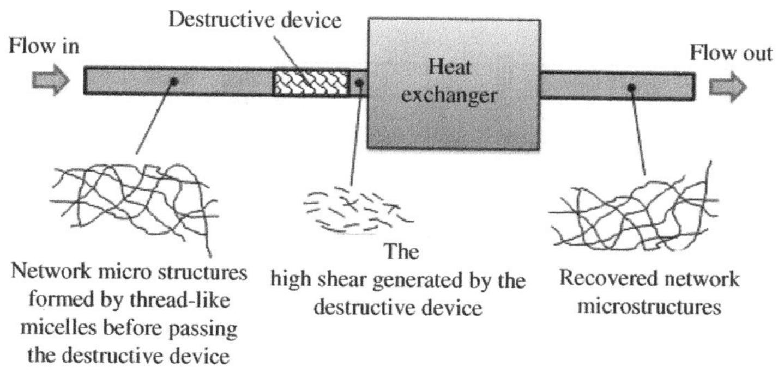

Figure 2.32 A conceptual schematic showing the heat transfer enhancement method for turbulent drag-reducing surfactant solution flows [48] (Reprinted from *Journal of Non-Newtonian Fluid Mechanics*, **116**, Qi, Y.Y., Weavers, L.K. and Zakin, J.L., "Enhancing heat-transfer ability of drag reducing surfactant solutions with ultrasonic energy," 71–93, 2003, with permission from Elsevier.)

of turbulent drag-reducing flows are made only for the methods in connection with that shown in Figure 2.32.

In order to destroy SIS structures in a drag reducer solution, researchers mainly adopt the mechanically destroying method, and the local energy injecting method can also be used. Three types of heat transfer enhancement methods are introduced below: the diameter effect method, local destroy device method, and ultrasonic energy injection method.

2.4.2.1 The Diameter Effect Method

In a surfactant drag reducer solution at a certain concentration and temperature, the SIS microstructures that can generate a turbulent drag-reducing effect are destroyed only when the wall shear stress is larger than a critical value (corresponding to Re_C, depicted in Figure 2.3). Beyond this critical wall shear stress, surfactant solution flow begins to lose its turbulent drag-reducing effect. This critical value of wall shear stress is unique. Under this constraint, the critical Reynolds number, Re_C, is uniquely corresponding to the hydraulic diameter of the flow passage, the so-called diameter effect of turbulent drag-reducing flows by additives.

Based on the diameter effect of turbulent DR by additives, it is possible to choose a specified flow rate for two different diameters of flow passage, so that the flow Reynolds number exceeds Re_C for the smaller flow passage but is still below Re_C for the larger one. This is because, at the same flow rate, the flow Reynolds number is larger for a smaller diameter of flow passage, while Re_C is smaller for the smaller flow passage case. If a proper flow rate range can be found for a surfactant drag reducer solution flow at a certain concentration, it can thus be realized that the turbulent

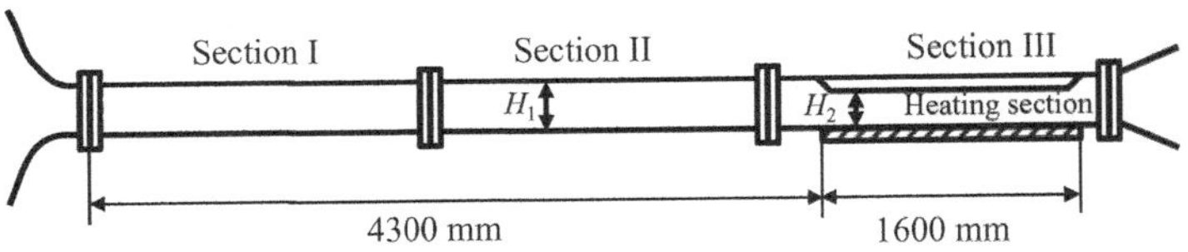

Figure 2.33 A schematic of a test section for heat transfer enhancement using a local abrupt contraction of a flow passage for turbulent drag-reducing surfactant solution flows. Designated in the figure, $H_1 = 40$ mm; $H_2 = 30$ mm; the width of the channel is 500 mm [14] (Reprinted from *Journal of Enhanced Heat Transfer*, **8**, Li, P.W., Kawaguchi, Y. and Yabe, A., "Transitional heat transfer and turbulent characteristics of drag-reducing flow through a contracted channel," 23–40, © 2001, with permission from Begell House, Inc.)

drag-reducing effect disappears or partly disappears in a flow passage with smaller diameter (and so convective heat transfer enhancement is obtained) and meanwhile the turbulent drag-reducing state is still maintained in a larger flow passage. Based on this idea, Li *et al.* [14] designed a heat transfer enhancement device for turbulent drag-reducing flow and experimentally investigated its performance. As shown in Figure 2.33, the rectangular cross-section dimensions of the 2D channel used in this experiment are the same as the flow channel dimensions shown in Figure 2.2, but only with a length of 6 m. The heating section is 1.6 m. Its front edge is 4.3 m downstream of the channel entrance. Opposing to the heating section, a 10 mm thick acrylic resin plate is attached to the channel wall, resulting in an abrupt contraction at the local position of the heating section, that is, from the upstream larger passage ($H_1 = 40$ mm) to a smaller passage ($H_2 = 30$ mm). The tested turbulent drag-reducing fluid is an aqueous solution of the CTAC/NaSal system.

The experimental result at first testifies to the diameter effect of turbulent DR in a drag reducer solution flow after the formation of the abruptly contracted channel, as shown in Figure 2.34. At temperature of 30°C, the critical Reynolds number, Re_C, is around 18 000 for a 30 ppm CTAC/NaSal aqueous solution flow in the contracted channel, while it is about 26 000 in the upstream larger channel. Therefore, it can be confirmed from the measurement results shown in Figure 2.34 that, for the tested flow cases, the flow Reynolds number range from 20 000 to 38 000 can satisfy the conditions for obtaining turbulent DR in the upstream larger channel, meanwhile obtaining a convective heat transfer enhancement in the small channel of the heating section.

In the heat transfer experiment, the heat flux for the heating plate is 5275 W/m^2 and the entrance temperature of the heating section is 30°C. Figure 2.35 shows the local Nusselt number along the heating section in the streamwise direction for a 30 ppm CTAC/NaSal aqueous solution at Re = 30 000. Here, $x/H_1 = 0$ corresponds to the front edge of the heating section. The experimental results indicated that, before $x/H_1 = 15$, the convective heat transfer coefficient has an invisible difference with and without the abrupt contracted part; however, in the downstream region of $x/H_1 > 15$, Nu at local points for the contracted-channel case is suddenly increased as compared

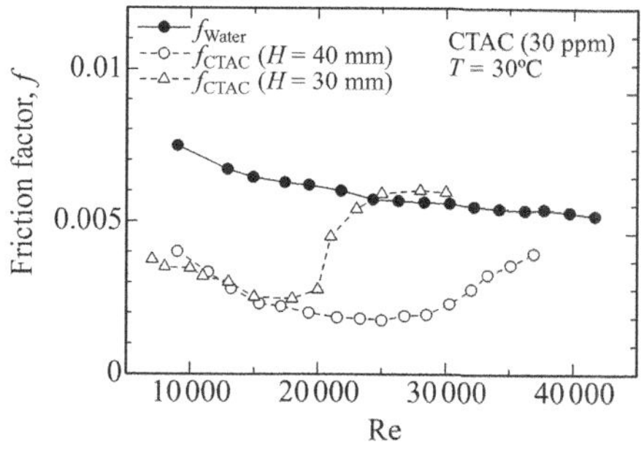

Figure 2.34 Influence of the diameter of the flow passage on the characteristics of turbulent DR of surfactant solution flow [14] (Reprinted from *Journal of Enhanced Heat Transfer*, **8**, Li, P.W., Kawaguchi, Y. and Yabe, A., "Transitional heat transfer and turbulent characteristics of drag-reducing flow through a contracted channel," 23–40, © 2001, with permission from Begell House, Inc.)

with its smooth-channel counterpart. It indicates a significant enhancement of convective heat transfer at the contracted heating section. This phenomenon cannot appear in any Newtonian fluid flows. In experimental studies, solution flows with six different concentrations (30 ppm, 35 ppm, 40 ppm, 45 ppm, 50 ppm, and 55 ppm) and at different Re have been tested. A significant heat transfer enhancement effect is obtained for all the tested cases.

Local heat transfer enhancement is obtained in the above-mentioned simple experimental facility; meanwhile, an obvious turbulent drag-reducing effect is maintained in the channel upstream of the contracted section. In the real application system, this corresponds to obtaining turbulent DR and energy-saving results and meanwhile local heat transfer augmentation in heat exchangers (the heat transfer enhancement is compared with the heat transfer ability of turbulent drag-reducing flow; if compared with the Newtonian fluid flow without any heat transfer enhancement efforts, it might

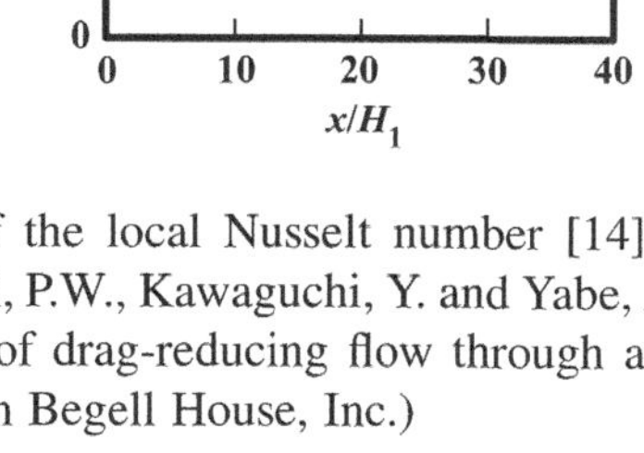

Figure 2.35 Distribution of the local Nusselt number [14] (Reprinted from *Journal of Enhanced Heat Transfer*, **8**, Li, P.W., Kawaguchi, Y. and Yabe, A., "Transitional heat transfer and turbulent characteristics of drag-reducing flow through a contracted channel," 23–40, © 2001, with permission from Begell House, Inc.)

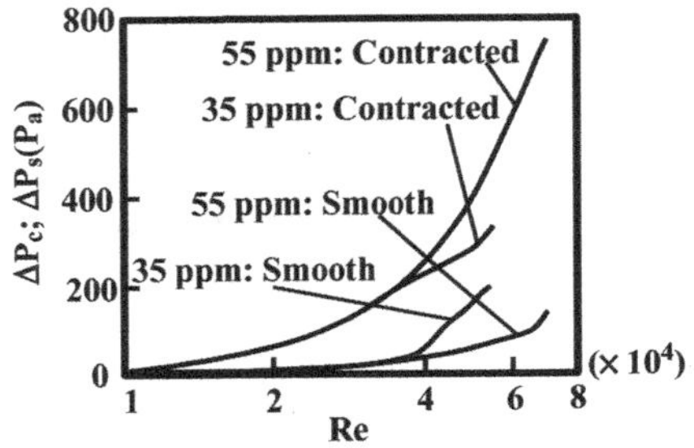

Figure 2.36 Increase of pressure difference due to the abrupt contraction in turbulent drag-reducing surfactant solution flows. ΔP_C designates the pressure difference of a contracted channel (1.6 m long), and ΔP_S is the pressure difference of a 1.6 m long smooth channel [14] (Reprinted from *Journal of Enhanced Heat Transfer*, **8**, Li, P.W., Kawaguchi, Y. and Yabe, A., "Transitional heat transfer and turbulent characteristics of drag-reducing flow through a contracted channel," 23–40, © 2001, with permission from Begell House, Inc.)

be still weaker). In general, accompanying the heat transfer enhancement is the payment of penalty for additional pressure loss. Figure 2.36 shows the measured increment of pressure loss between 1.6 m long distance before and after installation of the contracted section for the turbulent drag-reducing flow of an aqueous solution of CTAC/NaSal. The pressure difference of the contracted channel, ΔP_C, is five times larger than that for the smooth channel with the same measured distance, ΔP_S. Nevertheless, provided DR = 70%, the increment of local pressure loss together with the installation of contracted part is about two times larger than that in the smooth channel at the same flow rate. This is absolutely acceptable for the long-distance liquid transportation application system.

2.4.2.2 The Local Destroy Device Method

As implied by the name, the local destroy device method is to install a mechanically destroying device in the upstream edge of heat exchangers of a turbulent drag-reducing surfactant solution flow system, in order to locally destroy the SIS microstructures responsible for the generation of DR (as schematically shown in Figure 2.32) and make the solution lose its drag-reducing effect and recover or partly recover its convective heat transfer capability. The methods of installing the mesh, static mixer, and honeycomb at the entrance of the heat transfer element; changing smooth pipes to be spirally corrugated pipes; and so on all belong to this.

Figure 2.37 schematically shows the structures and dimensions of a mesh plug installed at the entrance of the heating section in a turbulent drag-reducing surfactant solution channel flow studied by Li *et al.* [13]. The typical results of heat transfer enhancement in drag-reducing flow by a mesh structure with different parameters are shown in Figure 2.38.

Mesh type	Mesh code	Opening width (mm)	Wire diameter (mm)	Opening rate (%)
A	16	1.3	0.2	76
B	50	0.33	0.2	35
C	150	0.098	0.07	33

Note:

$$R = \left(\frac{a}{a+d}\right)^2 \times 100$$

***R*: opening rate**
***a*: opening width**
***d*: diameter of wire**

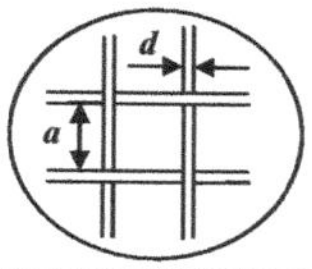

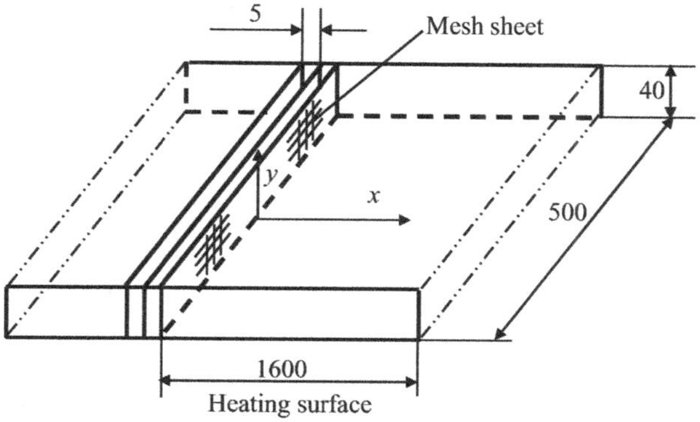

Figure 2.37 Schematics of a 2D channel and the mesh plug dimensions installed at the entrance of the heating section [13] (Reprinted with permission from Li, P.W., Kawaguchi, Y., Daisaka, H., Yabe, A., Hishida, K. and Maeda, M., "Heat transfer enhancement to the drag-reducing flow of surfactant solution in two-dimensional channel with mesh-screen inserts at the inlet," *Journal of Heat Transfer* **123**, 779–789, © 2001 ASME International.)

It is seen that the convective heat transfer ability of turbulent drag-reducing flow by surfactant additives is improved by the insertion of mesh structures at the entrance of the heating section. This is because mesh structures destroy the SIS microstructures formed in surfactant solution flow, making the flow lose its abilities to cause DR and HTR. The *y*-axis of Figure 2.38 is the average Nusselt number. Actually, the effect of heat transfer enhancement by mesh structures is relatively limited. The heat transfer rate is increased obviously only at the close downstream of the mesh structures. As it moves downstream, the heat transfer ability decreases fast. At a downstream distance of about ten times the channel height, the heat transfer enhancement effect nearly disappears [13]. It is because the SIS structures in surfactant solution can be immediately recovered after mechanical destruction. By comparing different measured cases, it can be found that the heat transfer enhancement capability becomes better for a smaller opening width and bigger number of the mesh, but at the same time, the increase of flow resistance is also more significant.

Figure 2.39 schematically shows the structures of a static mixer and honeycomb utilized by Qi *et al.* [49] for the study of heat transfer enhancement in turbulent

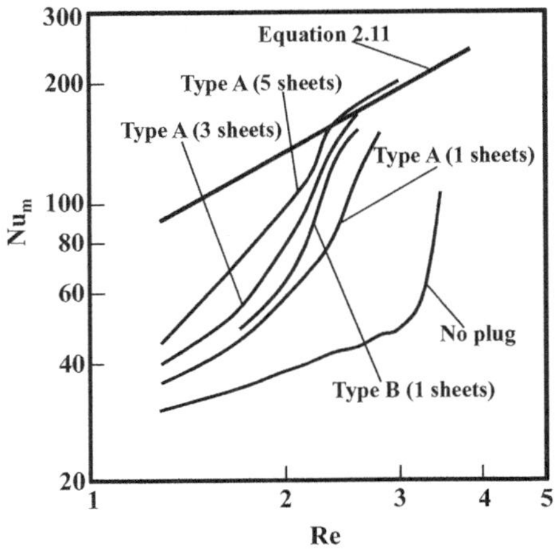

Figure 2.38 Average Nusselt number as a function of the Reynolds number for different mesh dimensions and piece numbers (the fluid inlet temperature is 30°C, and the concentration of aqueous solution of CTAC/NaSal is 30 ppm) [13] (Reprinted with permission from Li, P.W., Kawaguchi, Y., Daisaka, H., Yabe, A., Hishida, K. and Maeda, M., "Heat transfer enhancement to the drag-reducing flow of surfactant solution in two-dimensional channel with mesh-screen inserts at the inlet," *Journal of Heat Transfer* **123**, 779–789, © 2001 ASME International.)

drag-reducing surfactant solution flows. Their experimental results indicated that the honeycomb has no obvious effects on enhancing the heat transfer of turbulent drag-reducing surfactant solution flows. This is possibly because the honeycomb only divides the original flow passage into a series of smaller ones, but cannot destroy the microstructures in the solution. The static mixer, however, can obviously enhance convective heat transfer in drag-reducing surfactant solution flow, but at the same

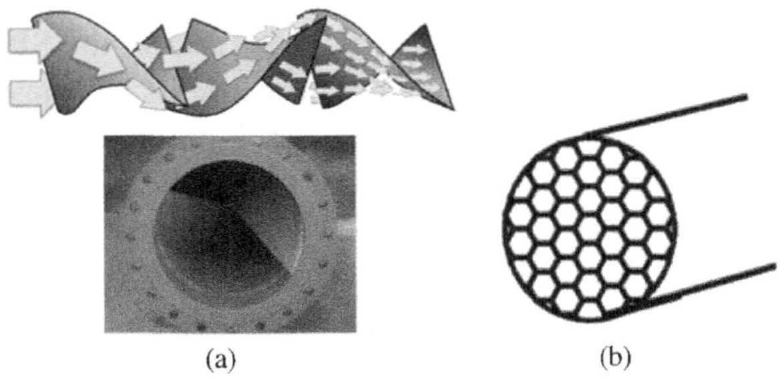

Figure 2.39 Schematics of mechanical destruction devices for microstructures in a drag reducer solution (a static mixer has two types: metal A and plastics B). (a) Static mixer B with 12 spiral units; and (b) honeycomb [50]. (Reprinted from *International Journal of Heat and Mass Transfer*, **46**, Qi, Y.Y., Kawaguchi, Y., Christensen, R.N. and Zakin, J.L., "Enhancing heat transfer ability of drag reducing surfactant solutions with static mixers and honeycombs," 5161–5173, 2003, with permission from Elsevier.)

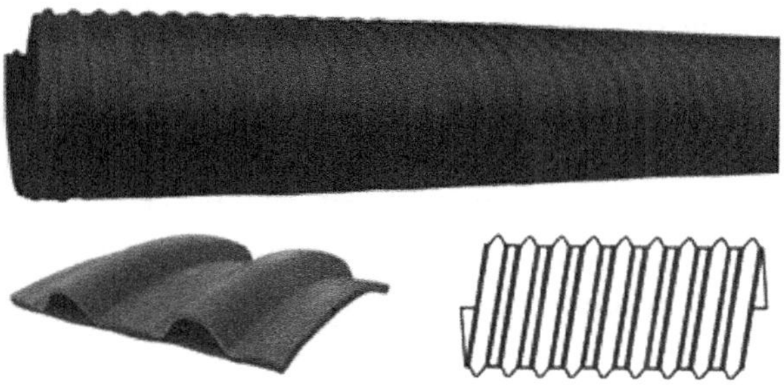

Figure 2.40 Schematic of a spiral corrugated pipe

time it needs to pay a penalty of significantly increased flow resistance. A series of experiments also show that the heat transfer enhancement capability is different for different materials of the static mixers. And if one considers the decrease of penalty coming from flow resistance increase, there is an optimum number of static mixer units.

Figure 2.40 is a schematic of spiral corrugated pipe used for heat transfer enhancement in turbulent drag-reducing surfactant solution flow by Qi *et al.* [50]. The test section they used is a tube–tube heat exchanger. The inner tubes are composed of four spiral corrugated pipes, in which surfactant solution flows as hot fluid. The outer tube is smooth. Cold water flows in between the inner and outer tubes. The measurement results show that the heat transfer coefficient for drag-reducing flow in the spiral corrugated pipes is significantly increased compared to that in smooth pipes under the measured range of the Reynolds number for the tested two kinds of surfactant solutions. The convective heat transfer coefficient can reach 1.2–1.4 times larger than that for water flows in a smooth pipe at the same Reynolds number range. This is undoubtedly inspiring to the heat transfer enhancement community for heat exchangers used in a liquid circulating system with the application of turbulent drag-reducing surfactant additives. However, experimental results also indicate that it has to pay for the penalty for the increase of local flow resistance at a spiral corrugated pipe together with obtaining heat transfer enhancement. Therefore, it needs to consider the overall characteristics of the liquid circulation system in order to determine which type of heat transfer enhancement method to use. For example, if the straight piping section is long enough for a liquid circulation system, the cost for the increase of local flow resistance at the spiral corrugated pipes might be only a very small portion. Significant economic benefits can still be obtained.

2.4.2.3 The Ultrasonic Energy Injection Method

The ultrasonic energy injection method is another approach that can temporarily destroy the microstructures in surfactant solution. Its mechanism is to induce vibration through the propagation of ultrasonic waves in solution flow media. The injected energy is mechanical energy, which cannot be absorbed by molecules of the flow media, but induce local cavitation in the liquid and generate effective mechanical energy releasing sources. The released mechanical energy may destroy large molecule or micelle structures formed by small molecules and make it lose its efficacy. The ultrasonic energy injection method was used as early as the 1940s to destroy the long-chain structures of high-polymer molecules and make the polymer degrade. When ultrasonic energy is used to destroy micellar microstructures in surfactant solution flows, its destructive effect is temporary. After ceasing to inject ultrasonic energy, the micellar structures can be recovered within a very short time interval. Therefore, this method can be used to enhance heat transfer; meanwhile, the surfactant solution flow will not lose its turbulent drag-reducing ability at the locations without ultrasonic energy injection.

Qi *et al.* [48] experimentally studied heat transfer enhancement in three surfactant solution flows (two of them display strong viscoelasticity and turbulent drag-reducing effect and another has no viscoelasticity, i.e., the first normal stress difference is zero, but also displays very good turbulent drag-reducing effect). The flow loop in their experimental set-up is mainly composed of two beaker-like containers. One container (hereinafter Container 1) serves as a liquid storage tank and is also covered with a water jacket, which is for controlling the temperature of the surfactant solution in this container. An ultrasonic emission probe is dipped into the liquid in this container to inject ultrasonic energy. The other container (Container 2) is set with one inlet and one outlet. A closed loop between these two containers is then connected with pump and piping lines. A magnet rod is set at the bottom inside Container 2. After driving this magnet rod to rotate by the electromagnetic mixer, a swirling flow can be generated in this container. Temperature and heat transfer sensors are also installed to measure the heat transfer coefficient of the solution flow. On the other hand, through visualizing the depth of the dip at the free surface of swirling flow in Container 2, the disappearance of viscoelasticity and the destruction of inner microstructures of the tested solution can be judged. Experimental results show that the injected ultrasonic energy can help the dip at the free surface of swirling flow of the two viscoelastic surfactant solutions to be formed; it increases its depth, and at the same time obviously enhances the convective heat transfer. However, for the surfactant solution without viscoelasticity, the influence of the ultrasonic energy injection is very weak. Furthermore, it is also indicated that the heat transfer enhancement effect on viscoelastic surfactant solution flows by the ultrasonic energy injection method is dependent on the surface tension, shear viscosity, temperature, geometrical characteristics of the inner micellar structures (long and threadlike, rod-like, or even spherical), and so on of the solution. It is also dependent on the ultrasonic energy intensity injected into the solution. Therefore, it still calls for further investigations before the ultrasonic energy injection method is used in real applications in heat transfer enhancement for turbulent drag-reducing surfactant solution flows.

References

1. Dean, R.B. (1978) Reynolds number dependence of skin friction and other bulk flow variables in two-dimensional rectangular duct flow. *Journal of Fluids Engineering*, **100**, 215–223.
2. Li, F-C., Kawaguchi, Y. and Hishida, K. (2004) Investigation on the characteristics of turbulence transport for momentum and heat in a drag-reducing surfactant solution flow. *Physics of Fluids*, **16**, 3281–3295.
3. Li, F-C., Wang, D-Z., Kawaguchi, Y. and Hishida, K. (2004) Simultaneous measurements of velocity and temperature fluctuations in thermal boundary layer in a drag-reducing surfactant solution flow. *Experiments in Fluids*, **36**, 131–140.
4. Li, F-C., Kawaguchi, Y., Segawa, T. and Hishida, K. (2005) Reynolds-number dependence of turbulence structures in a drag-reducing surfactant solution channel flow investigated by PIV. *Physics of Fluids*, **17**, 075104.
5. Li, F-C., Kawaguchi, Y. and Hishida, K. (2005) Structural analysis of turbulent transport in a heated drag-reducing channel flow with surfactant additives. *International Journal of Heat and Mass Transfer*, **48**, 965–973.
6. Li, F-C., Kawaguchi, Y., Hishida, K. and Oshima, M. (2006) On turbulent contribution to frictional drag in wall-bounded turbulent flow. *Chinese Physics Letters*, **23**, 1226–1229.
7. Li, F-C., Kawaguchi, Y., Hishida, K. and Oshima, M. (2006) Investigation of turbulence structures in a drag-reduced turbulent channel flow with surfactant additive using stereoscopic particle image velocimetry. *Experiments in Fluids*, **40**, 218–230.
8. Li, F-C., Yu, B., Wei, J-J. *et al.* (2008) Experimental study of drag-reduction mechanics for a dilute surfactant solution flow. *International Journal of Heat and Mass Transfer*, **51**, 835–843.
9. Virk, P.S. (1975) Drag reduction fundamentals. *AIChE Journal*, **21**, 625–656.
10. Zakin, J.L., Myska, J. and Chara, Z. (1996) New limiting drag reduction and velocity profile asymptotes for nonpolymeric additives systems. *AIChE Journal*, **42**, 3544–3546.
11. Gnielinski, V. (1976) New equation for heat and mass transfer in turbulent pipe and channel flow. *International of Chemical Engineering*, **16**, 359.
12. Cho, Y.I. and Hartnett, J.P. (1982) Non-Newtonian fluids in circular pipe flow. *Advances in Heat Transfer*, **15**, 59.
13. Li, P.W., Kawaguchi, Y., Daisaka, H. *et al.* (2001) Heat transfer enhancement to the drag-reducing flow of surfactant solution in two-dimensional channel with mesh-screen inserts at the inlet. *ASME Journal of Heat Transfer*, **123**, 779–789.
14. Li, P.W., Kawaguchi, Y. and Yabe, A. (2001) Transitional heat transfer and turbulent characteristics of drag-reducing flow through a contracted channel. *Journal of Enhanced Heat Transfer*, **8**, 23–40.
15. Lu, B., Zheng, Y., Davis, H.T. *et al.* (1998) Effect of variations in counterion to surfactant ratio on rheology and microstructures of drag reducing cationic surfactant systems. *Rheologica Acta*, **37**, 528.
16. Usui, H. and Saeki, T. (1993) Drag reduction and heat transfer reduction by cationic surfactants. *Journal of Chemical Engineering of Japan*, **26**, 103.
17. Hu, Y.T. and Matthys, E.F. (1997) Evaluation of micellar overlapping parameters for a drag-reducing cationic surfactant system: light scattering and viscometry. *Langmuir*, **13**, 4995.
18. Wei, J-J., Kawaguchi, Y., Yu, B. and Feng, Z-P. (2006) Rheological characteristics and turbulent friction drag and heat transfer reductions of a very dilute cationic surfactant solution. *ASME Journal of Heat Transfer*, **128**, 977–983.
19. Li, F-C., Dong, Y., Kawaguchi, Y. and Oshima, M. (2008) Experimental study on swirling flow of dilute surfactant solution with deformed free-surface. *Experimental Thermal and Fluid Science*, **33**, 161–168.
20. Kawaguchi, Y., Tawaraya, Y., Yabe, A. *et al.* (1996) Turbulent transport mechanism in a drag reducing flow with surfactant additive investigated by two component LDV. Proceedings of the

8th International Symposium on Application of Laser Techniques to Fluid Mechanics, Lisbon, 29.4.1–29.4.7.
21. Jiao, L-F., Li, F-C., Su, W-T. *et al.* (2008) Experimental study on surfactant drag-reducer applying to some district heating system. *Energy Conservation Technology*, **126**, 195–201 (in Chinese).
22. Usui, H., Kamada, T. and Suzuki, H. (2004) Surfactant drag reduction caused by a cationic surfactant with excess addition of counter-ions. *Journal of Chemical Engineering of Japan*, **37**, 1232–1237.
23. Zakin, J.L., Zhang, Y. and Ge, W. (2007) Drag reduction by surfactant giant micelles, in *Giant Micelles: Properties and Applications*, vol. **140**, Surfactant Science Series (eds R. Zana and E.W. Kaler), Taylor and Francis, New York.
24. Zakin, J.L., Lu, B. and Bewersdorff, H.W. (1998) Surfactant drag reduction. *Reviews in Chemical Engineering*, **14**, 253–318.
25. Hu, Y.T. and Matthys, E.F. (1997) Effect of metal ions and compounds on the rheological properties of a drag-reducing cationic surfactant solution exhibiting shear-induced structure formation. *Journal of Colloid and Interface Science*, **186**, 352–359.
26. Gasljevic, K., Aguilar, G. and Matthys, E.F. (1999) An improved diameter scaling correlation for turbulent flow of drag-reducing polymer solutions. *Journal of Non-Newtonian Fluid Mechanics*, **84** (2–3), 131–148.
27. Savins, J.G. (1964) Drag reduction characteristics of solutions of macromolecules in turbulent pipe flow. *Society of Petroleum Engineering Journal*, **4**, 203.
28. Whitsitt, N.F., Harrington, J.L. and Crawford, H.R. (1964) Effect of wall shear stress on drag reduction, in *Viscous Drag Reduction*, Plenum Press, NY, p. 265.
29. Gasljevic, K. and Matthys, E.F. (1995) On the diameter effect for turbulent flow of drag-reducing surfactant solugion. FED-Vol.231/MD-VoL.66, Developments and Applications of Non-Newtonian Flows, ASME, pp. 237–243.
30. Astarita, G., Greco, G. and Nicodemo, L. Jr. (1969) Excess pressure drop in laminar flow through sudden contraction of non-newtonian liquids. *AIChE Journal*, **15** (4), 564–567.
31. Lee, W.K., Vaseleski, R.C. and Metzner, A.B. (1974) Turbulent drag reduction in polymeric solutions containing suspended fibers. *AIChE Journal*, **20** (1), 128–133.
32. Savins, J.G. and Seyer, F.A. (1977) Drag reduction scale-up criteria. *Physics of Fluids*, **20** (10), s78–s84.
33. Taylor, D.D. and Sabersky, R.H. (1974) Extrapolation to various tube diameters of experimental data taken with dilute polymer solutions in a smooth pipe. *Letter in Heat and Mass Transfer*, **1**, 103.
34. Granville, P.S. (1977) Scaling-up of pipe-flow frictional data for drag-reducing polymer solution. Proceedings of the 2nd International Conference on Drag Reduction, B1.1.
35. Granville, P.S. (1984) A method for predicting additive drag reduction from small-diameter pipe flow. Proceedings of the 3rd International Conference on Drag Reduction, C3-1.
36. Matthys, E.F. and Sabersky, R.H. (1982) A method of predicting 'diameter effect' for heat transfer and friction of drag reducing fluids. *International Journal of Heat and Mass Transfer*, **25**, 1343.
37. Sellin, R.H.J. and Ollis, M. (1983) Effect of pipe diameter on polymer drag reduction. *Industrial and Engineering Chemistry Product Research and Development*, **22**, 445.
38. Hoyt, J.W. (2003) Scale-up from laboratory pipe-flow data to large flow. Proceedings of ASME FEDSM'03, p. 45656.
39. Schmitt, K., Brunn, P.O. and Durst, F. (1988) Scaling-up correlations for drag reducing surfactants. *Progress and Trends in Rheology II: Proceedings of the 2nd Conference of European Rheologists, Prague, June 17–20, 1986*, pp. 249–252.
40. Gasljevic, K., Aguilar, G. and Matthys, E.F. (2001) On two distinct types of drag-reducing fluids, diameter scaling, and turbulent profiles. *Journal of Non-Newtonian Fluid Mechanics*, **96**, 405–425.

41. De Loof, J.P., de Lagarde, B., Petry, M. and Simon, A. (1977) Proceedings of the 2nd International Conference on Drag Reduction, BHRA Pub. Cranfield, UK, paper B2 (eds H.S. Stevens and J.A. Clarke).
42. Wilson, K.C. (1989) Additives: theoretical models, in *Drag reduction in fluid flows: Techniques for Friction Control* (eds R.H.J. Sellin and R.T. Moses), Ellis Horwood Publishers, England.
43. Indartono, Y.S., Usui, H., Suzuki, H. and Komoda, Y. (2005) Temperature and diameter effect on hydrodynamic characteristic of surfactant drag-reducing flows. *Korea-Australia Rheology Journal*, **17** (4), 157–164.
44. Meyer, W. (1966) A correlation of the frictional characteristics for turbulent flow of dilute viscoelastic non-Newtonian fluids in pipes. *AIChE Journal*, **12**, 522–525.
45. Virk, P.S., Mickley, H.S. and Schmitt, K.A. (1970) The ultimate asymptote and mean flow structure in Toms' phenomenon. *ASME Journal of Applied Mechanics*, **37** (2), 488–493.
46. Bewersdorff, H.W. and Ohlendorf, D. (1988) The behavior of drag-reducing cationic surfactant solutions. *Colloid Polymer Science*, **266**, 941–953.
47. Usui, H., Itoh, T. and Saeki, T. (1998) On pipe diameter effects in surfactant drag-reducing pipe flows. *Rheologica Acta*, **37** (2), 122–128.
48. Qi, Y.Y., Weavers, L.K. and Zakin, J.L. (2003) Enhancing heat-transfer ability of drag reducing surfactant solutions with ultrasonic energy. *Journal of Non-Newtonian Fluid Mechanics*, **116**, 71–93.
49. Qi, Y.Y., Kawaguchi, Y., Christensen, R.N. and Zakin, J.L. (2003) Enhancing heat transfer ability of drag reducing surfactant solutions with static mixers and honeycombs. *International Journal of Heat and Mass Transfer*, **46**, 5161–5173.
50. Qi, Y.Y., Kawaguchi, Y., Lin, Z.Q. *et al.* (2001) Enhanced heat transfer of drag reducing surfactant solutions with fluted tube-in-tube heat exchanger. *International Journal of Heat and Mass Transfer*, **44**, 1495–1505.

3

Turbulence Structures in Drag-Reducing Surfactant Solution Flow

The main concern in this book is the turbulent drag-reducing effects of surfactant drag reducers in channel or pipe flows. Regarding research results on turbulent drag-reducing effects in the homogeneous isotropic turbulence of viscoelastic fluid [1–5], they will not be mentioned at this stage. The turbulent flows with solid wall(s), such as channel flows, pipe flows, or boundary layer flows, are all named "wall turbulence" or "wall-bounded turbulent flow." When studying wall turbulence, it is of particular importance to understand the near-wall turbulent bursting event and the consequent coherent structures of turbulent vortices due to the bursting event. For the mechanism of turbulent drag-reducing effect by surfactant drag reducer, it can be qualitatively explained as follows. In the turbulent flow of surfactant drag reducer solution, microstructures can be formed by shear (shear-induced structures). These microstructures impart viscoelasticity into the solution flow, and the viscoelasticity interacts with turbulent bursting events or turbulent vortex structures. In consequence, the frequency and intensity of turbulent bursting events are all greatly decreased; the turbulent vortex structures and turbulent kinetic energy spectrum are also changed. Therefore, studying the turbulence structural as well as statistical characteristics of drag-reducing surfactant solution flow may help researchers gain insight into the turbulent drag reduction (DR) mechanism. On the other hand, after fully recognizing the coupling information between velocity and temperature fields in drag-reduced flows, the mechanism of heat transfer reduction (HTR) accompanying turbulent DR can be deeply understood also, which may provide new ideas on heat transfer enhancement for drag-reducing flows. Hence, in this chapter, the state-of-the-art principal experimental approaches for measuring turbulence structures in turbulent

Turbulent Drag Reduction by Surfactant Additives, First Edition. Feng-Chen Li, Bo Yu, Jin-Jia Wei and Yasuo Kawaguchi.

drag-reducing flows by additives are at first introduced briefly, followed by a detailed elucidation of the statistical characteristics of turbulent velocity and temperature fields and the structural characteristics of turbulent vortices of drag-reducing flows. Finally, the mechanisms of turbulent DR by additives and HTR accompanying DR are revealed through quantitative analyses of variation features of the Reynolds shear stress and turbulent heat flux distributions.

3.1 Measurement Techniques for Turbulence Structures in Drag-Reducing Flow

For the measurement of turbulent velocity fields, point-wise measurement approaches were used in the early years via a Pitot tube, hot wire (or hot film) velocimetry, and so on. However, except for shortcomings coming from disturbances to the flow field for these measurement approaches, when used for measuring the turbulent velocity field of drag-reducing polymer solution or surfactant solution flows, the microstructures in the solutions may be entangled on the sensor probe, which makes the momentum and heat transfer characteristics of the probe change greatly [6]. For hot wire or hot film velocimetry, the viscoelastic rheological properties of drag-reducing fluids may generate wrong information to the measurement results [7,8], since the calibration of this kind of velocimetry is based on the heat transfer laws for Newtonian fluid flow. The calibrated results of velocimetry will not be valid anymore when applied to drag-reducing fluids [9,10]. The above-mentioned issues restrain such kinds of measurement methods with a probe (which disturbs the velocity field) from being used for the measurement of turbulent velocity fields of drag-reducing flows. With the applications and developments of laser techniques in the field of experimental fluid dynamics, laser Doppler velocimetry (LDV) and particle image velocimetry (PIV) have successively become the two key effective approaches for measuring turbulence structures in drag-reducing flows.

3.1.1 Laser Doppler Velocimetry

As mentioned in reference [11], the first systematically measured data for velocity fields in turbulent drag-reducing flow by means of LDV were those reported by Goldstein *et al.* in 1969, which covered point-wise velocity measurement in the center region of drag-reducing flow in a pipe [12]. From then on, a large amount of experimental results on drag-reducing flows using the LDV technique were published, and the understanding of turbulent DR changed essentially: from only flow drag (i.e., the macro or overall features) to the turbulent velocity fluctuation intensity, the Reynolds shear stress, turbulent bursting events, and so on (i.e., the microscopic characteristics). Great progress was therefore made for understanding the mechanisms

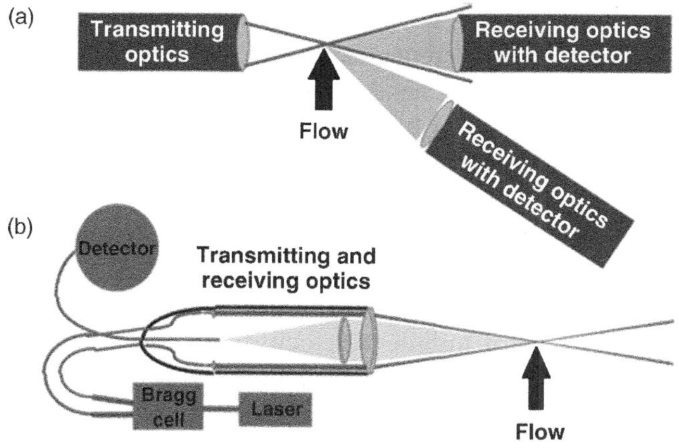

Figure 3.1 A standard LDV system. (a) Forward or side-scattering set-up; and (b) backward-scattering setup

of turbulent DR. Up to now, as a powerful supplementation to the PIV technique due to its high sampling rate, the LDV technique is still being utilized in experimental studies on the mechanism of turbulent drag-reducing flows.

The LDV system is mainly composed of a laser, transmitting optics, receiving optics, and a detector for Doppler signals (as shown in Figure 3.1). Depending on the different locations of receiving optics, LDV set-ups can be divided into two types: forward or side scattering, and backward scattering. Nowadays, the standard LDV system is mainly the double-beam backward-scattering type (Figure 3.1b), whose transmitting optics and receiving optics are unified. The difference of the Doppler frequency shift generated by the two laser beams is inspected with the detector, and so the velocity component of tracing particles in the flow field normal to the angle bisector of the two laser beams is obtained. Of course, for a standard LDV system, it also consists of the bursting spectrum analyzer (BSA) and computer installed with a data-sampling board and analytical software, which are not shown in Figure 3.1b. The direction of the velocity component measured with a double-beam LDV system is related to the plane determined by the double beams (or beam pair). If illuminating the flow field with multiple beam pairs can be realized, multidimensional measurement of the velocity field can thus be performed. Therefore, one-dimensional (1D), two-dimensional (2D), and three-dimensional (3D) LDV systems can be realized based on real requirement. For the measurement principle and other detailed information about LDV, one refers to the related references, for example reference [13].

With correct operation of LDV measurement (including adjustment of optics, selection and addition of tracing particles, option of gain factor of BSA signal, etc.), a relatively high data-sampling rate can be easily obtained and large enough sampling space can be saved, which is the principal advantage of the LDV technique. Through statistical analysis of the sampled data of a turbulent velocity field collected by LDV, the mean velocity distribution, velocity fluctuation intensity, spectrum characteristics, coupling information between different quantities (e.g., the Reynolds shear stress), and so on can be obtained. On the other hand, conditional resampling with set conditions for the original data sample taken by LDV can also be performed. By doing so, the physical procedures with some typical characteristics can be extracted and the quantitative information can be provided. The most typical advance is to extract the turbulent bursting event for wall turbulence and investigate its occurrence frequency, intensity, distribution to the Reynolds shear stress, and so on.

The LDV technique coupled with a fine-wire thermocouple can be used for simultaneous measurements of turbulent velocity field and fluctuating temperature field when heat transfer takes place in a turbulent flow. This experimental technique has been employed to study the statistical characteristics of turbulent velocity field and temperature field when turbulent DR and HTR co-occur in a heated drag-reducing surfactant solution flow. The turbulent heat flux normal to the heated wall has thus been obtained, through which the mechanism of HTR accompanying turbulent DR can be explored quantitatively [14–16]. The LDV measurement volume and geometrical information of a fine-wire thermocouple probe used in those experimental studies are shown in Figure 3.2.

3.1.2 PIV

Although LDV has its advantages such as non-intrusiveness to the flow field, fast response to the fluctuating velocity (high data-sampling rate), and so on, its drawback is that LDV can only measure the instantaneous velocity at a fixed point and it is unavailable to obtain more structural information of turbulent velocity field. However, captured by the PIV technique is the instantaneous velocity vectors distributed in a 2D plane in the turbulent flow field illuminated with a laser sheet (up to now, only the 2D PIV technique has been applied to the experimental investigation of turbulent

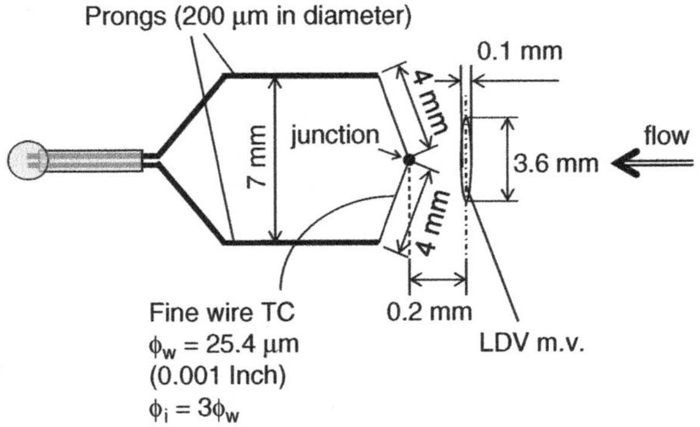

Figure 3.2 Schematics of a LDV measurement volume and fine-wire thermocouple probe

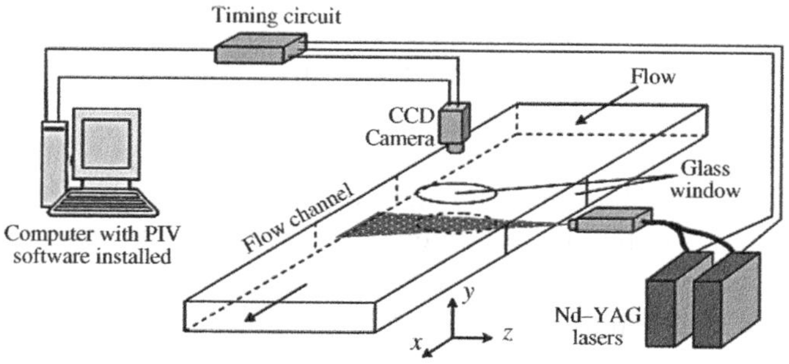

Figure 3.3 Schematic of the components of a conventional 2D PIV system

drag-reducing flows, and so the real 3D PIV technique will not be mentioned herein; Figure 3.3 shows the layout of a typical 2D PIV system). If the measured velocity vector at each point includes only two velocity components within the illuminated 2D plane, this kind of PIV technique is just the conventional 2D two-component (2D-2C) PIV; if the measured velocity vector at each point not only includes the two components within the illuminated 2D plane, but also includes the other component normal to the illuminated plane, then this kind of PIV technique is named 2D-3C PIV or the stereoscopic PIV (SPIV). In particular, note here that some researchers said that the measured results by SPIV were a 3D velocity field, or they named SPIV as 3D-PIV. In the strict sense, such a statement was wrong because the distribution of these three-component velocity vectors is still restricted within a 2D plane, which does not have real 3D characteristics, for example, we have no way to obtain the velocity gradient normal to this plane. Reports of the application of PIV technique to the experimental investigations on turbulent drag-reducing flow by additives was started around the beginning of the twenty-first century. Until now, there have been tens of typical papers associated with PIV measurement results about turbulent drag-reducing flows by additives published in international journals [17–31]. These publications have provided fruitful information on the statistical and structural characteristics of turbulent drag-reducing flows and on understanding the mechanism of turbulent DR.

A standard 2D digital PIV setup is shown in Figure 3.3. The measurement target is the tracing-particle seeded flow in a transparent test section. The principal components of a PIV system include two pulsed laser optics for generating a pair of pulsed laser sheets, which are for illuminating the flow field to be measured; a CCD camera for imaging the particles in the laser sheet-illuminated area of interest; a computer installed with PIV system software for controlling the operation of the PIV system and storing data; a synchronization device for generating signal pulses to control the double-pulsed laser emitting and for communicating with the camera and computer; and PIV software for setting up and controlling the major components, and for analyzing the images with tracing particles to derive vector representations of the measured flow field. For working principles, categorizations, application examples, and other detailed information about PIV system, one can refer to reference [32].

When analyzing a PIV-measured velocity field, except that the statistical characteristics of a turbulent velocity field can be obtained, such as the distributions of mean velocity, fluctuation intensities of different velocity components, Reynolds shear stress, and so on, some typical structural characteristics of the turbulent velocity field can also be obtained through spatial correlation analyses of instantaneous velocity vectors between different points in the measured 2D plane. For example, for the wall turbulence, if the measured 2D plane is close to the wall and parallel to the wall surface, the quantitative information of spatial intervals in the spanwise direction between the near-wall low-speed streaks can be attained by spatial correlation analysis of velocity vectors between different points in the spanwise direction. Such information reflects strength and occurrence frequency of the near-wall coherent structures and turbulent bursting events in wall turbulence [25]. On the other hand, for the PIV-measured velocity fields, some other analyses can also be performed, such as wavenumber-based spectrum analysis for the fluctuating velocity component in one direction, conditional resampling analysis in order to gain the vortex structure and its strength at the location where turbulent bursting event happens, utilizing specific turbulent vortex analytical methods (e.g., proper orthogonal decomposition, or POD) [30] to extract the characteristics of the vortex, and so on. From these different methods, very fruitful information about the characteristics of turbulent flows can be elaborated. Analyses of the differences and equivalences between the turbulent drag-reducing flow of additives solution and the Newtonian fluid turbulent flow by comparison are actually of great help in deeply understanding the mechanisms of turbulent DR.

3.2 Statistical Characteristics of Velocity and Temperature Fields in Drag-reducing Flow

Hereinafter, taking the channel flow as an example, the statistical characteristics of velocity and temperature fields in turbulent drag-reducing flows of surfactant solution are elucidated. For the other wall-bounded turbulent drag-reducing flows, such as pipe flows and boundary layer flows, there are similar statistical characteristics. The velocity and temperature fields shown below are mainly based on the simultaneous measurements by means of LDV coupling with a fine-wire thermocouple for a one-wall-heated turbulent drag-reducing channel flow of surfactant solution. The experimental system for this measurement is the same as that plotted in Figure 2.2 in Chapter 2. As for the detailed introductions to the experimental set-up and the surfactant drag reducer used, one refers to Section 2.2.1. For other referred experimental data, the publication resource will be indicated additionally. The statistical characteristics of turbulent velocity and temperature fields mainly include the distributions of mean velocity and temperature, fluctuation intensity of instantaneous velocity components in the different directions and fluctuation intensity of instantaneous temperature, correlations between different fluctuating quantities, high-order moments of different fluctuating quantities, and spectrum analyses of the fluctuating quantities.

3.2.1 *Distribution of Averaged Quantities*

For a turbulent channel flow, there is no net flow rate in the wall-normal direction (the y direction of the coordinate as shown in Figure 2.2). The mean velocity in this direction is thus zero. Hence, it needs to analyze only the distribution of mean streamwise (in the x direction) velocity, U, in the y direction. In the cases with heating, it also needs to analyze the distribution of mean temperature (generally, it is the temperature difference, Θ, between the heated wall and local point in the flow) in the direction normal to the heated wall. The mean velocity U and mean temperature difference Θ at a local point are obtained through ensemble averaging the sampled instantaneous data at this point, as follows:

$$U = \sum_{i=1}^{N} U_i \Big/ N, \quad \Theta = \sum_{i=1}^{N} \left(T_w - T_f\right)_i \Big/ N \tag{3.1}$$

where U_i is the measured instantaneous streamwise velocity component at the local point of the flow field, T_w is the temperature of the heated wall, T_f is the instantaneous temperature at the local point, and N is the sampling space.

3.2.1.1 Distribution of Mean Velocity

Figure 3.4 gives the comparison of mean velocity distributions between turbulent drag-reducing flow of surfactant solution and turbulent water flows. The abscissa is the distance away from the solid wall and is normalized with the half-height of the channel. The streamwise mean velocity of the ordinate is nondimensionalized with the average velocity within the cross-section of channel. This kind of nondimensionalization is also named "data analysis based on the outer scale". The two cases for water flow shown in Figure 3.4, at $\mathrm{Re} = 17700$ and $\mathrm{Re} = 704$, represent the typical mean velocity distribution in turbulent and laminar flows of the Newtonian fluid, respectively. For turbulent water flow, the mean velocity distribution complies with the $1/n$ power law, with the power exponent $1/n$ of 1/6 or 1/7; while for laminar water flow, the mean velocity shows parabolic distribution in the y direction, and the velocity at the center point of the channel is 1.5 times the average velocity within

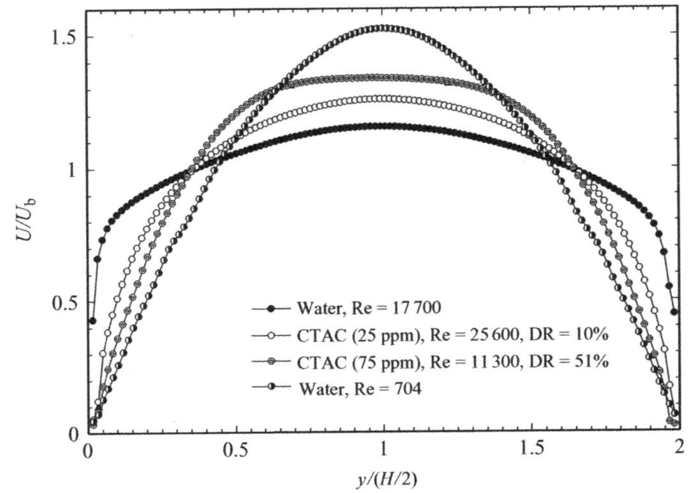

Figure 3.4 Mean velocity distribution in a channel flow based on an outer scale

the cross-section of the channel. For the turbulent drag-reducing surfactant solution flow, however, its mean velocity distribution is in between the $1/n$ power law of turbulent water flow and the parabolic distribution of laminar water flow. In the close vicinity of wall surface, the velocity gradient of drag-reduced flow is smaller than that of turbulent water flow and larger than that of laminar flow, while it is opposite in the bulk region and the velocity level at the center point is also in between the water turbulent flow and laminar flow. From the comparison between the two cases for drag-reducing flow with different DR rates as shown in Figure 3.4, it can be seen that, with increase of DR, the above-mentioned distribution of mean velocity in drag-reduced flow becomes further away from that of turbulent water flow and closer toward that of laminar flow. It can be known from other statistical information as provided later that the turbulent drag-reducing flow by additives is actually a flow state in between laminar flow and fully turbulent flow. But characteristics of drag-reducing flow are more like those of the normal turbulent flow of the Newtonian fluid, and so the addition of drag reducer cannot change the turbulent flow state to a laminar one.

When analyzing the characteristics of velocity field in wall turbulence, additional cases are to provide the distributing features of some statistical quantities based on the inner scale, since with this kind of expression the mean velocity profile within the near-wall inner layer of wall turbulence in the Newtonian fluid flow satisfies with the Prandtl law-of-the-wall, that is, the relationship between the streamwise mean velocity and the wall-normal distance within the linear substrate layer and log-law layer in the near-wall region of wall turbulence can be written as relatively regular functions. The term "based on the inner scale" represents the nondimensionalization method using the friction velocity and the wall-normal distance. Accordingly, the dimensionless quantities based on the friction velocity are designated with the superscript "+." The friction velocity has been defined in Chapter 2: $u_\tau = \sqrt{\tau_W/\rho}$. The dimensionless velocity and wall-normal distance based on inner scale read as

$$U^+ = U/u_\tau, \quad y^+ = yu_\tau/v \tag{3.2}$$

where$v$ is the kinematic viscosity of solvent. Since in the wall turbulence, the average size of the turbulent vortex is decreasing with a decrease of the distance away from the wall, the distance y can also be looked upon as the length scale of the turbulent vortex. Consequently, y^+ is the typical Reynolds number of the turbulent vortex at location y away from the wall. For the wall-bounded turbulent flow of the Newtonian fluid, the near-wall velocity distribution obeys,

$$\text{Viscous substrate layer}: \quad U^+ = y^+, \quad (0 < y^+ < 5) \tag{3.3}$$

$$\text{Log-law layer}: \quad U^+ = 2.5\ln y^+ + 5.5, \quad (y^+ > 40) \tag{3.4}$$

For the turbulent drag-reducing flow by additives, the velocity profile in the viscous substrate layer is still linearly distributed, as Equation 3.3 shows. The different point is that the effective y^+ region for linear velocity distribution is significantly larger for drag-reducing flow as compared with its Newtonian counterpart. This is often referred to as the thickening phenomenon of the linear substrate layer in turbulent drag-reducing flow. In the log-law layer, the velocity distribution in turbulent drag-reducing flow changes a lot. With an increase of DR, U^+ is increased in this region and its distribution departs more from that expressed by Equation 3.4. But no matter how DR is increased, the increase of U^+ in the log-law layer has a limitation. Virk summarized a large amount of experimental data for turbulent drag-reducing polymer solution flows and proposed the following ultimate curve for velocity distribution [33]:

$$U^+ = 11.7\ln y^+ - 17.0 \tag{3.5}$$

Although Equation 3.5 was empirically proposed based on turbulent drag-reducing flows of polymer solutions, it is still applicable to the overwhelming majority of the reported turbulent drag-reducing surfactant solution flows.

Figure 3.5 gives the typical mean streamwise velocity profiles normalized with the inner scale for the Newtonian fluid turbulent flow and turbulent drag-reducing flow of surfactant solution. It is shown that the three-layer (linear substrate layer, buffer layer, and log-law layer) model of velocity distribution for the Prandtl turbulent boundary layer flow agrees with the experimental results very well for the Newtonian fluid wall turbulence, whereas for the turbulent drag-reducing surfactant solution flow, although

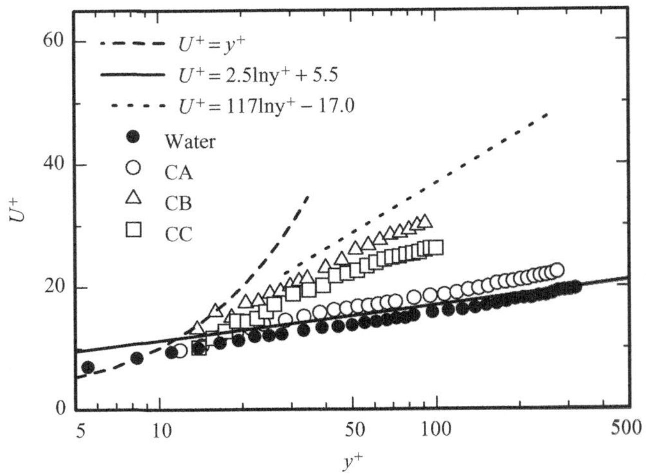

Figure 3.5 Mean velocity distribution in a channel flow based on an inner scale [15]. The parameters shown in the figure are as follows: water, Re = 25 000; CA, Re = 35 000, DR = 33.0%; CB, Re = 25 000, DR = 70.0%; and CC, Re = 15 000, DR = 65.1% (Reprinted with permission from Li, F.-C., Kawaguchi, Y., and Hishida, K., "Investigation on the characteristics of turbulence transport for momentum and heat in a drag-reducing surfactant solution flow," *Physics of Fluids* **16**, 3281–3295, 2004. © 2004 American Institute of Physics.)

there still exists the analogous three-layer structure of velocity distribution, as already described above, the viscous substrate layer becomes thicker, and the buffer layer and log-law layer move away from the wall obviously. The characteristics of velocity distribution of drag-reducing flow shown in Figure 3.5 are typical for turbulent drag-reducing flow by additives.

3.2.1.2 Distribution of Mean Temperature

Accompanying turbulent DR, the convective heat transfer ability is also greatly weakened. From the statistics of a turbulent flow field, the variations appearing in mean velocity distribution (and the other statistics of turbulent flow field, as elucidated later on) are bound to change the temperature distribution in the flow field. This phenomenon and its related occurrence mechanisms have been investigated in detail through elaborately designed experiments [14,15,34,35]. In order to make a comparison with mean velocity profiles, the distributive characteristics of mean temperature Θ with y can also be expressed with a dimensionless form normalized by inner scales, that is, nondimensionalization by friction temperature. The definition of friction temperature has been provided in Chapter 2: $T_\tau = \frac{q_W}{\rho c_p u_\tau}$. Thus, mean temperature can be nondimensionalized as

$$\Theta^+ = \Theta/T_\tau \tag{3.6}$$

The temperature profile for the Newtonian fluid flow in a channel or pipe was empirically compiled by Kader as follows [36]:

$$\Theta^+ = \mathrm{Pr}\cdot y^+ \exp(-\Gamma) + \left\{2.12\ln\left[(1+y^+)\cdot\frac{1.5(2-2y/H)}{1+2(1-2y/H)^2}\right] + \beta(\mathrm{Pr})\right\}\exp\left(-\frac{1}{\Gamma}\right) \tag{3.7}$$

where $\Gamma = 10^{-2}(\mathrm{Pr}\cdot y^+)^4/(1+5\mathrm{Pr}^3 y^+)$, $\beta(\mathrm{Pr}) = (3.85\mathrm{Pr}^{1/3} - 1.3)^2 + 2.12\ln(\mathrm{Pr})$, and Pr is the molecular Prandtl number. The applicable range of this equation covers the whole thermal boundary layer. On the other hand, for laminar flow, the mean temperature profile complies with the following theoretical equation:

$$\Theta^+ = \mathrm{Pr}\cdot y^+ \tag{3.8}$$

The profiles of the measured mean temperature nondimensionalized with inner scales for both water and turbulent drag-reducing surfactant solution flows are plotted in Figure 3.6. Kader's equation (Equation 3.7) and laminar flow equation (Equation 3.8) are also plotted in the figure for comparison. From Figure 3.6, it can be seen that the mean temperature profile for water flow agrees well with Equation 3.7. For the turbulent drag-reducing surfactant solution flow, however, the mean temperature profiles are significantly different from that of water. For the cases with relatively larger DR (CB and CC), it is obviously seen that a large temperature gradient exists at

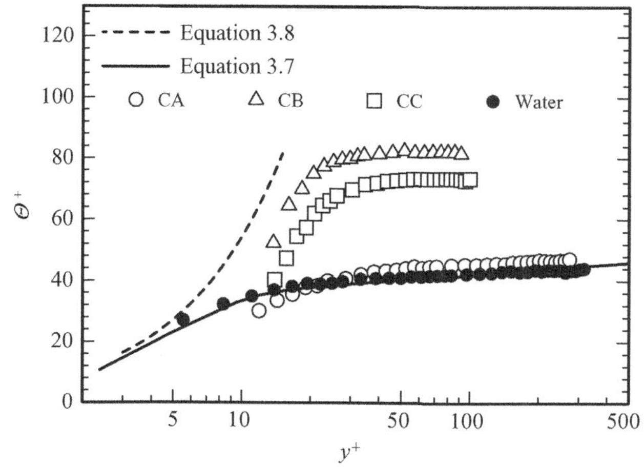

Figure 3.6 Mean temperature distributions for channel flows based on inner scales [15]. The parameters shown in the figure are as follows: water, Re = 25 000; CA, Re = 35 000, DR = 33.0%; CB, Re = 25 000, DR = 70.0%; and CC, Re = 15 000, DR = 65.1% (Reprinted with permission from Li, F.-C., Kawaguchi, Y., and Hishida, K., "Investigation on the characteristics of turbulence transport for momentum and heat in a drag-reducing surfactant solution flow," *Physics of Fluids* **16**, 3281–3295, 2004. © 2004 American Institute of Physics.)

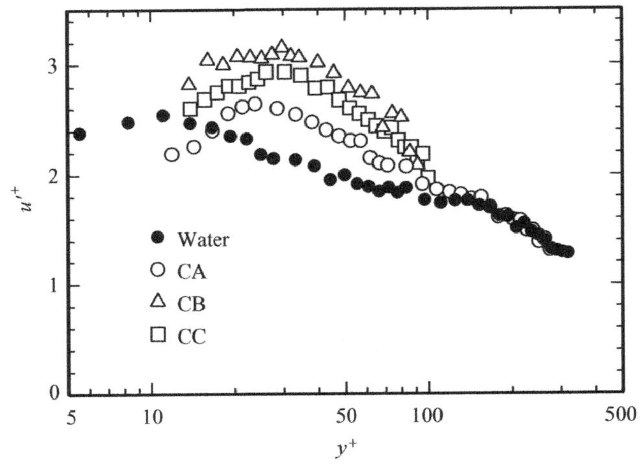

Figure 3.7 Turbulent intensity profiles for the streamwise velocity component nondimensionalized with inner scales [15]. The running parameters shown in the figure are as follows: water, Re = 25 000; CA, Re = 35 000, DR = 33.0%; CB, Re = 25 000, DR = 70.0%; and CC, Re = 15 000, DR = 65.1% (Reprinted with permission from Li, F.-C., Kawaguchi, Y., and Hishida, K., "Investigation on the characteristics of turbulence transport for momentum and heat in a drag-reducing surfactant solution flow," *Physics of Fluids* **16**, 3281–3295, 2004. © 2004 American Institute of Physics.)

the near-wall region ($y^+ < 50$ in the figure). This phenomenon was also observed in the experiments of turbulent drag-reducing pipe flows [34,35]. Gasljevic *et al.* [34] named this high-temperature-gradient layer as the elastic layer in order to make an analogy to the elastic layer of the velocity profile in Virk's three-layer model. In the outer region away from the heated wall, the gradient of temperature profiles becomes very small in drag-reducing flows. From the characteristics of mean temperature distribution, it is again manifested that the turbulent drag-reducing flow is actually a flow state in between the laminar and turbulent flows of the Newtonian fluid.

3.2.2 Distribution of Fluctuation Intensities

For a time-dependent fluctuating quantity, the root-mean-square value is usually statistically calculated to be its fluctuation intensity. In turbulent drag-reducing surfactant solution flows, the fluctuation intensity of every fluctuating quantity (every fluctuating velocity component and fluctuating temperature) changes obviously to a different extent as compared with a fully turbulent flow of Newtonian fluid. Figures 3.7–3.9 provide the fluctuation intensity profiles for the streamwise and wall-normal velocity components and temperature with nondimensionalization by inner scales.

Figure 3.7 shows the typical characteristics of turbulent intensity profiles of the streamwise velocity component for the turbulent drag-reducing surfactant solution flow, that is, the location of the maximum value of u'^+ shifts further away from the

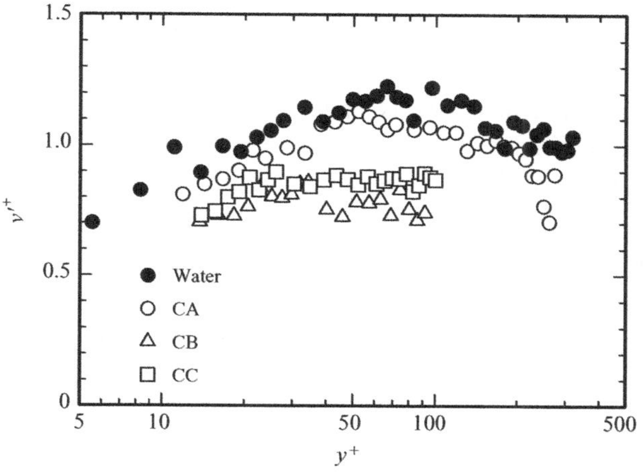

Figure 3.8 Turbulent intensity profiles for the wall-normal velocity component nondimensionalized with inner scales [15]. The running parameters shown in the figure are as follows: water, Re = 25 000; CA, Re = 35 000, DR = 33.0%; CB, Re = 25 000, DR = 70.0%; and CC, Re = 15 000, DR = 65.1% (Reprinted with permission from Li, F.-C., Kawaguchi, Y., and Hishida, K., "Investigation on the characteristics of turbulence transport for momentum and heat in a drag-reducing surfactant solution flow," *Physics of Fluids* **16**, 3281–3295, 2004. © 2004 American Institute of Physics.)

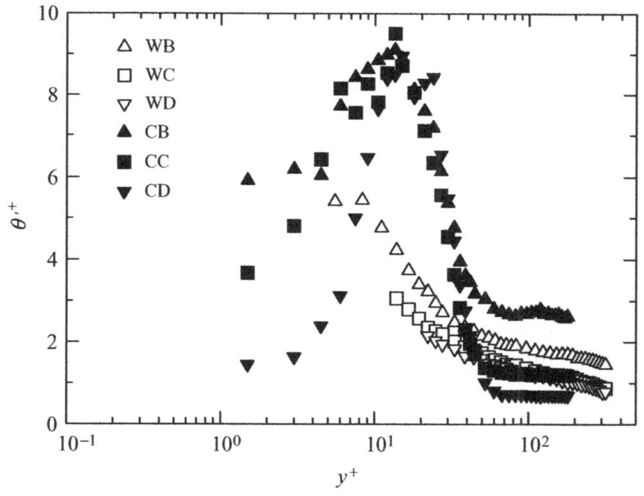

Figure 3.9 Temperature fluctuation intensity profiles nondimensionalized with inner scales [14]. The Reynolds number for all the flow cases shown in this figure is the same: Re = 25 000. For other parameters: WB, WC, and WD are turbulent water flows, and their friction temperatures are 0.023°C, 0.052°C, and 0.067°C, respectively; and CB, CC, and CD are drag-reducing surfactant solution flows with DR = 69.9%, and their friction temperatures are 0.011°C, 0.024°C, and 0.041°C, and HTR = 69.4%, 68.6%, and 69.5%, respectively (Reprinted with permission from Li, F.-C., Kawaguchi, Y., and Hishida, K., "Investigation on the characteristics of turbulence transport for momentum and heat in a drag-reducing surfactant solution flow," *Physics of Fluids* **16**, 3281–3295, 2004. © 2004 American Institute of Physics.)

wall as compared with turbulent water flow (as shown in Figure 3.7, from $y^+ = 12$ of turbulent water flow to $y^+ = 23 \sim 30$ of the turbulent drag-reducing flow case reported in [15]) and the peak levels of u'^+ profiles become larger. This indicates the thickening of a buffer layer in the turbulent boundary layer. This is a repeatedly observed phenomenon in turbulent drag-reducing flows by additives. Some researchers also named it "enhancement" of the streamwise turbulent intensity in drag-reducing flows. However, after detailed investigations, it is found that if the fluctuation intensity of u is not normalized with u_τ but with the average velocity within the cross-section of flow channel, U_b, then, in almost all the reported experimental results about turbulent drag-reducing flows by additives, the fluctuation intensity of the streamwise velocity component was smaller than, or at least at a similar level as, that in turbulent water flow. It testifies that, when transporting different liquids with the same flow rate, compensation for turbulent intensity in the streamwise direction is also reduced for turbulent drag-reducing flows. Thereafter, the sayings about enhancement of the streamwise turbulent intensity in drag-reducing flows are actually not accurate.

As shown in Figure 3.8, the turbulent intensity of the wall-normal velocity component v'^+ for turbulent drag-reducing flows is significantly depressed as compared with the Newtonian fluid case, and the depression becomes increasingly serious with the increase of DR level. This is the consequent phenomenon accompanying the occurrence of turbulent DR. Or, conversely to say, depression of the wall-normal velocity fluctuation intensity directly induces the weakening of momentum transport between the fluid and the wall, which is one of the reasons for turbulent DR.

Figure 3.9 shows the temperature fluctuation intensity profiles measured by fine-wire thermocouples. Since in the experimental study shown in Figure 3.9, not only a fine-wire thermocouple was used for temperature measurement but also LDV was used for velocity measurement simultaneously in order to obtain the coupling information of temperature and velocity fluctuations, the measurable distance approaching the wall was limited by the "acceptable measuring distance" of LDV. (Due to heating, a relatively strong convection was generated near the heated wall, which makes the refractive index of fluid in the proximity of the heated wall inhomogeneous. This influences the coherence between the two laser beams. The LDV measurement is thus constrained locally. With increase of the wall-heating flux, this influencing area becomes larger.) This prevents the temperature measurement for several turbulent water flow cases from approaching very near to the heated wall surface. As shown in Figure 3.9, the profiles of temperature fluctuation intensity θ'^+ do not show peaky structure at small y for the two water flow cases WC and WD. It can be seen that, similar to the velocity fluctuation intensity profiles in the streamwise direction, the value of the maximum θ'^+ is larger for the turbulent drag-reducing flows and the location of the maximum value shifts away from the heated wall surface. Meanwhile, this phenomenon manifests that the buffer layer in the thermal boundary layer of turbulent drag-reducing flow is thickened. This is one of the reasons causing HTR.

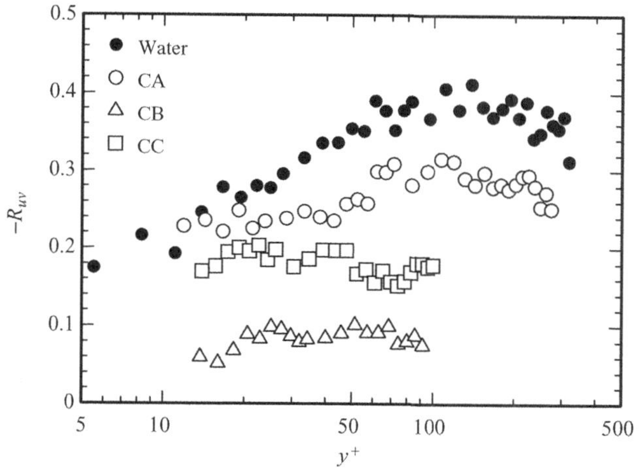

Figure 3.10 Distributions of a cross-correlation coefficient between u and v [15]. The case parameters shown in this figure are as follows: water, Re = 25 000; CA, Re = 35 000, DR = 33.0%; CB, Re = 25 000, DR = 70.0%; and CC, Re = 15 000, DR = 65.1% (Reprinted with permission from Li, F.-C., Kawaguchi, Y., and Hishida, K., "Investigation on the characteristics of turbulence transport for momentum and heat in a drag-reducing surfactant solution flow," *Physics of Fluids* **16**, 3281–3295, 2004. © 2004 American Institute of Physics.)

3.2.3 Correlation Analyses of Fluctuating Quantities

For drag-reducing surfactant solution flows, the occurrence of DR and HTR is caused by the weakening of momentum and heat transports between fluid flow and a solid wall. The main physical quantity responsible for accomplishing the momentum and heat transports is the fluctuation intensity of the wall-normal velocity component. As summarized previously, v' is greatly decreased through the whole profile. This actually reflects one aspect only for the occurrence of DR and HTR. If considering the turbulent momentum flux and turbulent heat flux normal to the wall transported by the fluctuating velocity component v, the quantities that influence its transfer rates are directly related to the correlations between v and u, and between v and θ, respectively.

Figures 3.10 and 3.11 plot the distributions of cross-correlation coefficients between (1) v and u, R_{uv} and (2) v and θ, $R_{v\theta}$, respectively. In wall-bounded turbulent flows, it is always negatively correlated between v and u, and between v and θ, in a statistically averaged sense. For turbulent drag-reducing surfactant solution flows, a salient feature is that the correlations between v and its transported two fluctuating quantities are all greatly depressed, and the depression is more pronounced with increase of DR and HTR. On the other hand, the profiles of R_{uv} and $R_{v\theta}$ are respectively very alike for all the cases, indicating that the turbulent momentum and heat-transporting processes are analogous to each other. This also qualitatively reveals that the surfactant drag reducer introduces similar effects to both the turbulent momentum transport and turbulent heat transport processes.

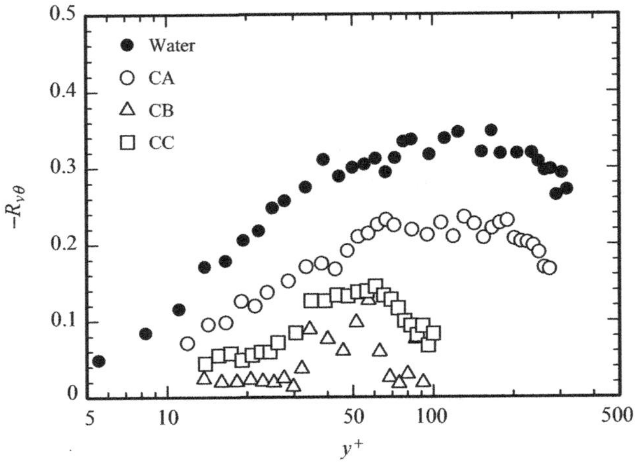

Figure 3.11 Distributions of a cross-correlation coefficient between v and θ [15]. The case parameters shown in this figure are as follows: water, Re = 25 000; CA, Re = 35 000, DR = 33.0%; CB, Re = 25 000, DR = 70.0%; and CC, Re = 15 000, DR = 65.1% (Reprinted with permission from Li, F.-C., Kawaguchi, Y., and Hishida, K., "Investigation on the characteristics of turbulence transport for momentum and heat in a drag-reducing surfactant solution flow," *Physics of Fluids* **16**, 3281–3295, 2004. © 2004 American Institute of Physics.)

3.2.4 Spectrum Analyses of Fluctuating Quantities

From the microscaled viewpoint of the inner flow structures, how turbulent flow differs from laminar flow is that turbulent vortices with different scales distribute in the whole flow space. The scale here not only refers to the spatial scale (reciprocal of wavenumber) but also includes the temporal scale (reciprocal of frequency). In turbulent flows, the vortex scale may have a very wide range of variations, and different leveled turbulent kinetic energy distributes locally among different scaled vortices; the turbulent kinetic energy then successively transfers from larger scaled vortices to smaller ones, and is finally dissipated at the smallest scaled (i.e., at the dissipation scale or Kolmogorov scale) vortex due to viscosity. In order to obtain the turbulent kinetic energy distributions at different vortex scales, it needs to carry out spectrum analysis for the turbulent velocity field. Through analysis of the differences between the power spectrum of velocity fluctuations in turbulent drag-reducing surfactant solution flow and turbulent flow of the Newtonian fluid, the influencing mechanism of drag reducer on the characteristics of inner microscaled turbulent vortices and turbulent kinetic energy transport might be derived straightforwardly.

Figure 3.12 shows the 1D power spectrum of the streamwise fluctuating velocity component at one point in both turbulent water flow and turbulent drag-reducing flow of surfactant solution, which was measured by Warholic *et al.* by means of LDV [37]. From the power spectrum plot, it can be seen that the following salient features appeared in drag-reducing surfactant solution flow: as compared with the spectrum distribution of

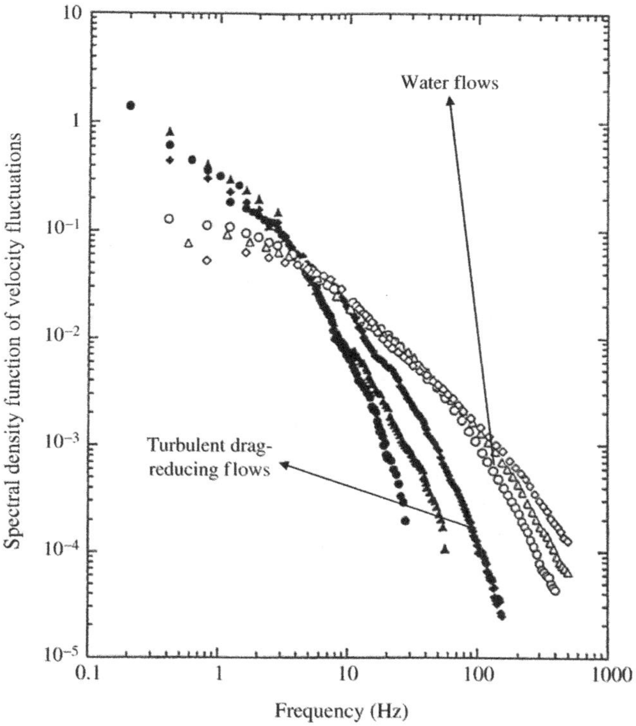

Figure 3.12 Power spectrum of the LDV measured velocity fluctuations in the streamwise direction [37]. W_u is the power spectrum of the streamwise fluctuating velocity component u, $\overline{u^2}$ is the square of fluctuation intensity of u, and H is half of the channel height (Reproduced with permission from Warholic, M.D., Schmidt, G.M. and Hanratty, T.J., "The influence of a drag-reducing surfactant on a turbulent velocity field," *Journal of Fluid Mechanics* **388**, 1–20, 1999, Cambridge University Press.

velocity fluctuations in the Newtonian fluid turbulent flow, the spectrum level is increased in the low-frequency range and decreased in the high-frequency range. This phenomenon is stressed in reference [6] as the redistribution of turbulent kinetic energy among the turbulent vortices in turbulent drag-reducing flows. The low-frequency range corresponds to vortices at larger temporal scales, and on the contrary the high-frequency range corresponds to vortices at smaller temporal scales. Meanwhile, the temporal scale and spatial scale are corresponding to each other, that is, the turbulent vortex at large temporal scale also has large spatial scale. Therefore, the variation appearing in the spectrum distribution for turbulent drag-reducing flow indicates that the turbulent kinetic energy is increased for vortices at a larger scale and decreased for vortices at a smaller scale. Particularly, due to the suppression effect by drag reducer, the small vortices around the dissipation scale in the Newtonian fluid turbulent flow disappeared, which changes the cascading process of turbulent kinetic energy transfer in the original turbulent flow. This phenomenon has been a consensus

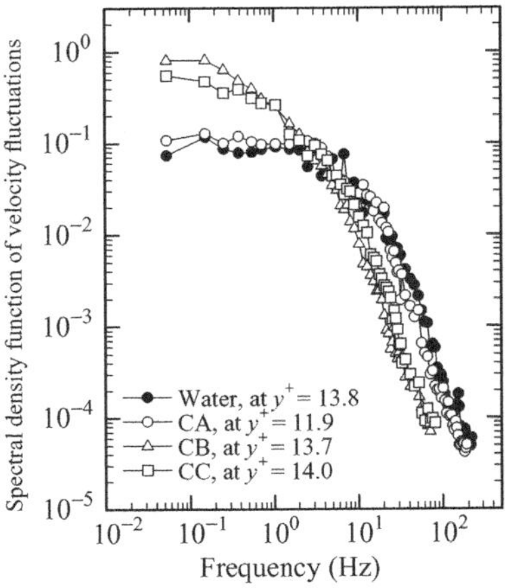

Figure 3.13 Power spectra of temperature fluctuations [15]. The case parameters are as follows: water, Re = 25 000; CA, Re = 35 000, DR = 33.0%; CB, Re = 25 000, DR = 70.0%; and CC, Re = 15 000, DR = 65.1% (Reprinted with permission from Li, F.-C., Kawaguchi, Y., and Hishida, K., "Investigation on the characteristics of turbulence transport for momentum and heat in a drag-reducing surfactant solution flow," *Physics of Fluids* **16**, 3281–3295, 2004. © 2004 American Institute of Physics.)

described in all the experimental studies measuring the fluctuating velocity field in turbulent drag-reducing flow by additives.

The features appearing in the power spectrum of velocity fluctuations in drag-reducing surfactant solution flows also emerge in the power spectrum of its temperature fluctuations, as shown in Figure. 3.13. In Figure 3.13, it shows spectra of temperature fluctuations at a point located in the buffer layer ($y^+ = 11.9 \sim 14.0$) of drag-reducing surfactant flow and turbulent water flow in a channel. It can be seen that at $F < 4$ Hz, the value of the power spectrum of temperature fluctuations is increased in drag-reducing flow as compared with turbulent water flow, indicating the increase of turbulent kinetic energy at large scales, whereas after about $F > 5$ Hz, the spectrum in drag-reducing flow is decreased, indicating the decrease of turbulent kinetic energy at small scales. Moreover, the truncated frequency of power spectrum curve is always small for the drag-reducing flow as compared with turbulent water flow, manifesting the disappearance of small turbulent vortices from some small scale.

However, it is worth specifically noting that, although it has been observed that both depression of turbulent vortices at high frequencies or at small scales and increase of the turbulent kinetic energy at low frequencies or large scales co-occur in a drag-reducing surfactant solution flow from the power spectra of streamwise velocity fluctuations and temperature fluctuations, it can be concluded here that this phenomenon is only associated with the DR phenomenon caused by polymeric and surfactant additives, but not with all kinds of turbulent drag-reducing flows. For example, reference [38] reported an experimental investigation of turbulent DR phenomenon in a wind tunnel with wall oscillation in the spanwise direction. From the estimated power spectra of streamwise velocity fluctuations, it was found that the turbulent kinetic energy was dramatically reduced at low frequencies, whereas the energy at higher frequencies was increased. This indicates a transfer of turbulent kinetic energy from large-scale to small-scale turbulent vortices. This phenomenon is right opposite to the transfer of turbulent energy from small-scale to large-scale turbulent vortices in drag-reducing flows by additives.

As mentioned previously, the turbulent velocity field measured with PIV may not have frequency as high as that measured with LDV. (Nowadays, the PIV technique has been developed to be able to realize high-frequency data sampling, that is, the so-called dynamic PIV or high-speed PIV. However, no high-speed PIV has been used for an experimental study on turbulent drag-reducing flows by additives yet.) For a conventional PIV system, its sampling frequency (here, it refers to the frequency of the PIV velocity field obtained from analyses of images) is normally 4 or 8 Hz. But most advantageous for the PIV technique is its capability of obtaining the distribution of a two-component velocity field in a plane at the same time. Within the permission of memory space of the data acquisition system, a PIV system can acquire multiple turbulent velocity fields (the number can reach hundreds or even thousands according to the requirement) at different instances within the same one plane. This enables the estimation of a wavenumber-based power spectrum of velocity fluctuations in the measured plane through statistical analysis.

Figure 3.14 provides the estimated wavenumber-based power spectra plots of velocity fluctuations in two directions, u and v, which are ensembles of spectra of 500 frames of PIV-measured velocity fields in the same x–y plane taken from a channel flow. This figure is plotted as a 2D fashion, which covers the wavenumber-based power spectra at all different locations away from the wall within half-channel height. In this way, the effects of drag-reducing surfactant additives on the power spectra of velocity fluctuations with respect to wavenumber can obviously be seen from the following two aspects: (1) how the turbulent kinetic energy can be redistributed through the wavenumbers, and (2) how much the additives affect the power spectra at different flow regions with respect to the location away from the wall surface. The two drag-reducing flow cases shown in Figure 3.14 correspond to the two drag-reducing characteristics regimes, Regime II and Regime III, that is, the developing and decreasing regimes for drag-reducing ability, respectively, as plotted in Figure 2.3 of Chapter 2. On the other hand, only one case of turbulent water flow is given here for comparison, since the characteristics of wavenumber-based spectra are similar for those of water flow at different Reynolds numbers.

Through comparison of turbulent water flow and turbulent drag-reducing surfactant solution flow, it can be seen that the important contributions to the power spectra of velocity fluctuations are shrunk to smaller wavenumbers (a small threshold value for a spectrum has been chosen below which the wavenumber is truncated) for the drag-reducing flow. It indicates that the addition of surfactant drag reducer increases the

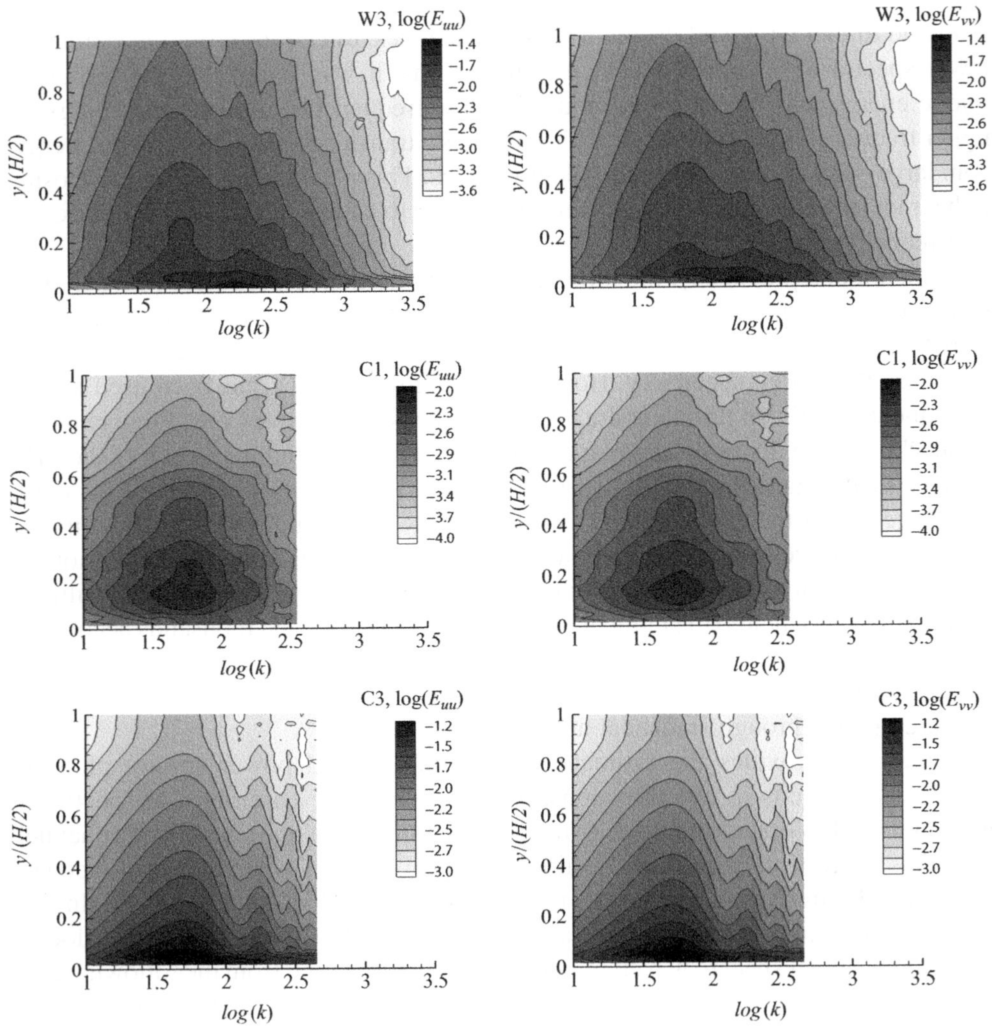

Figure 3.14 Distribution of wavenumber-based spectra of velocity fluctuations in the PIV-measured velocity fields [24]. Function designates the wavenumber-based spectrum of velocity fluctuation *u*. The case parameters are as follows: W3, water, Re = 20 100; C1, surfactant solution, Re = 6500, DR = 49%; and C3, surfactant solution, Re = 18 200, DR = 43% (Reprinted with permission from Li, F.-C., Kawaguchi, Y., Segawa, T. and Hishida, K., "Reynolds-number dependence of turbulence structures in a drag-reducing surfactant solution channel flow investigated by PIV," *Physics of Fluids* **17**, 075104, 2005. © 2005 American Institute of Physics.)

spectra at small wavenumbers characterizing large turbulent vortices, whereas those at large wavenumbers characterizing small vortices are decreased or negligible, or, in other words, the small vortices corresponding to large wavenumbers have disappeared due to the suppression by drag reducer. On the other hand, the 2D contour maps of wavenumber-based spectra show that drag-reducing additives also shrink the contour area with relative large power spectra value toward the walls, indicating that the center region in drag-reducing surfactant solution flow with a large DR rate is less important in turbulence production (and, consequently, the frictional drag) as compared with the water flow. Moreover, from comparison, it can be observed that the contour maps of power spectra in Regime II and Regime III are also different. Although the DR levels for the two cases are similar, in Regime II, the most important flow layer in turbulence production is apart from the wall, but in Regime III, it is much closer to the wall similar to the water flow (see the contour darkness and the shape of contour lines of the two flows). This implies that, when the drag-reducing ability of surfactant solution flow decreases (after the critical Reynolds number Re_C, shown in Figure 2.3 of Chapter 2), the drag-reducing surfactant solution flow may have exhibited water-like behavior in the proximity of the wall at which a large shear rate exists, regardless of the relatively large DR level. Many detailed discussions on this Reynolds-number dependent phenomenon have been made in reference [24].

3.3 Characteristics of Turbulent Vortex Structures in Drag-reducing Flow

In wall turbulence, the characteristics of turbulent flow are influenced mainly by two factors, that is, the turbulent bursting events and near-wall low-speed streaks, both of which are closely associated with the turbulent coherent structures. Therefore, in order to reduce the frictional drag for wall-bounded turbulent flow, the most effective approach, eventually, is to suppress the process of the turbulent bursting event, inducing the modification of turbulence coherent structures and near-wall low-speed streaks. Quantitatively, this results in a decrease of Reynolds shear stress and turbulent DR (the quantitative relationship between the Reynolds shear stress and friction factor will be discussed in the next section). Hence, through detailed investigation of the structural characteristics of turbulent vortices in drag-reducing surfactant solution flows, researchers can gain more intuitive help in fully understanding the occurrence mechanism of turbulent DR. Since the vortex structures in turbulence are fully 3D, the velocity vectors distributed in one plane as measured with conventional PIV have not been enough to capture the detailed structural information for 3D vortices. Although SPIV cannot measure fully a 3D turbulent velocity field, as done by the real 3D PIV technique, it can obtain the three velocity components within one plane taken in the flow field simultaneously. When some statistical analysis approaches, such as conditional resampling, supplement the research, this is also of great help in studying the 3D characteristics of turbulent vortex structures. In this section, the structural characteristics of turbulent vortices in drag-reducing flows will be elucidated in detail based on experimental results on turbulent drag-reducing surfactant solution channels by means of SPIV (conventional 2D PIV measurement results are also included, referring to the system shown in Figure 3.3). The applied SPIV system is schematically

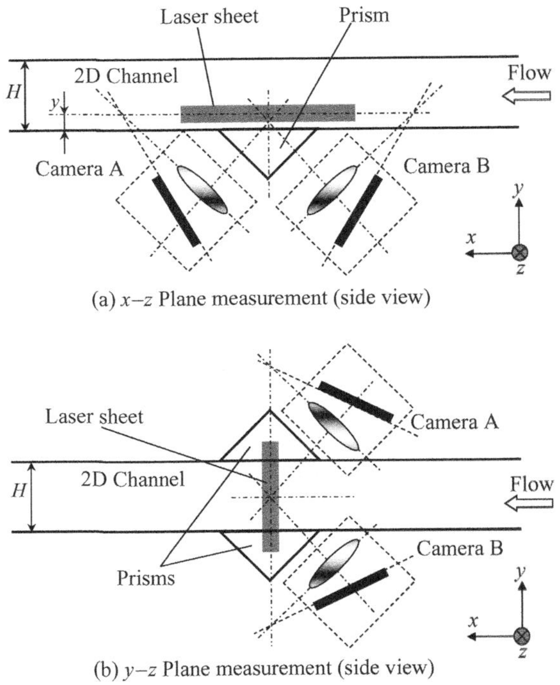

Figure 3.15 Schematic diagram of optical configurations for SPIV measurement, viewing for stereoscopic imaging with the Scheimpflug condition

shown in Figure 3.15. The coordinative system that will be used in the following descriptions is also plotted in this figure.

3.3.1 Identification Method of Turbulent Vortex by Swirling Strength

In order to investigate the structural information of turbulence, it needs to establish an effective approach for identifying the turbulence structure or turbulent vortex at first. Following the definition of Kline and Robinson [39], a vortex is defined as a region of concentrated vorticity around which the pattern of streamlines is roughly circular when viewed in a frame moving with the center of the vortex. The conventional quantity for identifying vortices is vorticity. However, vorticity is contributed not only by the swirling motion of vortex cores but also by any shearing motion present in the flow field. This inevitably confuses the identification of turbulence vortices. In connection with the 3D turbulent velocity field (e.g., a 3D velocity field obtained by direct numerical simulation, or DNS, of turbulent flow), there have been several different approaches for not only extraction of turbulent vortex structures, such as the D-criterion method [40], Q-criterion method [41], and so on. As for the analysis of the 2D velocity vector field in a plane measured with PIV, by utilizing the so-called swirling strength method [42–44], it can effectively identify the locations of vortex cores and their strength. This method is to characterize the location of vortex cores by the imaginary part of the eigenvalue of the local velocity gradient tensor. For a 2D velocity field (u_1, u_2) in a plane $(x_1 - x_2)$ measured by PIV, only the in-plane velocity gradients can be used to form a 2D tensor:

$$A = \begin{bmatrix} \partial u_1/\partial x_1 & \partial u_1/\partial x_2 \\ \partial u_2/\partial x_1 & \partial u_2/\partial x_2 \end{bmatrix} \tag{3.9}$$

The characteristic equation of the tensor A is

$$\lambda^2 - p\lambda + q = 0 \tag{3.10}$$

where $p = \partial u_1/\partial x_1 + \partial u_2/\partial x_2$, and $q = (\partial u_1/\partial x_1)(\partial u_2/\partial x_2) - (\partial u_1/\partial x_2)(\partial u_2/\partial x_1)$. Depending on the signal of the discriminant for this characteristic equation, $\Delta = p^2/4 - q$, Equation 3.10 either has two real eigenvalues ($\Delta > 0$) or has a pair of complex conjugate eigenvalues ($\Delta < 0$), $\lambda_{cr} \pm i\lambda_{ci}$. The magnitude of the imaginary part, λ_{ci}, was defined as the swirling strength of the vortex by Zhou *et al.* [43,44]. Thus, the turbulent vortices are identified by plotting contours of λ_{ci}. Within the contour map, the locations on which λ_{ci} have positive values represent the locations of vortex cores.

3.3.2 *Distribution Characteristics of Turbulent Vortex in the x-y Plane*

Although the turbulent velocity field in one plane taken by 2D PIV is not able to recover the 3D characteristics of turbulent vortex structures perfectly, through analysis of the vortex information in each PIV measurement plane for the drag-reducing surfactant solution flows, the influencing mechanisms of drag reducer on the turbulent vortex structures and its strength can also be explored from different aspects.

Figure 3.16 shows a snapshot velocity field in the x–y plane measured with 2D-2C PIV for both turbulent water flow and turbulent drag-reducing surfactant solution flow. The velocity vectors are composed of (u, v), where $u = U_{i,j} - U_j$ (the instantaneous velocity component in the x direction at a different y location minus its mean velocity at this location) and $v = V_{i,j}$. The contour map of swirling strength is also overlapped on to this figure, so that the cross-sections of the 3D vortical structures in the PIV measurement plane can be figured out from the velocity vectors. The vortex packets formed by a sequence of vortices can be observed from the velocity field and vortex distribution in the x–y plane. Each one in such identified vortex packets is actually the cross-section of the head of a 3D hairpin vortex in the wall-bounded turbulent flow. The heads of the vortex in a packet show a growth angle with respect to the wall surface, within which a triangular region (or ramp-shaped region) is formed. These ramp-shaped regions are the low-momentum regions caused by the penetration of low-speed fluid elements into a high-speed fluid region during the turbulent bursting events (see Figure 3.16).

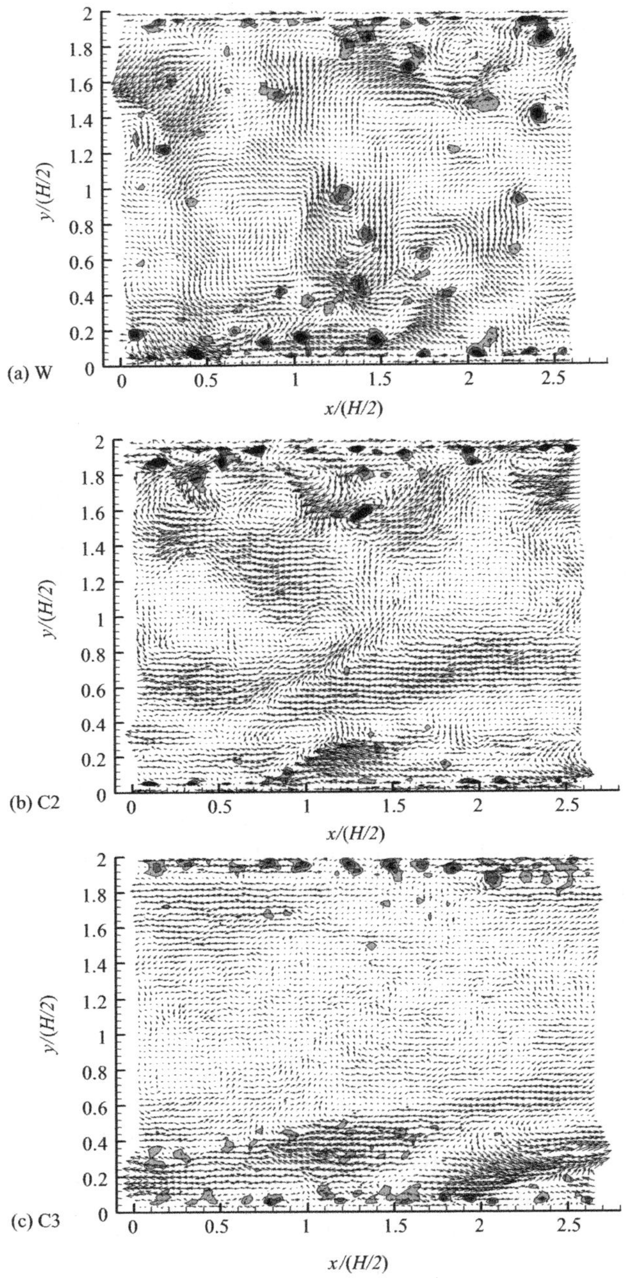

Figure 3.16 Instantaneous velocity field together with the vortex structures in the *x*–*y* plane of turbulent channel flow [27]. The flow is from left to right. The case parameters of this figure are as follows: (a) W, water, Re = 18 000; (b) C2, Re = 21 000, DR = 34%; and (c) C3, Re = 13 000, DR = 58% (Reprinted from *International Journal of Heat and Mass Transfer*, **51**, Li, F.-C., Yu, B., Wei, J.-J., Kawaguchi, Y., Hishida, K., "Experimental study of drag-reduction mechanism for a dilute surfactant solution flow," 835–843, 2008, with permission from Elsevier.)

As seen from Figure 3.16, the structures of vortex packets near the walls are closely associated with turbulent bursting events. A strong Q2 motion (this is the so-called quadrant method for analysis of the turbulent bursting events in wall turbulence, which categorizes the different flow patterns of fluid elements during the bursting events; Q2 motion represents those fluid elements whose velocity fluctuations satisfy $u < 0, v > 0$, that is, when categorizing the fluctuating velocity components at all points in the x–y plane onto a coordinate plane with u as its abscissa and v as its ordinate, the groups of u and v are gathered into the second quadrant) may penetrate into the core region at a high rate. In a DNS study [44], Zhou *et al.* found that the auto-generation process of the primary hairpin vortex forming secondary and tertiary hairpin vortices showed a threshold behavior, that is, only in the case of strong initial vortex structures was the auto-generation process possible. The weak vortices are either convected downstream by themselves or annihilated during convection, forming the ramp-shaped low-momentum region with a gentle angle. Inversely, from this phenomenon, it can be supposed that the inclination angle of the near-wall ramp-shaped low-momentum region (also known as the growth angle of the vortex packet) in wall turbulence may characterize the strength of turbulent bursting events. Compared with turbulent water flow (Figure 3.16a), it can be seen that the near-wall vortex structures near the walls in the drag-reducing surfactant solution flow are changed gradually with DR level: the inclination angle of the ramp-shaped low-momentum regions decreases, and the appearance of coherent vortex structures becomes less frequent.

3.3.3 *Distribution Characteristics of Turbulent Vortex in the y-z Plane*

For any wall-bounded turbulent flows, the velocity component in the streamwise direction (x direction herein) is always overwhelmingly predominant. Therefore, for measuring the distribution of velocity components in the y–z plane, the conventional 2D-2C PIV is not available. The reason is as follows. The time for the tracing particles passing through the laser sheet with limit thickness (order of 1 mm) in the x direction is very short, which requires the time interval between the two PIV images (used for calculating one velocity frame) to be short enough, otherwise the two sampled pictures with particle images may lose their correlation. On the other hand, since the statistical averaged value of the velocity components in the y and z directions are all 0 in wall flow, the time interval between the two PIV images should be long enough in order to obtain large enough relative displacement of the tracing particles in the two successive pictures. It is just this controversy that constrains the conventional 2D-2C PIV to be unable to measure the velocity components in the y–z plane. Thereafter, it is necessary to use the stereoscopic 2D-3C PIV technique for the measurement of the three velocity components in the y–z plane.

Figure 3.17 shows an overview of a 3D image of turbulent velocity fields measured in the y–z plane for water and CTAC solution flows, respectively. It shows that the

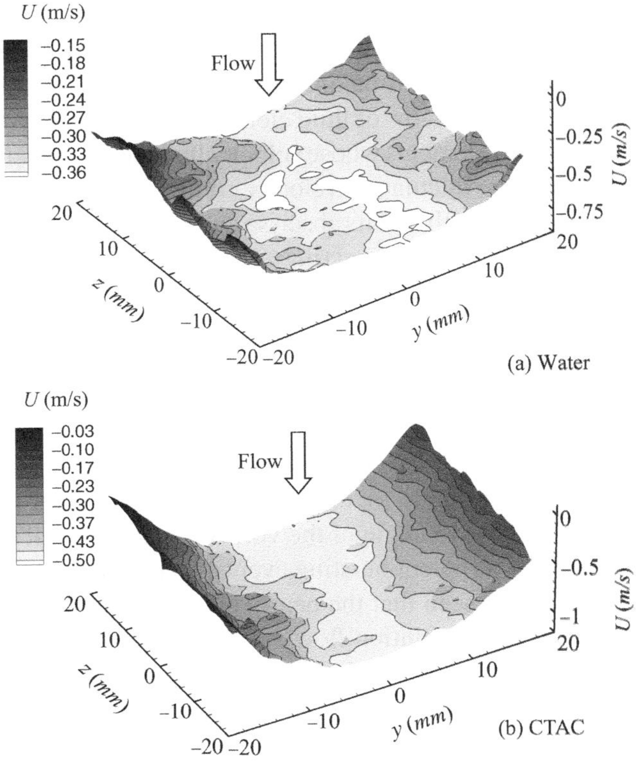

Figure 3.17 An example of the measured 2D-3C velocity contour map in the *y*–*z* plane for water and CTAC solution flows, respectively. The upper figure is for turbulent water flow at Re = 11 000, and the lower one is for turbulent drag-reducing surfactant solution flow at Re = 15 000 and with DR = 54%

surface of the 3D contour map of *U* becomes smoother for CTAC solution flow than for water flow, qualitatively indicating a depression of turbulence, and the shape of the *U*-contour profile shifts from the turbulent case toward the laminar one (a gentler gradient near the wall and an elevated amplitude in the center) for CTAC solution flow.

Through detailed investigation of a SPIV-measured velocity field in the *y*–*z* plane, the vortex structures in turbulent channel flow can be analyzed from another viewpoint. Figure 3.18 demonstrates the measured velocity field in the *y*–*z* plane for water and drag-reducing surfactant solution flows, respectively. The vectors shown in this figure are (*w*, *v*). The contours are contributions of conditioned vorticity: $\omega_x^+|_{\lambda_{ci}}$ ($\omega_x^+|_{\lambda_{ci}}$ means the streamwise vorticity, $\omega_x^+ = \omega_x(H/2)/u_\tau$, at locations where λ_{ci} is nonzero. This expression not only identifies the location of vortex cores, at which λ_{ci} is nonzero, but also provides the strength of the vortex, as seen in the upper part (i) in the figure, and the *u* component within a locally enlarged area, as

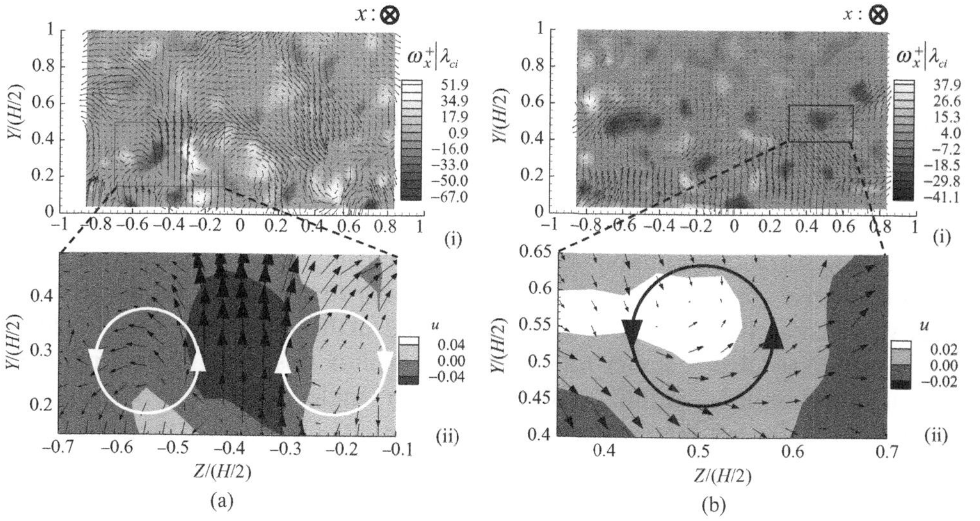

Figure 3.18 Instantaneous velocity field in the y–z plane measured with SPIV [25]. (i) Velocity vectors (w, v) with patches of $\omega_x^+|_{\lambda_{ci}}$ (positive and negative values represent clockwise and counter-clockwise rotations, respectively); and (ii) a close view to highlight a vortex or vortex pair and ejection motion. (a) Turbulent water flow at Re = 11 000; and (b) turbulent drag-reducing surfactant solution flow at Re = 15 000, DR = 54% (With kind permission from Springer Science + Business Media: *Experiments in Fluids*, "Investigation of turbulence structures in a drag-reduced turbulent channel flow with surfactant additive using stereoscopic particle image velocimetry," **40**, © 2006, 218–230, Li, F.-C., Kawaguchi, Y. and Hishida, K. and Oshima, M.)

shown in the lower part (ii) of the figure. Two obvious phenomena can be seen from Figure 3.18a: (1) many vortex cores distribute across the channel height (only half-channel height is shown here due to symmetry), and the number of counter-rotating vortex pairs is very small compared with the entire number of vortices. The counter-rotating vortex pairs observed in the velocity field in the y–z plane are actually the cross-sections of the two hairpin-vortex legs (only corresponding to the hairpin vortices with relatively symmetrical legs) inclined from the center flow region to the wall in wall turbulence. Note that although Figure 3.18a(ii) demonstrates an example of a clear counter-rotating vortex pair, after careful inspection, it was found that about 90% of all the measured velocity fields in the y–z plane show only a single-vortex core or pairs that are quite asymmetrical. This implies that most of the hairpin vortices have only one-leg or very asymmetrical two-leg structures in the wall-bounded turbulent flows. On the other hand, the close association between vortex structures in the wall turbulence and turbulent bursting events near the wall are clearly evident in Figure 3.18a(ii): between the two vortex legs of a counter-rotating vortex pair, negative u characterizing low-momentum fluid is always accompanied by positive v, which can be seen from the velocity contour map and vector directions as shown in

Figure 3.18a(ii). This phenomenon indicates one of the bursting events, that is, the turbulent ejection motions of low-momentum fluid from the wall to the flow center region in between the two legs of the hairpin vortex. A careful inspection of the instantaneous velocity fields confirms that this is also the case for a single vortex: the signature of ejection motions ($u < 0, v > 0$) appears at the left side of the vortex tube with respect to the direction of vorticity (as shown in Figure 3.18b).

Figure 3.18b plots an instantaneous 2D-3C velocity field in the *y–z* plane for the drag-reducing surfactant solution flow. Across the channel height, the distribution of many streamwise vortex cores is also obvious, similar to the water flow, whereas the vorticity is decreased by the additives. After viewing all the measured velocity fields in the *y–z* plane, no clear counter-rotating vortex pair is observed in drag-reducing surfactant solution flow. Figure 3.18b highlights a single vortex core. The ejection motion occurs at the left side (with respect to the vorticity direction) of the vortex tube, as mentioned above.

3.3.4 Distribution Characteristics of Turbulent Vortex in the x-z Plane

From the instantaneous velocity field in the *x–y* plane of turbulent channel flow, the cross-section of the hairpin vortex head and the near-wall ramp-shaped low-momentum region below the vortex packet can be visualized, while from the instantaneous velocity field in the *y–z* plane, the cross-section of hairpin vortex leg(s) and the turbulent bursting event close to the vortex leg(s) can be observed. On the other hand, from the instantaneous velocity field in the *x–z* plane, the other important information of the 3D turbulent vortical coherent structures can be analyzed from another aspect.

Figure 3.19a(i) shows a contour map of u^+ (the superscript "+" indicates normalization with friction velocity) in the *x–z* plane at $y^+ = 16.5$ for water flow. Shown in the figure are also the contour patches of λ_{ci} plotted in solid and dashed lines. The values of λ_{ci} within the areas enclosed by the solid lines are positive, representing the counter-clockwise rotating fluid motions, and those within the areas enclosed by the dashed lines are negative, representing the clockwise rotating motions. Figure 3.19a(ii) plots an enlarged part of the low-speed streak, on which the velocity vectors are (u^+, w^+) vectors and the contour is v^+. From the figure, it can be clearly seen that the near-wall turbulence structures are closely associated with the turbulent bursting events, that is, the near-wall low-speed streaks and coherent vortex packets in the wall-bounded turbulent flow. In Figure 3.19a(ii), the velocity vectors opposite to the flow direction represent low-speed fluid and form the near-wall low-speed streaks. From Figure 3.19a(i), it can be observed that in water flow the low-speed streaks are irregularly meandering in roughly the streamwise direction with relatively narrow width. From the contour map of λ_{ci}, it can also be seen that a series of wall-normal vortex cores align with the low-speed streaks with opposite signals of

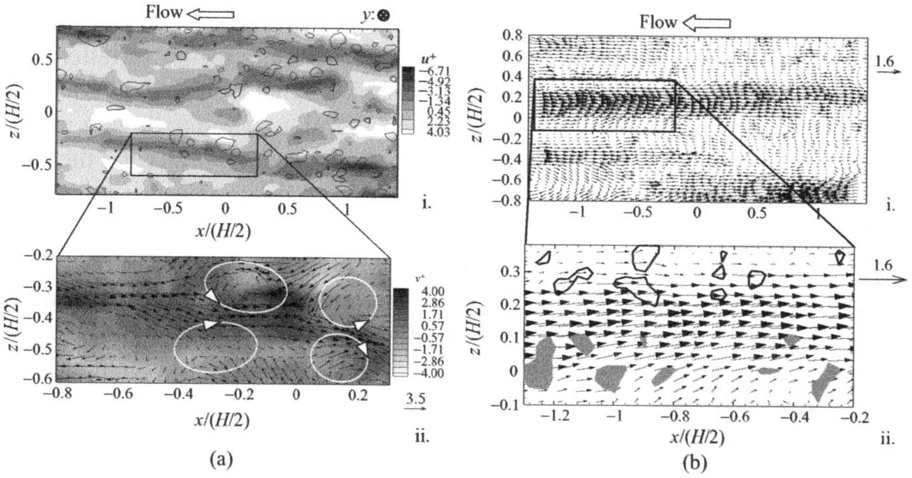

Figure 3.19 Instantaneous velocity field in the *x–z* plane measured with SPIV [25]. (a) Turbulent water flow at Re = 11 000, and the measurement plane locates at $y^+ = 16.5$; and (b) turbulent drag-reducing surfactant solution flow at Re = 15 000 and with DR = 54%, and the measurement plane locates at $y^+ = 13.8$. (i) The near-wall low-speed streaks in a large field of view; and (ii) a close view of the flow structures (With kind permission from Springer Science + Business Media: *Experiments in Fluids*, "Investigation of turbulence structures in a drag-reduced turbulent channel flow with surfactant additive using stereoscopic particle image velocimetry," **40**, © 2006, 218–230, Li, F.-C., Kawaguchi, Y. and Hishida, K. and Oshima, M.)

vorticity or an opposite swirling direction, as designated in Figure 3.19a(ii) at both sides of each streak. This reveals two essential facts: (1) the appearance of hairpin vortex packets, and (2) the formation of near-wall low-speed streaks. After inspection of all the 200 velocity fields in the *x–z* plane measured with SPIV, the number of paired counter-rotating vortex cores with comparable strength as marked in Figure 3.19a(ii) is actually very small; most are single ones or paired ones with quite non-equivalent strength, indicating that the asymmetrical hairpin vortices are the dominant structure. This is consistent with the phenomenon observed from the instantaneous velocity in the *y–z* plane. On the other hand, the SPIV technique also measures the wall-normal velocity component, *v*, in the *x–z* plane. From the contour map shown in Figure 3.19a(ii), it can be seen that at the low-speed region, the wall-normal velocity component, *v*, is positive (from the wall and toward the bulk region), indicating again the ejection motions of low-momentum fluid from the wall due to the pumping effect of a group of hairpin vortices.

Figure 3.19b shows the instantaneous velocity field in the *x–z* plane at $y^+ = 13.8$ for the turbulent drag-reducing surfactant solution flow. In the close view shown in Figure 3.19b(ii), the contour patches of λ_{ci} around the low-speed streak are also plotted. The shaded and circled areas on the contour of λ_{ci} represent fluids with counter-clockwise (positive vorticity) and clockwise (negative vorticity) rotating fluid motions, respectively. Through comparing Figure 3.19b with Figure 3.19a, it can be seen that the alignment of the near-wall low-speed streaks in the turbulent drag-reduced flow becomes relatively regular compared with that of turbulent water flow, which results from the depression of turbulence. Similar to water flow, a series of vortex cores distribute at both sides of the near-wall low-speed streaks and the signals of vorticity at both sides are opposite to each other. It is harder, however, to extract an obvious counter-rotating vortex pair from the instantaneous velocity field in the x–z plane in turbulent drag-reduced flow, and it is even harder to extract a swirling motion from the (u^+, w^+) field due to a rather weak strength of the vortices (quantitative analysis shows that the wall-normal vorticity at the conditionally captured vortex cores is greatly reduced in the drag-reducing flow: about ten times smaller than that in turbulent water flow).

In order to obtain the detailed local features and evolution of the vortex structures associated with the low-speed fluids from the buffer layer to the bulk region, conditional averaging technique can be utilized for all the SPIV-measured velocity fields in the x–z planes. Now, since the vortical features to be investigated are related to the low-speed streaks, the conditionally chosen event is a local minimum streamwise velocity, $U_{\min}$, that is, a local point where U is lower than any of the neighboring points. Meanwhile, in order to eliminate the contributions to the average from the local minima in high-speed regions, a threshold value of $0.75\ \bar{U}_{ave}$ is also set for $U_{\min}$, that is, $U_{\min} < 0.75\ \bar{U}_{ave}$, where $\bar{U}_{ave}$ designates the averaged value of the streamwise velocity components at all the points in the measured x–z plane. After extracting all the events satisfying the above two conditions (200 PIV-measured velocity fields in the x–z plane are inspected, and there are several events in each field), overlapping and then averaging all the extracted events with the "event point" as the center, the conditionally averaged velocity field is then obtained, which reflects the statistically averaged typical characteristics within the low-momentum region. For both turbulent water flow and drag-reducing surfactant solution flow, conditional averages are performed for velocity fields in the x–z plane at three different locations away from the wall. The results are shown in Figures 3.20–3.22, respectively.

Figure 3.20a and 3.20b show the conditionally averaged velocity field in the x–z plane in the buffer layer for water flow at $y^+ = 16.5$ and drag-reducing surfactant solution flow at $y^+ = 13.8$, respectively. It can be seen that the contour map of u^+ elongates in the stream-wise direction, as shown in Figure 3.20a(i) and 3.20b(i). This reveals the stream-wise elongation feature of the near-wall low-speed streaks. The swirling motions neighboring the conditional event in the span-wise direction are expected to be the dominant contributions to the wall-normal vorticity. Therefore, ω_y^+ is used here to show the strength of the vortical structure as well as location of the vortex core, which is for the purpose of comparison between drag-reducing flow and water flow.

Figure 3.20a(ii) and 3.20b(ii) clearly shows that a counter-rotating vortex pair, with almost the same size and vorticity, appears and almost symmetrically rides on

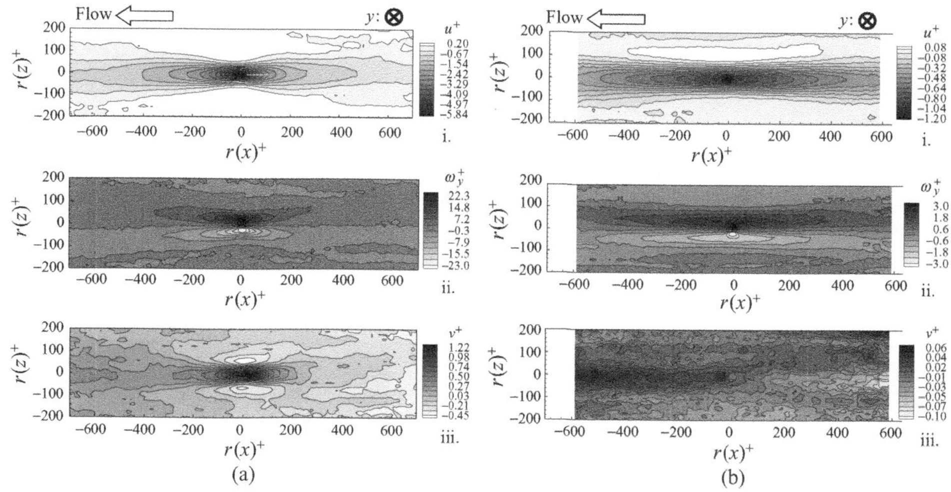

Figure 3.20 Conditionally averaged vortex structure in the *x–z* plane [25]. (a) Turbulent water flow at Re = 11 000, and the measurement plane is at $y^+ = 16.5$; and (b) drag-reducing surfactant solution flow at Re = 15 000, DR = 54%, and the measurement plane is at $y^+ = 13.8$. (i) A contour map of u^+; (ii) a contour of vorticity in the *y* direction; and (iii) a contour of v^+ (With kind permission from Springer Science + Business Media: *Experiments in Fluids*, "Investigation of turbulence structures in a drag-reduced turbulent channel flow with surfactant additive using stereoscopic particle image velocimetry," **40**, © 2006, 218–230, Li, F.-C., Kawaguchi, Y. and Hishida, K. and Oshima, M.)

the conditional event point in the spanwise direction. The two vortices greatly elongate in the streamwise direction, resulting in narrow elliptical shapes of the vortex cores. Influenced by the drag-reducing surfactant additives, the symmetrical counter-rotating vortex pair is elongated in the streamwise direction to a much greater extent in drag-reducing surfactant solution flow than in water flow. With consideration of the 3D characteristics of a hairpin vortex structure, it can be imagined that the inclination angle of vortex tubes or leg(s) of the hairpin vortex in drag-reduced flow in the buffer layer is on average much smaller than that in the Newtonian fluid flow. This is in accordance with the phenomenon of the inclination angle of the ramp-shaped low-momentum region in the turbulent drag-reducing flow becoming small as observed in the instantaneous velocity field in the *x–y* plane. Furthermore, the magnitude of ω_y^+ in the buffer layer of the drag-reducing flow is decreased by about one order of magnitude compared with water flow, indicating from another aspect the suppression of turbulence by drag reducer.

Figure 3.7a(iii) and 3.7b(iii) shows the contour of averaged v^+ based on the local minimum-*U* event in turbulent water and drag-reducing flows, respectively. For water flow, it appears that in the center area around the conditional event point, v^+ has a large positive value, and next to this positive *v* region there are two areas in the span-wise direction with negative v^+ nearly symmetrical about the center. This kind of v^+

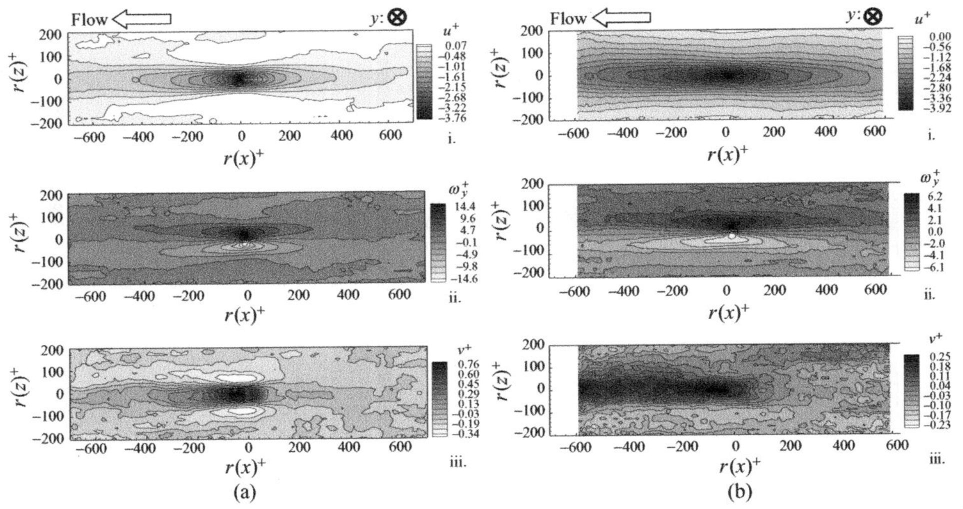

Figure 3.21 Conditionally averaged vortex structure in the x–z plane [25]. (a) Turbulent water flow at Re = 11 000, and the measurement plane is at $y^+ = 65.8$; and (b) drag-reducing surfactant solution flow at Re = 15 000, DR = 54%, and the measurement plane is at $y^+ = 55.2$. (i) A contour map of u^+; (ii) a contour of vorticity in the y direction; and (iii) a contour of v^+ (With kind permission from Springer Science + Business Media: *Experiments in Fluids*, "Investigation of turbulence structures in a drag-reduced turbulent channel flow with surfactant additive using stereoscopic particle image velocimetry," **40**, © 2006, 218–230, Li, F.-C., Kawaguchi, Y. and Hishida, K. and Oshima, M.)

distribution is a clear signature of counter-rotating motion of two stream-wise vortices. The aforementioned three regions are also elongated in the stream-wise direction, which allows one to view two counter-rotating vortex tubes or leg(s) of a hairpin vortex. However, in turbulent drag-reducing surfactant solution flow, the stream-wise vortex structures implied by the contour map of v^+ are less organized, as compared with water flow. Comparing the full range of contour levels, a significant decrease in v^+ caused by drag-reducing additives can be known. All these phenomena provide further evidence of depression of turbulence, that is, the weakening of the vortices and turbulent events by drag-reducing surfactant additives.

On the other hand, the contour map of v^+ shows that, in the buffer layer, negative v occurs upstream of the conditional event, and downstream v is positive. This implies that the vortex tubes or leg(s) of a hairpin vortex are inclined with respect to the wall with a positive inclination angle, and the PIV-measured x–z plane is obliquely cutting the vortex tubes. Implications of this phenomenon agree with the well-documented coherent vortex structures and turbulent events for a wall flow: (1) ejection motions ($v > 0$) occur due to the pumping effect of the hairpin vortex, and a ramp-shaped low-momentum region is formed between the wall and the hairpin vortices; and (2) sweeping motions ($v < 0$) of high-speed fluid toward the wall occur at the upstream of and above the hairpin vortices to compensate for the fluid ejected from the wall.

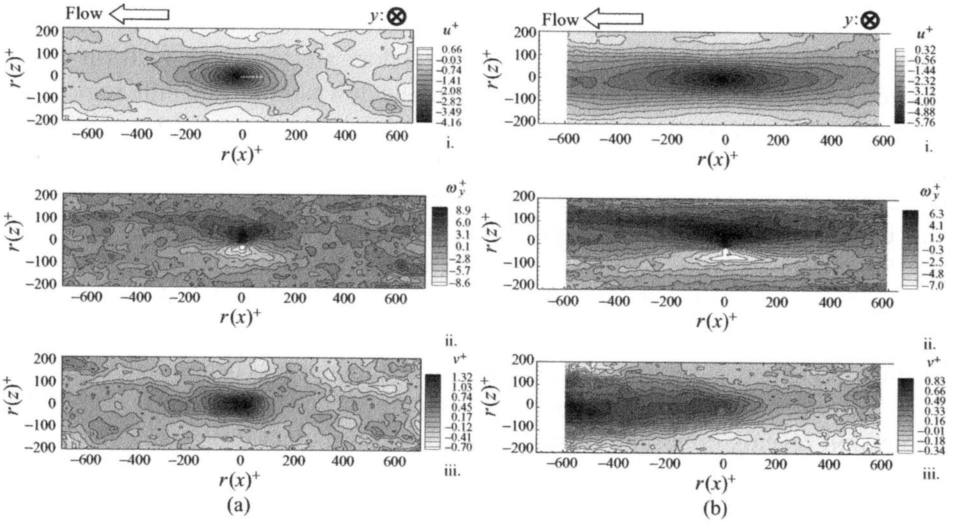

Figure 3.22 Conditionally averaged vortex structure in the *x–z* plane [25]. (a) Turbulent water flow at Re = 11 000, and the measurement plane is at $y^+ = 131.6$; and (b) drag-reducing surfactant solution flow at Re = 15 000, DR = 54%, and the measurement plane is at $y^+ = 110.4$. (i) A contour map of u^+; (ii) a contour of vorticity in the *y* direction; and (iii) a contour of v^+ (With kind permission from Springer Science + Business Media: *Experiments in Fluids*, "Investigation of turbulence structures in a drag-reduced turbulent channel flow with surfactant additive using stereoscopic particle image velocimetry," **40**, © 2006, 218–230, Li, F.-C., Kawaguchi, Y. and Hishida, K. and Oshima, M.)

The conditionally averaged results for *x–z* plane velocity fields at the other two *y*-positions (the log-law layer and bulk region) are shown in Figures 3.21 and 3.22, respectively, for the two kinds of channel flows. It can be seen that the contour of u^+ is broadened in the spanwise direction as the measured location changes from the buffer layer to the log-law layer and then the outer region in the two tested flows, showing the evolution of hairpin vortices with convection to the downstream and the outer region. For turbulent water flow, at $y^+ = 131.6$ in the *x–z* plane, the affected distance of the minimum-*U* event is shortened in the stream-wise direction; the elliptical shape of the cross-section of vortex tubes indicated by ω_y^+ becomes closer to a round shape as compared with those at other two locations; and the streamwise vortical motion indicated by the contour of v^+ becomes quite ambiguous, implying that, with the evolution of a hairpin vortex (which is strong enough to be convected far away from the wall and whose organized structure is maintained at least roughly) up to $y^+ = 131.6$, the inclination angle of leg(s) of a hairpin vortex might have become a near-vertical angle.

For the turbulent drag-reducing surfactant solution flow, however, the averaged *u*-contour does not show a shortened feature in the stream-wise direction even at the furthest PIV measurement location, $y^+ = 110.4$. There are two possible reasons for this phenomenon: (1) the wall-normal distance of $y^+ = 110.4$ is still not large enough, and so the bending of the hairpin vortex head to the bulk region has not occurred locally; or (2) the vortices in this drag-reduced flow are too weak to form clearly organized vortical structures with evolvement to the downstream and bulk region. The latter is more likely since the clearly organized streamwise swirling motions as appeared in the water flow in the buffer layer and log-law layer (see Figures 3.21aiii and 3.22aiii) do not appear in the drag-reduced flow.

The contour levels of ω_y^+ and v^+, plotted in Figures 3.20–3.22, are given in their full range. Through careful inspection, it is seen that, in the buffer layer, ω_y^+ and v^+ for turbulent drag-reducing surfactant solution flow are decreased by about one order compared with water flow; with the x–z plane moving away from the wall, the difference of ω_y^+ and v^+ between water and drag-reducing flows becomes less and less, and in the outer region ω_y^+ and v^+ for both flows are in the same order. This phenomenon reveals important information for the turbulent drag-reducing flow by additives: to reduce the overall skin friction for a wall-bounded turbulent flow, it could be the most effective one to suppress turbulence in the buffer layer.

3.4 Reynolds Shear Stress and Wall-Normal Turbulent Heat Flux

The suppression of turbulence by drag reducer in turbulent drag-reducing flows is reflected by the relative variations in distributions of mean velocity and temperature fields; the fluctuation intensities of velocity and temperature; and the correlations between different fluctuating quantities, among which the most salient is that the turbulent intensity of the wall-normal velocity fluctuations is greatly decreased and the correlations between the wall-normal and streamwise instantaneous velocity fluctuations and between the wall-normal velocity and temperature fluctuations are greatly reduced. These changes can be quantitated from the statistical viewpoint with the Reynolds shear stress ($-\overline{uv}$ or $-\overline{u^+v^+} = -\overline{uv}/u_\tau^2$) and the wall-normal turbulent heat flux ($-\overline{v\theta}$ or $-\overline{v^+\theta^+} = -\overline{v\theta}/(u_\tau T_\tau)$), respectively.

Figures 3.23 and 3.24 show the distributions in the channel height direction of the measured $-\overline{u^+v^+}$ and $-\overline{v^+\theta^+}$, respectively, in turbulent drag-reducing surfactant solution flow. From these two figures, it can be seen that the larger the DR and HTR are, the more the $-\overline{u^+v^+}$ and $-\overline{v^+\theta^+}$ are depressed in turbulent drag-reducing surfactant solution flow as compared with turbulent water flow. This experimental phenomenon indicates that, in turbulent flows, distributions of the Reynolds shear stress and wall-normal turbulent heat flux should have direct functionalized relationships with the momentum and heat exchanges between the flow and the wall, respectively.

In fact, for the wall-bounded turbulent flows, it is just because one additional term, that is, the Reynolds stress (or the turbulent heat flux), appears in the momentum equation (or energy equation), as compared with the laminar flow,

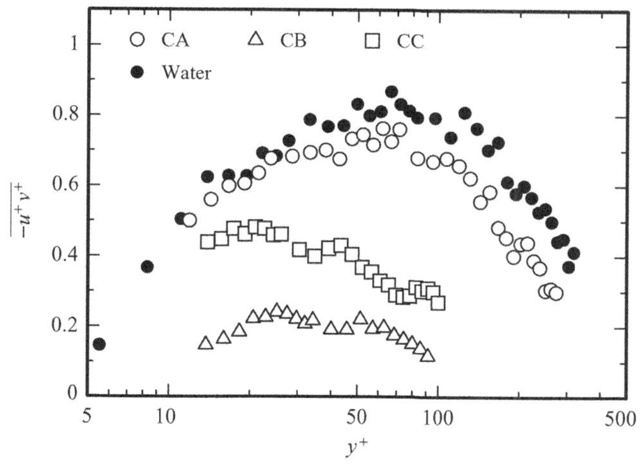

Figure 3.23 Distribution of the Reynolds shear stress [15]. The case parameters shown in the figure are as follows: water, Re = 25 000; CA, Re = 35 000, DR = 33.0%; CB, Re = 25 000, DR = 70.0%; and CC, Re = 15 000, DR = 65.1% (Reprinted with permission from Li, F.-C., Kawaguchi, Y., and Hishida, K., "Investigation on the characteristics of turbulence transport for momentum and heat in a drag-reducing surfactant solution flow," *Physics of Fluids* **16**, 3281–3295, 2004. © 2004 American Institute of Physics.)

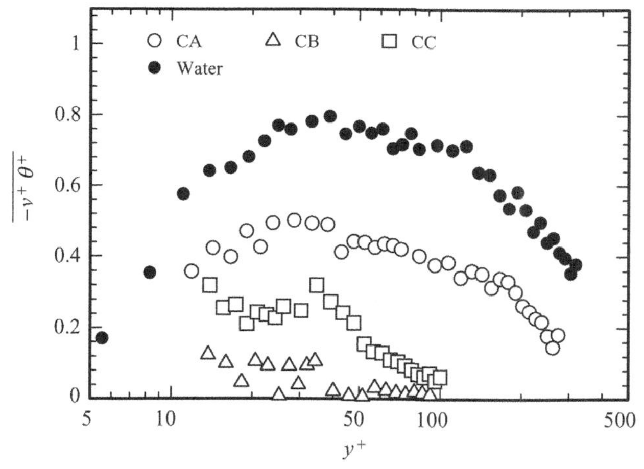

Figure 3.24 Distribution of the turbulent heat flux normal to the heated wall [15]. The case parameters shown in the figure are as follows: water, Re = 25 000; CA, Re = 35 000, DR = 33.0%, HTR = 20.2%; CB, Re = 25 000, DR = 70.0%, HTR = 77.3%; and CC, Re = 15 000, DR = 65.1%, HTR = 77.0% (Reprinted with permission from Li, F.-C., Kawaguchi, Y., and Hishida, K., "Investigation on the characteristics of turbulence transport for momentum and heat in a drag-reducing surfactant solution flow," *Physics of Fluids* **16**, 3281–3295, 2004. © 2004 American Institute of Physics.)

the frictional drag of flow (or convective heat transfer rate) becomes greatly increased compared with its assumed laminar counterpart at the same Reynolds number. Consequently, it is natural to imagine that the augmentations of momentum and heat transports are bound to be coming from the occurrences of the Reynolds shear stress and turbulent heat flux in turbulent flow. Furthermore, the frictional drag and convective heat transfer are the momentum and heat transports between flow and solid wall; hence, within the Reynolds stress and turbulent heat flux, the most important contributions are the Reynolds shear stress and the wall-normal component of turbulent heat flux. In 2002, Fukagata *et al.* for the first time derived the quantitative relationship between the Reynolds shear stress and the turbulent contribution to the frictional drag coefficient [45] for the Newtonian fluid wall-bounded turbulent flows (boundary layer flow, channel flow, and pipe flow). For the turbulent channel flow of the Newtonian fluid, the frictional drag coefficient can be expressed by the following equation:

$$f = \frac{12}{\mathrm{Re}} + \frac{6}{U_b^{+2}} \int_0^1 (1 - y^*)\left(-\overline{u^+v^+}\right) dy^* \tag{3.11}$$

where Re is the Reynolds number based on channel height, bulk velocity (U_b), and the kinematic viscosity of the fluid; and the superscript "*" designates normalization of length using the half-height of the channel. The first term in the right-hand side of Equation 3.11 attributes to viscous force (or laminar flow part), and the second to turbulent force (or turbulent flow part). Applying the same idea, one can derive the friction factor for turbulent drag-reducing flow by additives in a channel as follows [15–20]:

$$f = \frac{12}{\mathrm{Re}} + \frac{6}{U_b^{+2}} \int_0^1 (1 - y^*)\left(-\overline{u^+v^+}\right) dy^* + \frac{6}{U_b^{+2}} \int_0^1 (1 - y^*)\tau_e^+ dy^* \tag{3.12}$$

In this equation, the third term in the right-hand side is attributed to some additional stress after the addition of drag reducer. Since the aqueous solution flow of surfactant drag reducer or high polymer displays the rheological viscoelasticity, τ_e^+ in the above-mentioned additional term is also named "elastic stress." The quantitative equation for friction factor (Equation 3.12) indicates that, in the turbulent drag-reducing surfactant solution flow, the influence of drag reducer on the turbulent vortex coherent structures, vortex intensity, turbulent bursting events, and so on is eventually to decrease the Reynolds shear stress, which results in a significant decrease of the turbulent contribution in Equation 3.12 and consequently the turbulent drag-reducing effect. The Reynolds shear stress can be easily obtained through measurement of the instantaneous velocity field. However, since the elastic stress (which positively contributes to the friction factor) includes very complicated information on microstructures in the viscoelastic fluid, it cannot be measured directly from experiment. The addition of surfactant drag reducer into turbulent water flow directly results in a great decrease in the Reynolds shear stress. Although the positive elastic stress is introduced at the same time, compared with the negative contribution to the friction factor due to the decrease of the Reynolds shear stress, the positive contribution due to the elastic stress is much smaller. Thus, an obvious turbulent drag-reducing effect is eventually brought about.

With application of the similar idea for deriving Equation 3.11, the quantitative relationship between the turbulent contribution to the convective heat transfer coefficient (or the Nusselt number, Nu) and the wall-normal turbulent heat flux in a wall-bounded turbulent flow is as follows [15]:

$$\frac{1}{Nu} = \frac{\langle\Theta\rangle_m}{\langle\Theta\rangle}\left[\left(\frac{1}{2} - A\right) - \int_0^{\delta*} (1 - y')\left(-\overline{v^+\theta^+}\right)dy'\right] \tag{3.13}$$

where y' is the dimensionless coordinate normalized with channel height, $A = \int_0^1 \left[\frac{(1-y')}{\mathrm{Re}_{\tau\delta}} \int_0^{y^+} \left(\frac{U^+}{\langle U\rangle_\delta^+}\right) dy^+\right] dy'$, δ is the thickness of the thermal boundary layer, $\delta*$ is the dimensionless thermal boundary layer normalized with channel height (when one wall is heated for channel flow and the thermal boundary is fully developed, its thickness should be the same as channel height, that is, $\delta* = 1$), $\langle\Theta\rangle_m = \frac{\int_0^1 U\Theta dy'}{\int_0^1 U dy'}$ is the mixing temperature of the flow, and $\langle\Theta\rangle$is the mean temperature over the channel height (generally different from the mixing temperature).

With a constant thermal boundary layer thickness and at a certain flow rate, the Nusselt number has its minimum value ($\mathrm{Nu}_{\mathrm{min}}$) when the flow is assumed to be fully laminar corresponding to $-\overline{v\theta} = 0$, and its maximum value ($\mathrm{Nu}_{\mathrm{max}}$) when the flow is fully turbulent corresponding to the maximum value of $q_T = \int_0^1 (1 - y')(-\overline{v^+\theta^+})dy'$. For turbulent drag-reducing flow, since its flow state is in between fully laminar and fully turbulent, at the same flow rate, the Nusselt number satisfies with $\mathrm{Nu}_{\mathrm{min}} < \mathrm{Nu} < \mathrm{Nu}_{\mathrm{max}}$, and the turbulent contribution satisfies with $0 < \int_0^{\delta*} (1 - y')(-\overline{v^+\theta^+})dy' < q_T$. Equation 3.13 indicates that the depression of the wall-normal turbulent heat flux $-\overline{v^+\theta^+}$ (especially in the near-wall region because of the local large weight $1 - y'$) caused by the drag-reducing additives results in the decrease of the heat transfer coefficient.

It has been known that the heat transfer coefficient in a drag-reducing surfactant or polymer solution flow is usually reduced more than the friction factor, that is, HTR > DR, as compared with a water flow at the same Reynolds number. From the comparison of Equation 3.12 and Equation 3.13, it can be seen that addition of the drag-reducing additives causes a positive elastic contribution to the friction factor in addition to a highly reduction in turbulent contribution, whereas this is not the case for the Nusselt number, that is, the additives cause only a high reduction in turbulent contribution to the heat transfer rate without an additional positive elastic contribution. This might be a reason why HTR is higher than DR, and the Colburn analogy is no longer valid.

References

1. Friehe, C.A. and Schwarz, W.H. (1970) Grid-generated turbulence in dilute polymer solutions. *Journal of Fluid Mechanics*, **44**, 173–193.
2. De Angelis, E., Casciola, C.M., Benzi, R. and Piva, R. (2005) Homogeneous isotropic turbulence in dilute polymers. *Journal of Fluid Mechanics*, **531**, 1–10.
3. Kalelkar, C., Govindarajan, R. and Pandit, R. (2005) DR by polymer additives in decaying turbulence. *Physical Review E*, **72** (017301), 1–4.
4. Perlekar, P., Mitra, D. and Pandit, R. (2006) Manifestations of DR by polymer additives in decaying, homogenous, isotropic turbulence. *Physical Review Letters*, **97** (264501), 1–4.
5. Cai, W-H., Li, F-C. and Zhang, H-N. (2010) DNS study of decaying homogeneous isotropic turbulence with polymer additives. *Journal of Fluid Mechanics*, **665**, 334–356.
6. Wei, T. and Willmarth, W.W. (1992) Modifying turbulent structure with drag-reducing polymer additives in turbulent channel flows. *Journal of Fluid Mechanics*, **245**, 619–641.
7. Metzner, A.B. and Astarita, G. (1967) External flows of viscoelastic materials: Fluid property restrictions on the use of velocity-sensitive probes. *AIChE Journal*, **13**, 550.
8. Walker, D.T. and Tiederman, W.G. (1990) Turbulent structure in a channel flow with polymer injection at the wall. *Journal of Fluid Mechanics*, **218**, 377–403.
9. Smith, K.A., Merrill, E.W., Mickley, H.S. and Virk, P.S. (1967) Anomalous Pitot tube and hot film measurements in dilute polymer solutions. *Chemical Engineering Science*, **22**, 619.
10. Hetsroni, G., Zakin, J.L. and Mosyak, A. (1997) Low-speed streaks in drag-reduced turbulent flow. *Physics of Fluids*, **9**, 2397–2404.
11. Allan, J.J., Greated, C.A. and McComb, W.D. (1984) Laser-Doppler anemometer measurements of turbulent structure in non-Newtonian fluids. *Journal of Physics D: Applied Physics*, **17**, 533–549.
12. Goldstein, R.J., Adrian, R.J. and Kried, D.K. (1969) Turbulent and transition pipe flow of dilute aqueous polymer solutions. *Industrial Engineering Chemistry Fundamentals*, **8**, 498–502.
13. Fan, B-C. and Ye, J-F. (2007) *Measurement of Instantaneous Flow Field Parameters*, Harbin Engineering University Press (in Chinese).
14. Li, F-C., Wang, D-Z., Kawaguchi, Y. and Hishida, K. (2004) Simultaneous measurements of velocity and temperature fluctuations in thermal boundary layer in a drag-reducing surfactant solution flow. *Experiments in Fluids*, **36**, 131–140.
15. Li, F-C., Kawaguchi, Y. and Hishida, K. (2004) Investigation on the characteristics of turbulence transport for momentum and heat in a drag-reducing surfactant solution flow. *Physics of Fluids*, **16**, 3281–3295.
16. Li, F-C., Kawaguchi, Y. and Hishida, K. (2005) Structural analysis of turbulent transport in a heated drag-reducing channel flow with surfactant additives. *International Journal of Heat and Mass Transfer*, **48**, 965–973.
17. Warholic, M.D., Heist, D.K., Katcher, M. and Hanratty, T.J. (2001) A study with particle-image velocimetry of the influence of drag-reducing polymers on the structure of turbulence. *Experiments in Fluids*, **31**, 474–483.

18. Kawaguchi, Y., Segawa, T., Feng, Z-P. and Li, P-W. (2002) Experimental study on drag-reducing channel flow with surfactant additives – spatial structure of turbulence investigated by PIV system. *International Journal of Heat and Fluid Flow*, **23**, 700–109.
19. White, C.M., Somandepalli, V.S.R. and Mungal, M.G. (2004) The turbulence structure of drag reduced boundary layer flow. *Experiments in Fluids*, **36**, 62–69.
20. Yu, B., Li, F-C. and Kawaguchi, Y. (2004) Numerical and experimental investigation of turbulent characteristics in a drag-reducing flow with surfactant additives. *International Journal of Heat and Fluid Flow*, **25**, 961–974.
21. Gurka, R., Liberzon, A. and Hetsroni, G. (2004) Characterization of turbulent flow in a flume with surfactant. *ASME Journal of Fluids Engineering*, **126**, 1054–1057.
22. Liberatore, M.W., Baik, S., McHugh, A.J. and Hanratty, T.J. (2004) Turbulent DR of polyacrylamide solutions: effect of degradation on molecular weight distribution. *Journal of Non-Newtonian Fluid Mechanics*, **123**, 175–183.
23. Baik, S., Vlachogiannis, M. and Hanratty, T.J. (2005) Use of particle image velocimetry to study heterogeneous DR. *Experiments in Fluids*, **39**, 637–650.
24. Li, F-C., Kawaguchi, Y., Segawa, T. and Hishida, K. (2005) Reynolds-number dependence of turbulence structures in a drag-reducing surfactant solution channel flow investigated by PIV. *Physics of Fluids*, **17**, 075104.
25. Li, F-C., Kawaguchi, Y., Hishida, K. and Oshima, M. (2006) Investigation of turbulence structures in a drag-reduced turbulent channel flow with surfactant additive using stereoscopic particle image velocimetry. *Experiments in Fluids*, **40**, 218–230.
26. Li, F-C., Kawaguchi, Y., Segawa, T. and Hishida, K. (2006) On turbulent contribution to frictional drag in wall-bounded turbulent flow. *Chinese Physics Letters*, **23**, 1226–1229.
27. Li, F-C., Yu, B., Wei, J-J. *et al.* (2008) Experimental study of drag-reduction mechanism for a dilute surfactant solution flow. *International Journal of Heat and Mass Transfer*, **51**, 835–843.
28. Hou, Y.X., Somandepalli, V.S.R. and Mungal, M.G. (2008) Streamwise development of turbulent boundary-layer DR with polymer injection. *Journal of Fluid Mechanics*, **597**, 31–66.
29. Wei, J.J., Kawaguchi, Y., Li, F-C. *et al.* (2009) DR and turbulence characteristics in sub-zero temperature range of cationic and zwitterionic surfactants in EG/water solvent. *Journal of Turbulence*, **10**, 1–15.
30. Cai, W-H., Li, F-C., Zhang, H-N. *et al.* (2009) Study on the characteristics of turbulent drag-reducing channel flow by PIV combining with POD analysis. *Physics of Fluids*, **21**, 115103.
31. Tamano, S., Itoh, M., Inoue, T. *et al.* (2009) Turbulence statistics and structures of drag-reducing turbulent boundary layer in homogeneous aqueous surfactant solutions. *Physics of Fluids*, **21**, 045101.
32. Li, F-C. and Kishida, K. (2009) Particle image velocimetry techniques and its applications in multiphase systems, Chapter 3, in *Advances in Chemical Engineering – Characterization of Flow, Particles and Interfaces* (ed. J-H. Li), Springer, pp. 87–147.
33. Virk, P.S. (1975) DR fundamentals. *AIChE Journal*, **21**, 625–656.
34. Gasljevic, K., Aguilar, G. and Matthys, E.F. (2001) On two distinct types of drag-reducing fluids, diameter scaling, and turbulent profiles. *Journal of Non-Newtonian Fluid Mechanics*, **96**, 405–425.
35. Gasljevic, K., Aguilar, G. and Matthys, E.F. (2007) Measurement of temperature profiles in turbulent pipe flow of polymer and surfactant drag-reducing solutions. *Physics of Fluids*, **19**, 083105.
36. Kader, B.A. (1981) Temperature and concentration profiles in fully turbulent boundary layers. *International Journal of Heat and Mass Transfer*, **24**, 1541–1544.
37. Warholic, M.D., Schmidt, G.M. and Hanratty, T.J. (1999) The influence of a drag-reducing surfactant on a turbulent velocity field. *Journal of Fluid Mechanics*, **388**, 1–20.
38. Choi, K.S. and Clayton, B.R. (2001) The mechanism of turbulent DR with wall oscillation. *International Journal of Heat and Fluid Flow*, **22**, 1–9.

39. Kline, S.J. and Robinson, S.K. (1989) Quasi-coherent structures in the turbulent boundary layer, in *Part I: Status Report on a Community-Wide of the Data in Neat Wall Turbulence*, Hemisphere, pp. 218–247.
40. Perry, A.E. and Chong, M.S. (1987) A description of edding motions and flow patterns using critical-point concepts. *Annular Review of Fluid Mechanics*, **19**, 125–155.
41. Zhong, J., Huang, T.S. and Adrian, R.J. (1998) Extracting 3D vortices in turbulent fluid flow. *IEEE Transactions on Pattern Analysis and Machine Intelligence*, **20**, 193–199.
42. Chong, M.S., Perry, A.E. and Cantwell, B.J. (1990) A general classification of three-dimensional flow fields. *Physics of Fluids*, **2**, 765–777.
43. Zhou, J., Adrian, R.J. and Balachandar, S. (1996) Auto generation of near-wall vortical structures in channel flow. *Physics of Fluids*, **8**, 288–290.
44. Zhou, J., Adrian, R.J., Balachandar, S. and Kendall, T.M. (1999) Mechanisms for generating coherent packets of hairpin vortices in channel flow. *Journal of Fluid Mechanics*, **387**, 353–396.
45. Fukagata, K., Iwamoto, K. and Kasagi, N. (2002) Contribution of Reynolds stress distribution to the skin friction in wall-bounded flows. *Physics of Fluids*, **14**, L73–L76.

4

Numerical Simulation of Surfactant Drag Reduction

Drag reduction by additives generally covers theoretical, experimental, and numerical research. Detailed analyses have been made in the previous chapters, which show that the mechanism study of drag reduction has been much developed and that many mechanism theories have been proposed. However, there is no theory that can explain all phenomena in drag reduction by additives. Viscoelastic theory [1], being widely accepted, can explain most drag reduction phenomena, but it is challenged by the experimental fact of nonviscoelastic drag reduction [2]. Thus drag reduction theory is still in exploration. In experimental aspects, from the early detector measurements to the recent laser Doppler velocimetry (LDV) and particle image velocimetry (PIV), the precision of equipments is higher and higher and measurable parameters extend from macrocharacteristics such as flow rate, pressure drop, and so on to microcharacteristics such as fluctuating intensity, burst event, and so on, deepening the understanding of the drag reduction mechanism. However, even the most advanced PIV technology cannot accomplish three-dimensional tests and obtain instantaneous information in the whole field. Experimental measurements have been restricted in two-dimensional (or quasi-three-dimensional) and local characteristics. It has advantages such as quantification, easy control, convenient scale-up, and good repetitiveness and has been an increasingly important tool of mechanism study. This chapter will introduce the research advancements of numerical simulation in the field of drag-reducing flow by additives.

The numerical simulation of drag-reducing flow can be categorized by three classes: direct numerical simulation (DNS), large eddy simulation (LES), and Reynolds-averaged Navier–Stokes simulation (RANS). DNS resolves all scale turbulence directly from an exact flow governing equation, without any modeling hypotheses [3]. This is the most accurate numerical simulation method for turbulence and can be seen as a "numerical experiment" [4]. Through this "numerical experiment," one can obtain

Turbulent Drag Reduction by Surfactant Additives, First Edition. Feng-Chen Li, Bo Yu, Jin-Jia Wei and Yasuo Kawaguchi.

the turbulent information that can hardly be obtained, or even not be obtained at all, in experimental apparatus. For example, the fluctuating pressure has had no accurate measuring result until now, and the vortex distribution is also hard to test. These parameters can be easily and accurately obtained by DNS. It can also capture the evolution of all instantaneous variables in 3D space [5]. LES decomposes the turbulent instantaneous variables to large-scale and small scale motions by a sort of wave-filtering method. This method, however, resolves only large-scale motion. The relation between small- and large-scale motions can be modeled by a subgrid model [6]. RANS gets time-averaged governing equations of drag reduction and models the high-order fluctuations to close the average equations. The three classes of simulations have different applications due to their own characteristics: DNS has the highest precision and longest computing time, mainly applied to the mechanism study; RANS has the shortest computing time, mainly applied to engineering forecasting; and LES is between DNS and RANS on the aspects of precision and computing time. It has obtained more and more attention. There have been many works about DNS and RANS of drag reduction by additives, but to the best knowledge of the authors, only Lee begins LES [7]. Due to the few works on LES, the following sections will focus on DNS and RANS.

4.1 Direct Numerical Simulation of Drag-reducing Flow

Drag-reducing fluid has complex non-Newtonian fluid properties, for example, viscoelasticity, extension, shear thinning and shear thickening, and so on. Its flow characteristics are more complicated than those of Newtonian fluid. Thus, the DNS of drag-reducing flow has two main problems, that is, a governing equation (mathematical model) that can correctly describe the flow characteristics [8–42] and a corresponding robust and accurate numerical method [10–28]. Hence, these two difficulties will be discussed below.

4.1.1 A Mathematical Model of Drag-reducing Flow

Due to the complexity and uncertainty of drag reduction, as stated above, no commonly recognized mathematical model has been used to describe drag-reducing flow like the Navier–Stokes (N–S) equation describing Newtonian fluid flow. But numerous researchers proposed different mathematical models according to the rheological properties of drag-reducing fluid. They are mainly categorized into two classes based on the distribution of additives in solvent: homogeneous and inhomogeneous models. Most references focused on homogeneous models, established according to the characteristics of polymer, mainly having extensional models and viscoelastic models. Surfactant drag reduction is more complicated than polymer drag reduction, but their rheological properties are similar. Researchers have also applied the homogeneous models based on polymer to the study of surfactant drag reduction until now. The extensional and viscoelastic models will be introduced below.

Orlandi [8] obtained an anisotropic extensional viscosity model from experiments and theoretical analyses, introduced the extensional viscosity into the N–S equation, and obtained momentum equations for polymer drag reduction. Then the drag-reducing channel flow of dilute polymer solution was simulated by this model. It was shown that turbulent stress in the near-wall region apparently does not grow larger, and its peak value position shifts to the center of the channel. With increasing concentration, the intensity of longitudinal vortexes decreases and their spacing increases. These results all agreed with the experimental conclusions qualitatively, although they were not quantitatively compared with experimental data. Toonder [9] proposed another extensional viscosity model based on the anisotropic stress hypothesis to simulate fully developed drag-reducing pipe flow and qualitatively obtained statistical turbulent law. Some experimental phenomena were reproduced by this extensional viscosity model, but drag reduction rate is small and only increase of extensional viscosity is inadequate for apparent drag reduction. Additionally, the DNS did not re-appear the in important feature-onset Reynolds number of drag reduction because polymer long-chain structures were not considered.

Due to the shortcomings of the extensional viscosity model, most researchers have proposed and applied viscoelastic models, mainly the constitutive models such as FENE-P, Giesekus, Oldroyd-B, and Maxwell. Of these, FENE-P model gets the most attention. Sureshkumar [10] used the FENE-P model to do the DNS of fully developed channel flow. The study reproduced many experimental phenomena such as the expansion of a buffer layer, the decrease of Reynolds stress, the increase of streak spacing, and so on, and it proposed the onset Weissenberg number for drag reduction. Dimitropoulos *et al.* [11] studied the parameter effects of FENE-P and Giesekus models on drag reduction and verified that drag reduction is mainly attributed to depressing the generation of turbulence. Angelis *et al.* [12] further confirmed the ability of the FENE-P model to reproduce most of the essential effects of polymers in dilute solutions on the wall turbulence. They also observed that polymers become highly oriented parallel to the mean flow direction [13]. Ptasinski *et al.* [14] studied the turbulent statistics at the maximum drag reduction and the energy redistribution by high elasticity of polymer molecules. Housiadas and Beris [15] applied the FENE-P model to the DNS to study the elasticity and inertia of turbulent drag reduction systematically. They extended the low Weissenberg number (<50) research of Sureshkumar [10] and Dimitropoulos [11] to high Weissenberg number research (125). Their findings were that in the high-elasticity region, the mean Reynolds number and drag reduction rate continue to increase although at a smaller rate; and in the high inertia region, drag reduction rate is almost unchanged with increasing Reynolds number at the fixed Weissenberg number. Vaithianathan *et al.* [16] tested some physics of stirred and decaying turbulent polymer solutions by the FENE-P model in DNS. They conjectured that the traditional energy cascade in a Newtonian fluid was bifurcated into two directions. Dimitropoulos *et al.* [17] observed an initial development length, which was followed by a quasi-steady region where variations in drag reduction were weak. High drag reduction behavior can be present at short downstream distances from the inflow plane. Sibilla *et al.* [18] added a source term, which is due to the stress state generated by polymer elongation, into the Navier–Stokes equation to directly simulate drag-reducing channel flow. The polymer stress was presented by the FENE-P model. Near-wall turbulent coherent structures, obtained by applying Jeong and Hussain's vortex identification criterion to DNS flow fields, were compared with those observed in a Newtonian flow at the same Reynolds number. The mean vortex in the polymer flow was weaker, less tilted, and less inclined, while its length and radius are higher. Li *et al.* [19] performed DNS to turbulent channel flow by the FENE-P model and developed relationships between the flow and fluid rheological parameters. They studied different features at different drag reduction rates. Three different regimes of drag reduction, namely, low drag reduction (LDR: 0~20%), high drag reduction (HDR: 20%~52%), and maximum drag reduction (MDR: 52%~74%), were identified. It was found that in both LDR and HDR regimes the eddy viscosity varies with the distance from the channel wall, but in the MDR regime the ratio of the eddy viscosity to the Newtonian one tends to be a very small value of around 0.1 within the channel.

Other researchers applied the Maxwell, Oldroyd-B, and Giesekus models to DNS. Sureshkumar [20] utilized the upper convected Maxwell (UCM) model to investigate the local linear stability characteristics of viscoelastic periodic channel flow in the inertial and purely elastic flow regimes. The elasticity was presented by the ratio of the Weissenberg number and Reynolds number. Min *et al.* [21,22] investigated the role of elastic energy in the polymer drag reduction by the Oldroyd-B model. They obtained the maximum drag reduction as high as 44%, which was in good agreement with Virk's asymptote at the mean Reynolds number 4000. For large drag reduction, the variations of turbulence statistics such as the mean stream-wise velocity and velocity fluctuations were quite different from those of small drag reduction. For example, for small drag reduction, the stream-wise velocity fluctuations decrease in the viscous sublayer but increase in the buffer and log layers with increasing the Weissenberg number, but they decrease in the whole channel for large drag reduction. Eckhardt *et al.* [23] applied the Oldroyd-B model to the direct numerical simulation of shear turbulence for the coil–stretch transition in the polymers. They found that in the coiled state, the lengths of polymers were distributed algebraically with an exponent relation with the Deborah number; and in the stretched state, the potentially infinite stretching of the polymers was limited by the feedback to the flow. Min and Choi [24] studied the combined effect of polymer additives and active blowing and suction at the wall on drag reduction using direct numerical simulation by the Oldroyd-B model. They concluded that the combined drag reduction obtained by both the polymer additives and the blowing and suction was larger than the individual drag reduction either by polymers or by blowing and suction, but was smaller than the sum of the two separate drag reductions, indicating that the polymer solution is less efficient in producing drag reduction in the absence of strong near-wall stream-wise vortices. Li *et al.* [25] performed high-fidelity DNS of polymer-induced drag reduction in turbulent channel flows by FENE-P and Oldroyd-B constitutive equations and studied the effects of the Weissenberg number, polymer chain extensibility, and the Reynolds number on drag reduction. They found that large polymer extensibility and high Weissenberg numbers were required to obtain significant levels of drag reduction and the viscoelastic force performed positive work on the velocity fluctuations. The polymer chains extracted energy via stretching from the flow and hence reduced turbulent fluctuations. Tamano *et al.* [26] directly numerically simulated a zero-pressure gradient drag-reducing turbulent boundary layer using constitutive equation models such as the Oldroyd-B and Giesekus models. Budgets of turbulent kinetic energy and Reynolds normal stresses were firstly investigated for a drag-reducing turbulent boundary layer. They discovered that the maximum drag reduction rate for the Oldroyd-B model (42%), which has the higher elongational viscosity, is larger than that for the Giesekus model (16%). Correspondingly, the contribution of the polymer stress work to the turbulent kinetic energy budget for the Oldroyd-B model is larger.

Yu *et al.* [27–35] and Kawaguchi et al. [36–38] have made several detailed works for surfactant drag reduction, including systematic DNS using the Giesekus model, investigation of the effect of model parameters on drag reduction, a proposal of the efficient finite difference method, a mechanism study of drag reduction and heat transfer reduction, and so on. Gu [39] proposed a parallel Giesekus model to avoid the limitation of the Giesekus model that it has only a single time scale. The proposed model was established by dynamic response analysis that a double relaxation time can overcome the drawback of incompatibility of high and low frequencies at a single relaxation time. Moreover, the parallel Giesekus model can improve the description ability of two-step shear shinning for surfactant solution. This model also resolved the divergence problems at a high Weissenberg number and achieved a drag reduction rate as high as 86% with stronger coherent structures than those of the Giesekus model.

The above homogeneous models all assume that the polymers distribute uniformly in space and time, which means that the concentration of the polymer is constant everywhere. But this hypothesis does not agree with the experimental finding by White *et al.* of Stanford University [40] that the concentration of polymer in the low-speed streaks is much higher than that in the high-speed streaks and the spatial distribution of the polymer is related to the turbulent structure. This indicates that the homogeneous models have a relatively large declination from real physical problems, at least without the spatial and temporal distribution of drag-reducing additives which may affect further mechanism study. Hence, some researchers proposed inhomogeneous models for drag-reducing flow, that is, inhomogeneous models based on constitutive equations and discrete element models.

Dimitropoulos *et al.* [41] first performed similar DNS according to the above experimental fact [40] to better understand the effect of an inhomogeneous introduction of polymer additives in a turbulent boundary layer. They applied a two-fluid Hamiltonian model to distinguish the solvent and additive parts from the solution. The solvent part was presented by the momentum equation, while the additive part was presented by the coupled concentration and conformation evolution equations. The relation between polymer additional stress and conformation was presented by the FENE-P constitutive equation to finally close the whole model. The simulation predictions for the interaction of polymer transport and drag reduction presented in the work were in qualitative agreement with recent experimental observations. The development of drag reduction for inhomogeneous flows depends on the evolution of the near-wall concentration.

Yu *et al.* [30,34] established a bilayer model with coexisting drag-reducing and Newtonian fluids. The drag-reducing fluid part was presented by the Giesekus model. Newtonian and viscoelastic fluids separately flow at different layers with the interface parallel to the walls between them. The different features of drag reduction in different regions can be modeled by changing the positions and thicknesses of the layers. Simulation results reappeared the experimental phenomenon that heat transfer reduction rate is greater than the drag reduction rate and showed that the network structure induced by surfactant additives plays an important role in drag reduction and reduction at the buffer layer, and it imposes more effect on heat transfer reduction than on drag reduction, especially in the layer next to the wall. The DNS by the model also identified that the wall-normal turbulent heat flux decreases as viscoelastic flow layer expands, which is a primary reason for heat transfer reduction. Moreover, the close relation between speed streaks and thermal streaks indicated a close correlation between velocity and temperature.

On the other hand, Kajishima *et al.* [42] established a discrete-element model for drag-reducing channel flow, in which the polymer chain is treated as a spring–dumbbell element, that is, the polymer macromolecule is modeled as two beads linked by a spring and a damper. The mass of the polymer is assumed to gather on the beads, and the spring is assumed to not have any mass. The important model parameters are the spring coefficient representing the elasticity, the damping coefficient representing the viscosity, and the natural length of the polymer chain. Additionally, the force on polymer molecules by fluid was derived using the Stokes equation. Then the motion equation of polymer molecules was established according to Newton's Second Law of Motion. Finally, the additional stress on fluid by polymer was included into the N–S equation to relate all forces, according to Newton's Third Law of Motion. As a first step of research, they studied the effects of some important model parameters such as chain length, concentration, and viscoelasticity on drag reduction and obtained the mobility, gather, elongation, and rotation of polymer in flow. A quantitative relation between elongation and turbulent elastic energy was also discussed. Wang *et al.* [43,44] applied Kajishima's model to simulate the spatial distribution of polymer molecules and proved the phenomenon of a high concentration of polymer near-walls and a low concentration far from walls. The larger drag reduction, the further the non-uniform distribution of concentration. The main function of additives in the buffer layer was also proved. The near diminishment of the Reynolds shear stress found in the experiments was first reproduced in the simulation when the spring coefficient is relatively small. Hagiwara *et al.* [45] established a cluster model, considering multiple spring–dumbbells entanglement, to simulate the elongation of polymers, network structures, and their breaks.

4.1.2 The DNS Method of Drag-reducing Flow

Numerical methods of drag reduction include the spectrum methods and the finite difference method. Spectral DNSs for drag-reducing flow include the works of Sureshkumar *et al.* [10]; Dimitropoulus *et al.* [11]; Angelis *et al.* [12,13]; Vaithianathan *et al.* [16]; Li *et al.* [19]; and Li *et al.* [25]. Among these researches, Fourier or Chebyshev expansions were usually used for spatial discretization. The Crank–Nicolson or Runge–Kutta scheme was usually applied for temporal advancement. A representative work is what Vaithianathan *et al.* [16] did by using the FENE-P model. They addressed two bottle-neck difficulties in DNS by Cholesky decomposition: the polymer extension can numerically exceed the finite extensible length, causing the calculation to rapidly diverge; and the accumulation of numerical errors can cause a loss of symmetric and positive definites, leading to the growth of Hadamard instabilities. Numerical tests of the new algorithms showed significant departures from other approaches that rely on filtering to remove the instabilities.

Classical DNSs by the finite difference method are as follows. Hagiwara *et al.* [45] used a second-order central difference scheme to treat the convective and diffusion terms. Sibilla *et al.* [18] combined the second-order finite difference scheme with the projection method to ensure the continuity condition. Tamano *et al.* [26] utilized a second-order difference scheme on a staggered grid. The coupling algorithm of velocity and pressure was the second-order splitting method. The pressure Poisson equation was solved using the successive over-relaxation (SOR) method after fast Fourier transformation (FFT). The second-order upwind difference scheme was used for the polymer–stress convection term. Min *et al.* [22,44] treated the stress terms in a momentum equation by a fourth-order compact difference scheme and the convective terms in a constitutive equation by a corrected upwind compact scheme, using a second-order central difference scheme for other terms. Yu *et al.* applied second-order finite difference schemes for the momentum equation and the MINMOD scheme for the constitutive equation [27,28]. Dimitropoulos *et al.* [17] applied a second-order difference scheme to velocity and pressure terms and a fourth-order compact difference scheme to the gradient of stress terms in the momentum equation. A third-order upwind compact scheme was used for conformation tensors in the constitutive equation. A multigrid method was introduced into the pressure Poisson equation to accelerate the iteration. Their later research applied the finite volume method on a staggered grid and high-order compact schemes to the stress terms in the momentum and conformation evolution equations [41].

For temporal advancement, Sibilla *et al.* [18] utilized a third-order Runge–Kutta scheme in the drag reduction DNS. A semifractional step method was also widely used in the DNS of drag reduction [17,24,41]. Min *et al.* [22] advanced diffusion terms and additional stress terms in a momentum equation by the Crank–Nicolson scheme and convective terms by the third order Runge–Kutta scheme. Tamano *et al.* [26] selected a semi-implicit scheme for time advancement, that is, the Crank–Nicolson scheme for diffusion terms in the wall-normal direction and the third-order Runge–Kutta scheme for other terms.

Constitutive equations are hyperbolic equations without diffusion (viscous) terms. They therefore easily generate numerical instability to cause computation not converged. Strategies to enhance their numerical stability need to be studied. To solve this problem, the artificial viscous method has been widely used. Sureshkumar and Beris [46] noted that it was necessary to introduce artificial viscous terms in the constitutive equation to ensure numerical stability. Dimitropoulus *et al.* [11] introduced an artificial viscous term into the FENE-P and Giesekus models. Their numerical simulation first succeeded in viscoelastic channel flow. Min *et al.* [47] specially treated the convective terms in the FENE-P model and obtained a converged solution of channel turbulence at a relatively high Weissenberg number by the modified artificial viscous scheme. Other representative research using the artificial viscous term method can be found in references [10,12,17,19,25,26, 41]. Yu and his coworkers introduced the MINMOD scheme to the Giesekus constitutive equation for convective discretization. They obtained converged solutions at a high Weissenberg number ($We_\tau = 54$) and compared drag-reducing results by the MINMOD scheme with those by the artificial viscous scheme. An important conclusion was drawn that the artificial viscous scheme enhanced numerical stability but brought large artificial diffusion effect, causing a higher onset and lower drag reduction rate [27,28]. Hence, they noted that DNS should avoid artificial viscous terms. Based on this opinion, Yu *et al.* further proposed a faithful finite difference method for direct numerical simulation of drag-reducing flow [28]. This method dealt with the convective terms in constitutive equation using a high-resolution scheme, MINMOD, which has good stability and accuracy. This measurement can not only ensure high accuracy and avoid large artificial diffusion but also enhance numerical stability. Other spatial terms were discretized by a second-order central difference scheme. Temporal advancement used the Adams–Bashforth scheme. To fit dense near-wall mesh for DNS and avoid a zigzag pressure field, non-uniform staggered grids were used. The multigrid method was used to accelerate the iteration of the implicit pressure equation. After a lot of case verification and comparison with experiments, the proposed method can ensure both the accuracy and stability, and the computational works were comparable with the artificial method. The largest shortcoming of DNS for turbulence is huge computational consumptions. It is very necessary to reduce the consumptions and improve computational efficiency on the basis of the same accuracy and stability. One reason of this shortcoming is primarily that turbulence has a wide wave spectrum and frequency spectrum. Thus, DNS requires high spatial and time resolutions, that is, computational mesh must be dense enough to resolve the smallest scale turbulence. Moreover, enough samples of flow field should be captured to obtain the statistic characteristics of turbulence. Another reason for the huge computational works is the complexity of the DNS algorithm. For a single computer, DNS can only use sparse grid systems and low Reynolds numbers due to the limitation of computing speed and storage. Even in this case, computation is still time consuming to obtain a statistic converged result. This is not acceptable for engineering application and an important reason for not applying DNS to engineering.

4.2 RANS of Drag-reducing Flow

According to the previous section, direct numerical simulation can obtain all detailed instantaneous solutions of turbulence, but the computational consumptions needed are huge. Thus, it is not economic, or not acceptable, to apply DNS to engineering problems. DNS has been limited within the mechanism study of turbulence and drag reduction. In the opinion on real application, the details about the generation and development of turbulence, even large-scale turbulence, are not concerns in engineering. Those concerns followed more intensively engineering are whether mean velocity, mean friction, and fluctuation intensity can be accurately calculated. Therefore, cloture of Reynolds averaged equations will be an engineering method with commonality and economy in the latest future, that is, the turbulence model. Its general idea is to establish the relations between high- and low-order statistics. The Newtonian turbulence model will be summarized first to introduce the origin and development of this kind of method.

The study of the Newtonian turbulence model is more mature than that of drag-reducing flow. There have been successful high- and low-order turbulent models. Compared to turbulence models for Newtonian fluid flow, the turbulence model for drag-reducing flow develop slowly due to their complexity. Poreh *et al.* [48] and Edwards *et al.* [49], and Azouz *et al.* [50] applied a zero-equation model (an eddy viscosity model) and expressed the additive effect by modifying the Van Driest damping function. Durst *et al.* [51], Hassid *et al.* [52,53], and Poreh *et al.* [54] developed one-equation and two-equation turbulence models. Hassid *et al.* found that the turbulent length scale of the one-equation model was hard to determine. Hassid and Poreh modified the low Reynolds number Newtonian turbulence model by Jones and Launder [55]. They concluded that the damping function must be adjusted by fluid, the Reynolds number, and the pipe to ensure accurate prediction. Although all these methods showed that the modification of damping function can predict some flows precisely, they all have no commonality. Due to the complexity of drag-reducing flow, its turbulence model stopped to develop until the 1990s. The recent development of flow measurement and DNS led researchers to realize that the lack of the commonality of the above turbulence models relies on the lack of rheological properties.

Cruz and Pinho [56] established a low Reynolds number k–ε turbulence model for drag-reducing pipe flow. The model used a nonlinear molecular viscosity that is affected by the turbulent fluctuations, and a new damping function was introduced to account for near-wall effects. Flow predictions compared favorably with results from experiments with several viscoelastic fluids, especially the friction factor and the mean velocity. Comparisons of the turbulence kinetic energy were less good, but the model was able to capture the shift of the peak turbulence kinetic energy and rate of dissipation away from the wall, the decrease in those peak values and of the production of the turbulence kinetic energy, and the Reynolds shear stress deficit across the pipe. An advantage of the turbulence cloture is the fact that the input is only the mean velocity and fluid properties, and no other modifications of the turbulence model are required to deal with different fluids and pipes. The drawback of this model is large gaps between the prediction and the experiment for turbulent kinetic energy and its peak value position. The authors pointed out the two different paths for further investigations: (1) improvements to the current turbulence model including the development of more advanced models to predict the full Reynolds stress tensor, but based on the same or closely related rheological constitutive equation of the generalized Newtonian type; and (2) the adoption of more advanced and realistic viscoelastic constitutive equations that are able to predict shear thinning, Trouton ratio thickening, and the existence of normal stresses.

According to the first improvement, Pinho [57] expanded the model in reference [58] by the generalized Newtonian fluid (GNF) framework to account for strain thickening of the extensional viscosity and derived a new low Reynolds number k–ε model. A closed form for the average molecular viscosity was also considered. The turbulence model was qualitatively shown to increase the rate of decay of turbulent kinetic energy in isotropic grid turbulence for certain rheological conditions and accurately predicted shear thinning. It was also shown that the new k–ε model can predict a slower rate of decay of grid-generated turbulence than that of Newtonian fluids for fluids having a Trouton-thickening viscosity variation, in agreement with results from the literature. Thus, this work largely promoted the advancement of a drag-reducing turbulence model. For the second improvement above, Cruz *et al.* [58] included the new stress term previously neglected by Cruz and Pinho [56] in the momentum and turbulent kinetic energy equations for drag-reducing viscolastic fluids. The new stress accounted for the correlation between fluctuations of viscosity and fluctuations of the rate of deformation. With the inclusion of this new stress, the predictions of turbulent kinetic energy in drag-reducing pipe flow were significantly improved at the cost of a slight deterioration in the prediction of other quantities for some of the fluids. To further improve Cruz's model [58], Resende *et al.* [59] extended the viscoelastic isotropic low Reynolds number k–ε model of Cruz *et al.* [58] to anisotropy, based on the model of Park *et al.* [60] for the turbulent flow of Newtonian fluids. The isotropic part of the model has also been modified and shown to improve the predictions of the friction factor, the mean velocity, and the turbulent kinetic energy by the results comparison with four different solutions in pipe measurements.

The above drag-reducing viscoelastic turbulence models developed gradually with increasing predicting function and precision. However, the turbulence models proposed by Pinho and his coworkers considered viscoelasticity much simpler than the constitutive equations such as the FENE-P or Giesekus relation. These models were also not combined with the DNS database, so the results were not quite satisfied. Hence, Pinho *et al.* [61] used the FENE-P constitutive equation and obtained the primary relation between flow and additives by analysis of DNS data. They developed a single-point k–ε model for drag-reducing channel flow by introducing two viscoelastic terms, based on the Newtonian model of Nagano and Hishida [62]. The Reynolds stress was modeled by Boussinesq's eddy viscosity hypothesis, including the dissipation rate of turbulent kinetic energy of drag-reducing solution, not only the dissipation of solvent. The terms in derived equations of turbulent kinetic energy and dissipation were modeled using physical arguments, order-of-magnitude estimations, dimensional analysis, and comparison with the Newtonian model, that is, neglecting secondary and small order of magnitude variables or using Newtonian part modeling. Predictions were made for a drag reduction rate of 18%, showing good prediction of mean velocity, Reynolds stress, turbulent kinetic energy, dissipation, and onset drag reduction. The methodology from this model is extensible to the more reliable second-order closures for the Reynolds stresses and higher drag reduction. The model does not succeed in capturing the region of negative nonlinear terms close to the wall. As the Weissenberg number tends to 0, the model does not appear to asymptote to its Newtonian limits and the frictions predicted are larger than those of Newtonian flow. Recently, Resende *et al.* [63] improved the above model and modified the model parameters by DNS. The application sphere of this model extended from low drag reduction region to medium drag reduction region (up to 50%). Three main improvements were incorporated: (1) a modified eddy viscosity cloture; (2) the inclusion of direct viscoelastic contributions into the transport equations for turbulent kinetic energy and its dissipation rate; and (3) a new cloture for the cross-correlations between the fluctuating components of the polymer conformation and rate of strain tensors. The predicting results by the new model show improvements of mean velocity, turbulent kinetic energy, dissipation, conformation tensor, viscoelastic stress, and Reynolds stress compared with the previous model. The continuous works by Pinho and his coworkers is encouraging and is gradually improving the viscoelastic k–ε model for drag-reducing flow.

A few researchers attempted a second-order model. Leighton [64] established a Reynolds stress model according to the FENE-P constitutive equation. This model is suitable for the flows in boundary layer, duct, and pipe. The model parameters were determined by DNS data. The predicted mean velocity and strain are in good agreement with DNS data, but other variables are not.

4.3 Governing Equation and DNS Method of Drag-reducing Flow

Numerical simulation of drag-reducing flow usually uses channel or pipe geometry. Due to the simplicity of channel flow for the study and wide application of plane channels as the research domain in most references, drag-reducing channel flow is selected here to introduce the governing equation and numerical method.

The computational domain is shown in Figure 4.1. The x, y, and z represent the stream-wise, wall-normal, and span-wise directions, respectively, and the corresponding velocity components are u, v, and w (abbreviated by u_i [$i = 1,2,3$ represents the x, y, and z directions]). Periodic velocity boundary conditions are imposed in both the stream-wise and span-wise direction, while the nonslip condition is adopted for the top and bottom walls. For an energy equation, constant wall temperatures or constant heat flux boundary conditions are usually selected. Domain discretization and numerical method in DNS will be summarized briefly.

4.3.1 Governing Equation

For Newtonian fluid flow, the governing equations are the continuity equation, Navier–Stokes equation, and energy equation as follows:

Continuity equation:

$$\frac{\partial u_i}{\partial x_i} = 0 \tag{4.1}$$

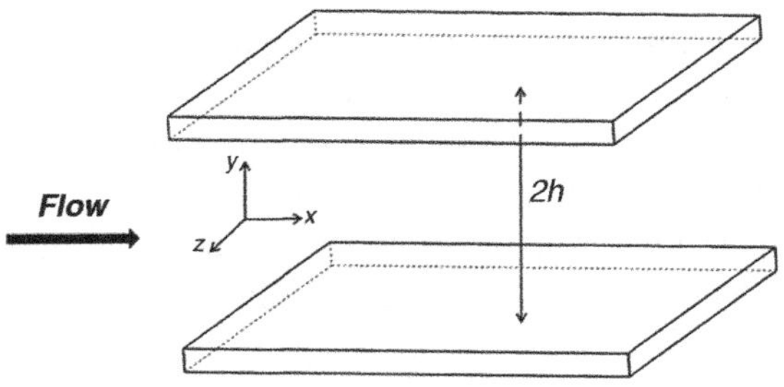

Figure 4.1 Computational domain of channel flow

Navier–Stokes equation:

$$\frac{\partial u_i}{\partial t} + u_j \frac{\partial u_i}{\partial x_j} = -\frac{1}{\rho}\frac{\partial p}{\partial x_i} + \frac{\mu}{\rho}\frac{\partial}{\partial x_j}\left(\frac{\partial u_i}{\partial x_j}\right) \tag{4.2}$$

Energy equation:

$$\frac{\partial T}{\partial t} + u_j \frac{\partial T}{\partial x_j} = \frac{\lambda}{\rho c_p}\frac{\partial}{\partial x_j}\left(\frac{\partial T}{\partial x_j}\right) \tag{4.3}$$

where p is pressure, T is temperature, ρ is density, μ is dynamic viscosity, and λ is relaxation time. For drag-reducing flow by additives, the additional stress generated by the interaction between additives and solvent cannot be described by the N–S equation. It needs to be represented by a constitutive equation, as which the Giesekus model is selected. The governing equations of drag-reducing flow are as follows (the continuity and energy equations are neglected because they are the same as those of Newtonian flow):

Momentum equation:

$$\frac{\partial u_i}{\partial t} + u_j \frac{\partial u_i}{\partial x_j} = -\frac{1}{\rho}\frac{\partial p}{\partial x_j} + \frac{\mu}{\rho}\frac{\partial}{\partial x_j}\left(\frac{\partial u_i}{\partial x_j}\right) + \frac{1}{\rho}\frac{\partial \tau_{ij}}{\partial x_j} \tag{4.4}$$

Giesekus constitutive equation:

$$\tau_{ij} + \lambda\left[\frac{\partial \tau_{ij}}{\partial t} + u_m \frac{\partial \tau_{ij}}{\partial x_m} - \tau_{im}\frac{\partial u_j}{\partial x_m} - \tau_{mj}\frac{\partial u_i}{\partial x_m} + \frac{\alpha}{\eta}\left(\tau_{im}\tau_{mj}\right)\right] = \eta\left(\frac{\partial u_j}{\partial x_i} + \frac{\partial u_i}{\partial x_j}\right) \tag{4.5}$$

where the additional stress τ_{ij} is induced by the conformation of network c_{ij}.They have the relation:

$$\tau_{ij} = \frac{c_{ij} - \eta\delta_{ij}}{\lambda} \tag{4.6}$$

where δ_{ij} is the Kronecker symbol, and η is the zero shear viscosity of additives. Substituting Equation 4.6 to the governing equations, the latter can be transformed into the following forms. Therefore, the calculation of additional stress is transformed to the calculation of conformation.

Momentum equation:

$$\frac{\partial u_i}{\partial t} + u_j \frac{\partial u_i}{\partial x_j} = -\frac{1}{\rho}\frac{\partial p}{\partial x_j} + \frac{\mu}{\rho}\frac{\partial}{\partial x_j}\left(\frac{\partial u_i}{\partial x_j}\right) + \frac{1}{\rho}\frac{\partial c_{ij}}{\partial x_j} \tag{4.7}$$

Giesekus constitutive equation:

$$\frac{1}{\lambda}\left(c_{ij}-\delta_{ij}\eta\right)+\frac{\partial c_{ij}}{\partial t}+u_m\frac{\partial c_{ij}}{\partial x_m}-\left(c_{im}-\delta_{im}\eta\right)\frac{\partial u_j}{\partial x_m}-\left(c_{mj}-\delta_{mj}\eta\right)\frac{\partial u_i}{\partial x_m}$$
$$+\frac{\alpha}{\lambda\eta}\left(c_{im}-\delta_{im}\eta\right)\left(c_{mj}-\delta_{mj}\eta\right)=\eta\left(\frac{\partial u_j}{\partial x_i}+\frac{\partial u_i}{\partial x_j}\right) \tag{4.8}$$

The half height of the channel h is used as the characteristic length, and the friction velocity u_τ as the characteristic velocity, and normalized time, pressure, additional stress, and temperature by h/u_τ, ρu_τ^2, $\eta u_\tau/h$ and $q_w/\rho c_p u_\tau$ respectively. The dimensionless variables are as follows:

$$x_i^*=\frac{x_i}{h},t^*=\frac{t}{h/u_\tau},u_i^*=\frac{u_i}{u_\tau},p^+=\frac{p}{\rho u_\tau^2}=\frac{\bar{p}+p'}{\rho u_\tau^2}=\overline{p^+}+p^{+^*},$$

$$\tau_{ij}^+=\frac{\tau_{ij}}{\eta u_\tau/h}=\frac{1}{\eta u_\tau/h}\begin{pmatrix}\tau_{xx} & \tau_{xy} & \tau_{xz}\\ \tau_{xy} & \tau_{yy} & \tau_{yz}\\ \tau_{xz} & \tau_{yz} & \tau_{zz}\end{pmatrix},$$

$$T_\tau=\frac{q_w}{\rho c_p u_\tau},\theta^+=\frac{\overline{T_w}-T}{T_\tau}$$

According to the definitions above, the dimensionless governing equations for drag-reducing flow are as follows:

Continuity equation:

$$\frac{\partial u_i^+}{\partial x_i^*}=0 \tag{4.9}$$

Momentum equation:

$$\frac{\partial u_i^+}{\partial t^*}+u_j^+\frac{\partial u_i^+}{\partial x_j^*}=-\frac{\partial p^{+^*}}{\partial x_i^*}+\frac{1}{\mathrm{Re}_\tau}\frac{\partial}{\partial x_j^*}\left(\frac{\partial u_i^+}{\partial x_j^*}\right)+\frac{\beta}{We_\tau}\frac{\partial c_{ij}^+}{\partial x_j^*}+\delta_{1i} \tag{4.10}$$

where $\delta_{1i}=\begin{cases}1 & (i=1)\\ 0 & (other)\end{cases}$.

Giesekus constitutive equation:

$$\frac{\partial c_{ij}^+}{\partial t^*}+\frac{\partial u_m^+c_{ij}^+}{\partial x_m^*}-\frac{\partial u_i^+}{\partial x_m^*}c_{mj}^+-\frac{\partial u_j^+}{\partial x_m^*}c_{mi}^+$$
$$+\frac{\mathrm{Re}_\tau}{We_\tau}\left[-\delta_{ij}+c_{ij}^++\alpha\left(c_{im}^+-\delta_{im}\right)\left(c_{mj}^+-\delta_{mj}\right)\right]=0 \tag{4.11}$$

Energy equation:

$$\frac{\partial \theta^+}{\partial t^*} + u_j^+ \frac{\partial \theta^+}{\partial x_j^*} = \frac{2u^+}{\displaystyle\int_{-1}^{1} U^+ dy^*} + \frac{1}{\mathrm{Re}_\tau \mathrm{Pr}} \frac{\partial}{\partial x_j^*}\left(\frac{\partial \theta^+}{\partial x_j^*}\right) \tag{4.12}$$

where Re_τ is the Reynolds number based on the friction velocity:

$$\mathrm{Re}_\tau = \frac{\rho u_\tau h}{\mu} \tag{4.13}$$

where We_τ is the Weissenberg number, representing dimensionless relaxation time. The larger the Weissenberg number, the stronger the elasticity of drag-reducing solution. Its definition is

$$We_\tau = \lambda \frac{\rho u_\tau^2}{\mu} \tag{4.14}$$

where λ is the relaxation time of the viscoelastic solution, μ is the zero shear-rate viscosity of solvent, and β is the zero shear-rate viscosity ratio of the additive contribution to the solvent contribution:

$$\beta = \frac{\eta}{\mu} \tag{4.15}$$

Another definition of β is the solvent contribution to the total zero shear-rate viscosity of the solution:

$$\beta = \frac{\mu}{\mu + \eta} \tag{4.16}$$

The dimensionless momentum equation based on the second definition of β has the following form:

$$\frac{\partial u_i^+}{\partial t^*} + u_j^+ \frac{\partial u_i^+}{\partial x_j^*} = -\frac{\partial p^+}{\partial x_i^*} + \frac{\beta}{\mathrm{Re}_\tau} \frac{\partial}{\partial x_j^*}\left(\frac{\partial u_i^+}{\partial x_j^*}\right) + \frac{1-\beta}{We_\tau} \frac{\partial c_{ij}^+}{\partial x_j^*} + \delta_{1i} \tag{4.17}$$

4.3.2 Numerical Method

Due to the very important role of the wall to the generation and development of turbulence, the velocity and temperature gradients along the wall-normal direction are large. To capture the severe variations of the near-wall flow and heat transfer information, non-uniform grids are used in the normal direction with denser mesh near the wall. However, the increase of grid numbers inevitably causes a greater increase of computational works. Moreover, the mesh density far from the walls is not necessarily the same as that of the near-wall region. Hence, local denser non-uniform mesh near walls is applied to ensure capturing near-wall flow and reducing computational works efficiently.

The transformation of the non-uniform grid points along the y is shown as follows:

$$y_j = \frac{h}{a}\tanh\left(\frac{1}{2}\zeta_j \ln\frac{1+a}{1-a}\right) \tag{4.18}$$

$$\zeta_j = -1 + 2\frac{j}{N} \tag{4.19}$$

where j represents the jth grid point in the y direction, N is the total grid numbers in the y direction, and a is an adjustable parameter of the transformation. A constant value of $a = 0.95$ is adopted herein (Figure 4.2). The periodic stream-wise and span-wise directions can use uniform grids with larger grid spacing than the wall-normal direction [65].

If one put the pressure and velocity on the same grid point without any special treatment, the pressure field obtained by DNS will be zigzag, violating the physical meaning. To avoid this phenomenon, a staggered grid is applied. A staggered grid is a system that stores the velocity components u, v, and w and the scalar field, including pressure p in four different grids as shown in Figure 4.3. Pressure p stores in the cell center (main cell). Velocity component u stores at the east and west borders of the main cell. Velocity component v stores at the south and north borders of the main cell, while

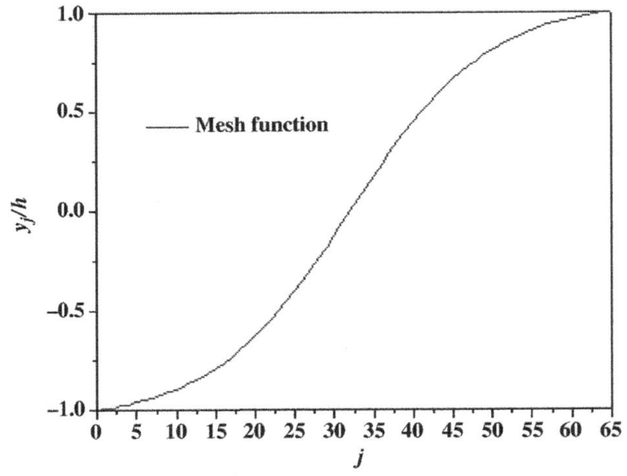

Figure 4.2 Mesh function

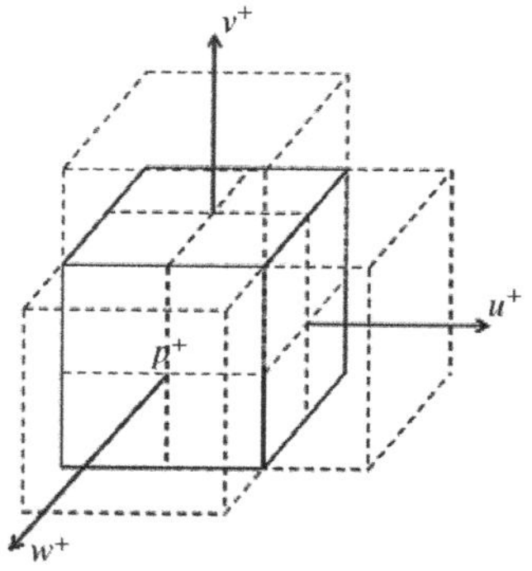

Figure 4.3 Schematic of a staggered grid

velocity component w stores at the front and back borders. The cells of u, v, and w determine the center as the corresponding locations of the velocity components and have half grid step overlap with the main cell. This is the name origin of the staggered grid, which was first proposed by Harlow and Welch in 1965 [66]. The method can not only avoid the unreal zigzag pressure field but also avoid the difficulty of the pressure boundary condition (no boundary pressure is needed in this case) and effectively control calculation errors.

Frequently used temporal advancement schemes in most references are the Adams–Bashforth and Runge–Kutta schemes. They will be introduced as follows where ϕ represents universe variables and L_h represents the discretization form of spatial terms, which will be introduced in the next section. The superscript represents time level.

1. Adams–Bashforth scheme

$$\phi^{n+1} = \phi^n + \Delta t \cdot \left[\frac{3}{2}L_h(\phi^n) - \frac{1}{2}L_h(\phi^{n-1})\right] \tag{4.20}$$

2. Third-order Runge–Kutta scheme

$$\begin{aligned} \phi^1 &= \phi^n + \Delta t \cdot L_h(\phi^n) \\ \phi^2 &= \frac{3}{4}\phi^n + \frac{1}{4}\left[\phi^1 + \Delta t \cdot L_h(\phi^1)\right] \\ \phi^{n+1} &= \frac{1}{3}\phi^n + \frac{2}{3}\left[\phi^2 + \Delta t \cdot L_h(\phi^2)\right] \end{aligned} \tag{4.21}$$

Uniform grids are taken as an example to introduce commonly used schemes for spatial discretization. The discretization of convective terms relates to the numerical stability so that only they are introduced here.

1. Second-order upwind difference scheme

$$\left.\frac{\partial \phi}{\partial x}\right|_i = \begin{cases} \dfrac{3\phi_i - 4\phi_{i-1} + \phi_{i-2}}{2\Delta x} & (u \geq 0) \\ \dfrac{-3\phi_i + 4\phi_{i+1} - \phi_{i+2}}{2\Delta x} & (u < 0) \end{cases} \tag{4.22}$$

2. Third-order upwind difference scheme

$$\left.\frac{\partial \phi}{\partial x}\right|_i = \begin{cases} \dfrac{2\phi_{i+1} + 3\phi_i - 6\phi_{i-1} + \phi_{i-2}}{6\Delta x} & (u \geq 0) \\ \dfrac{-\phi_{i+2} + 6\phi_{i+1} - 3\phi_i - 2\phi_{i-1}}{6\Delta x} & (u < 0) \end{cases} \tag{4.23}$$

3. Second-order central difference scheme

$$\left.\frac{\partial \phi}{\partial x}\right|_i = \frac{\phi_{i+1} - \phi_{i-1}}{2\Delta x} \tag{4.24}$$

4. Fourth-order central difference scheme

$$\left.\frac{\partial \phi}{\partial x}\right|_i = \frac{\phi_{i-2} - 8\phi_{i-1} + 8\phi_{i+1} - \phi_{i+2}}{12\Delta x} \tag{4.25}$$

5. Compact difference scheme

$$\Delta x \cdot \sum_{i=k_1}^{k_2} \alpha_i \left.\frac{\partial \phi}{\partial x}\right|_i = \sum_{l=l_1}^{l_2} \beta_l \phi_l \tag{4.26}$$

6. MINMOD scheme

$$\tilde{\phi}_f = \begin{cases} \dfrac{3}{2}\tilde{\phi}_C & 0 \leq \tilde{\phi}_C \leq \dfrac{1}{2} \\ \dfrac{1}{2}\left(1 + \tilde{\phi}_C\right) & \dfrac{1}{2} < \tilde{\phi}_C \leq 1 \\ \tilde{\phi}_C & other \end{cases} \tag{4.27}$$

where $\tilde{\phi} = \frac{\phi - \phi_U}{\phi_D - \phi_U}$, and $_{C,\ U,\ and\ D}$ represent the center, upward, and downward of the grid point i.

The momentum equations of drag-reducing flow were discretized using the above numerical method:

$$\frac{u_{i,j,k}^{+(n+1)} - u_{i,j,k}^{+(n)}}{\Delta t^*} = -\left(\frac{\partial p^{+^*}}{\partial x^*}\right)^{(n+1)} + (u\_disc)^* \tag{4.28}$$

$$u_{i,j,k}^{+(n+1)} = u_{i,j,k}^{+(n)} + \Delta t^* (u\_disc)^* - \Delta t^* \left(\frac{\partial p^{+^*}}{\partial x^*}\right)^{(n+1)} \tag{4.29}$$

where $(u\_disc)^*$ is the discrete form of the convective and diffusion terms. The Adams–Bashforth scheme is used for temporal advancement:

$$(u\_disc)^* = \frac{3}{2} u\_disc^{(n)} - \frac{1}{2} u\_disc^{(n-1)} \tag{4.30}$$

The discrete momentum equations are

$$u_{i,j,k}^{+(n+1)} = u_{i,j,k}^{+(n)} + \Delta t^* \left(\frac{3}{2} u\_disc^{(n)} - \frac{1}{2} u\_disc^{(n-1)}\right) - \Delta t^* \left(\frac{\partial p^{+^*}}{\partial x^*}\right)^{(n+1)} \tag{4.31}$$

Set $\langle u^+ \rangle_{i,j,k} = u_{i,j,k}^{+(n)} + \Delta t^* \left(\frac{3}{2} u\_disc^{(n)} - \frac{1}{2} u\_disc^{(n-1)}\right)$. Then the discrete momentum equations transform to

$$u_{i,j,k}^{+(n+1)} = \langle u^+ \rangle_{i,j,k} - \Delta t^* \left(\frac{\partial p^{+^*}}{\partial x^*}\right)^{(n+1)} \tag{4.32}$$

$$v_{i,j,k}^{+(n+1)} = \langle v^+ \rangle_{i,j,k} - \Delta t^* \left(\frac{\partial p^{+^*}}{\partial y^*}\right)^{(n+1)} \tag{4.33}$$

$$w_{i,j,k}^{+(n+1)} = \langle w^+ \rangle_{i,j,k} - \Delta t^* \left(\frac{\partial p^{+^*}}{\partial z^*}\right)^{(n+1)} \tag{4.34}$$

The above equations are substituted into the continuity equation on the time level $n + 1$, and then the pressure Poisson equation can be obtained:

$$\begin{aligned} &\left(\frac{\partial^2 p^{+^{*(n+1)}}}{\partial x^{*2}} + \frac{\partial^2 p^{+^{*(n+1)}}}{\partial y^{*2}} + \frac{\partial^2 p^{+^{*(n+1)}}}{\partial z^{*2}}\right)_{i,j,k} \\ &= \frac{1}{\Delta t^*}\left(\frac{\partial \langle u^+ \rangle_{i,j,k}}{\partial x^*} + \frac{\partial \langle v^+ \rangle_{i,j,k}}{\partial y^*} + \frac{\partial \langle w^+ \rangle_{i,j,k}}{\partial z^*}\right) \end{aligned} \tag{4.35}$$

The discrete form of the Poisson equation is

$$\frac{p_{i+1,j,k}^{+*(n+1)} - 2p_{i,j,k}^{+*(n+1)} + p_{i-1,j,k}^{+*(n+1)}}{\Delta x^{*2}} + \frac{p_{i,j,k+1}^{+*(n+1)} - 2p_{i,j,k}^{+*(n+1)} + p_{i,j,k-1}^{+*(n+1)}}{\Delta z^{*2}}$$

$$+ \frac{p_{i,j+1,k}^{+*(n+1)} - p_{i,j,k}^{+*(n+1)}}{\left(y_{j+1}^{*} - y_{j}^{*}\right)\left(y_{j+1/2}^{*} - y_{j-1/2}^{*}\right)} - \frac{p_{i,j,k}^{+*(n+1)} - p_{i,j-1,k}^{+*(n+1)}}{\left(y_{j}^{*} - y_{j-1}^{*}\right)\left(y_{j+1/2}^{*} - y_{j-1/2}^{*}\right)}$$

$$= \frac{1}{\Delta t^{*}}\left(\frac{\langle u^{+}\rangle_{i,j,k} - \langle u^{+}\rangle_{i-1,j,k}}{\Delta x^{*}} + \frac{\langle v^{+}\rangle_{i,j,k} - \langle v^{+}\rangle_{i,j-1,k}}{y_{j+1/2}^{*} - y_{j-1/2}^{*}} + \frac{\langle w^{+}\rangle_{i,j,k} - \langle w^{+}\rangle_{i,j,k-1}}{\Delta z^{*}}\right) \quad (4.36)$$

A multigrid method is applied to accelerate the iteration of the Poisson equation to obtain the pressure on the $n+1$ time level. Then, the velocity on the time level $n+1$ can be solved by substituting the pressure on the time level $n+1$. Finally, a time step advancement is fulfilled.

4.4 DNS Results and Discussion for Drag-reducing Flow and Heat Transfer

Numerical methods presented in Section 4.1.2 are applied to the surfactant drag-reducing DNS using the Giesekus constitutive relation. The simulation works mentioned in Section 4.1.1 [27–35] are summarized here. First, DNS for the drag-reducing flow induced by surfactant needs rheological parameters in a constitutive equation determined by experiments, which can also verify the reliability of the computing coding, and then the DNS for drag-reducing flow and heat transfer can be made, and the characteristic structures can be extracted to further study the mechanisms of drag reduction and heat transfer reduction. Second, many references show that the Weissenberg number plays the most important role in drag reduction so that it is studied in detail. Other parameters in the Giesekus constitutive equation are also fully inspected. Third, an inhomogeneous numerical model for surfactant drag-reducing flow, the bilayer model, will be introduced. The results for the above three groups will be summarized one by one as follows.

4.4.1 The Overall Study on Surfactant Drag Reduction and Heat Transfer by DNS

In the present study, we performed both numerical and experimental tests for a 75 ppm CTAC surfactant drag-reducing channel flow. The experimental measurements were

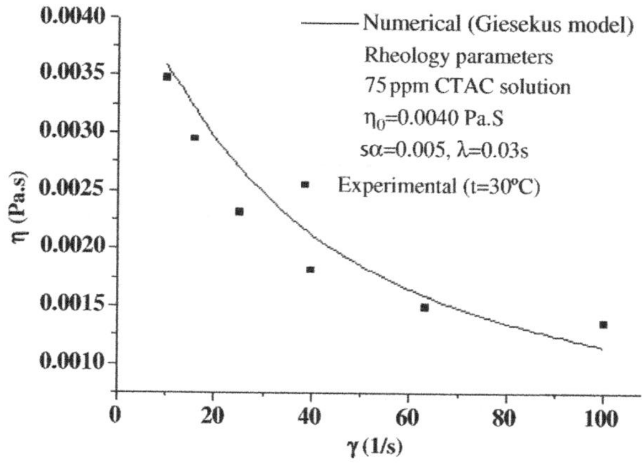

Figure 4.4 Shear viscosity versus shear rate (Reprinted from *International Journal of Heat and Fluid Flow*, **25**, Yu, B., Li, F.-C. and Kawaguchi, Y., "Numerical and experimental investigation on turbulence characteristics in a drag-reducing flow with surfactant additives," 961–974, 2004, with permission from Elsevier.)

carried out at a constant solution temperature of 30°C. Experiments were carried out at $\mathrm{Re}_b = 11350$ for the CTAC solution. Hence, the shear thinning in those studies was slight. From the present calculations, the contribution of surfactant additives is dominant in the zero shear-rate viscosity, and the solution is apparently shear thinning, as shown in Figure 4.4. A Giesekus constitutive equation was used to calculate the extra stress caused by the surfactant additives, and model parameters were obtained by well fitting the measured shear viscosities of the 75 ppm CTAC surfactant solution. Hence, in the numerical part of the present investigation, we directly connect realistic rheological properties (such as relaxation time, and the extensibility of the network structure in the surfactant solution) with the drag reduction rate. The relaxation time λ, mobility factor α, and additive contribution to zero-shear-rate viscosity are 0.3 s, 0.005, and 3.2×10^{-3} Pa · s, respectively.

The parameters are shown in Table 4.1. A calculation was carried out for surfactant solutions at $\mathrm{Re}_b = 12\,080$ ($\mathrm{Re}_\tau = 300$). The bulk Reynolds number is similar to that in the experiment ($\mathrm{Re}_b = 11\,350$). A computational box $10h \times 2h \times 5h$ in the x, y, and z directions was chosen. The computational domain in wall units (η_s, U_τ and ρ) is $3000 \times 600 \times 1500$. A grid system of $96 \times 128 \times 96$ (in x, y, and z) meshes was adopted. Non-uniform grids in the wall-normal direction were used with grid spacing

Table 4.1 Computational parameters of 75 ppm CTAC

	Re_τ	Re_m	We_τ	β	α	$L_x \times L_y \times L_z$	$N_x \times N_y \times N_z$	Δx^+	Δy^+	Δz^+
Surfactant	300	12 080	54	4	0.005	$10h \times 2h \times 5h$	$96 \times 128 \times 96$	31.3	0.4~9	15.6
Newtonian	380	13 070	×	0	×	$3h \times 2h \times 1.5h$	$128 \times 128 \times 128$	8.9	0.5~11	4.5

Table 4.2 Results of DNS (Reprinted from *International Journal of Heat and Fluid Flow*, **25**, Yu, B., Li, F.-C. and Kawaguchi, Y., "Numerical and experimental investigation on turbulence characteristics in a drag-reducing flow with surfactant additives," 961–974, 2004, with permission from Elsevier.)

Re_b	C_f	C_f^D	C_f^V	$DR\%$
12 080 (Numer.)	0.00493	0.00696	0.00248	29%
11 350 (Exp.)	0.00346	0.00707	0.00258	51%

of Δy^+ varying from about 0.45 next to the wall to 9 at the center. Uniform grids were used in the x and z directions, and the corresponding grid spacings were $\Delta x^+ = 31.3$ and $\Delta z^+ = 15.6$, respectively.

Table 4.2 shows that the measured drag reduction rate of 51% is higher than the predicted value of 29%. Both the numerical and the experimental friction factors were greater than Virk's asymptotic friction factor [67]. Figure 4.5 compares the mean stream-wise velocity profiles of the fully developed turbulence flow of a Newtonian fluid with that of the CTAC solution, for which the well-known relationships of Newtonian fluids – linear law in the viscous sublayer $u^+ = y^+$, logarithmic law $u^+ = 2.5 \ln y^+ + 5.0$ in the inner layer, and Virk's asymptote velocity profile for the drag-reducing flow – are given for comparison. The numerical and measured mean velocity profiles of the Newtonian fluid are in good agreement with $u^+ = y^+$ in the viscous sublayer and with $u^+ = 2.5 \ln y^+ + 5.0$ in the logarithmic layer, as shown in Figures 4.5a and 4.5b. Both calculated and experimental results show that the addition of surfactant additives dramatically decreases the mean velocity values as well as their gradient in the viscous sublayer. The mean velocity profile of surfactant solution is also shifted upward in the logarithmic layer with a steeper gradient than that of the Newtonian fluid, and a narrower logarithmic layer (a wider buffer layer) is established with the additives. Interestingly, the velocity profile of the surfactant solution (Figure 4.5b) more closely approaches Virk's asymptote velocity profile, compared with that shown in Figure 4.5a, which corresponds to a higher drag reduction rate.

The root-mean-square velocity fluctuations obtained from the numerical simulations and the experiments are presented in Figures 4.6a and 4.6b respectively. In Figure 4.6b, only two components (u^+_{rms} and v^+_{rms}) are given because our PIV measurement is two-dimensional. Both the numerical and experimental results show that, with the addition of surfactant additives, the peak positions of the stream-wise velocity fluctuation intensity u^+_{rms} are located wider apart from the wall than those for the Newtonian flows. In the experiment, surfactant additives decrease u^+_{rms}, but in the numerical calculations, they increase it. For larger Reynolds numbers $Re_b = 3 \times 10^4$, and the experimental results show that u^+_{rms} is enhanced compared with the Newtonian fluid. Thus, the decrease or increase of u^+_{rms} is not an essential feature of the drag-reducing flow caused by surfactant additives, while the shift of its peak-value position

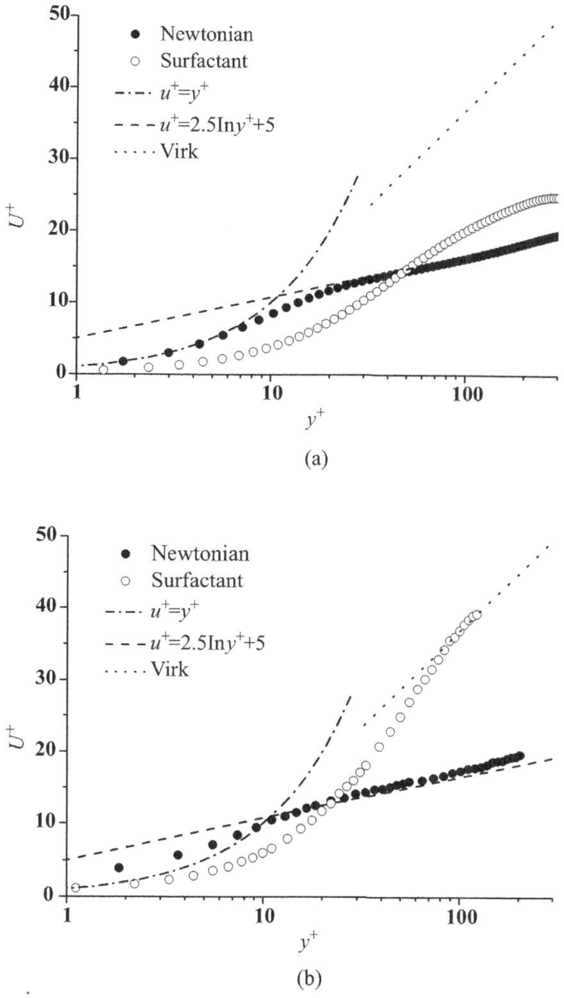

Figure 4.5 Mean velocity profiles: (a) numerical, and (b) experimental (Reprinted from *International Journal of Heat and Fluid Flow*, **25**, Yu, B., Li, F.-C. and Kawaguchi, Y., "Numerical and experimental investigation on turbulence characteristics in a drag-reducing flow with surfactant additives," 961–974, 2004, with permission from Elsevier.)

to the bulk flow is an essential feature, which is consistent with a wider buffer layer. Both the numerical calculations and the experiments show that v_{rms}^+ decreases appreciably for the surfactant solution. A decrease in w_{rms}^+, by adding surfactant additives, is also seen in Figure 4.6a. The decrease in the velocity fluctuation intensities in the wall-normal and span-wise directions, as to be shown later, is due to the energy redistribution.

The total shear stress can be used to confirm if the calculation has reached a statistically steady state. When a steady state is reached, the following balance

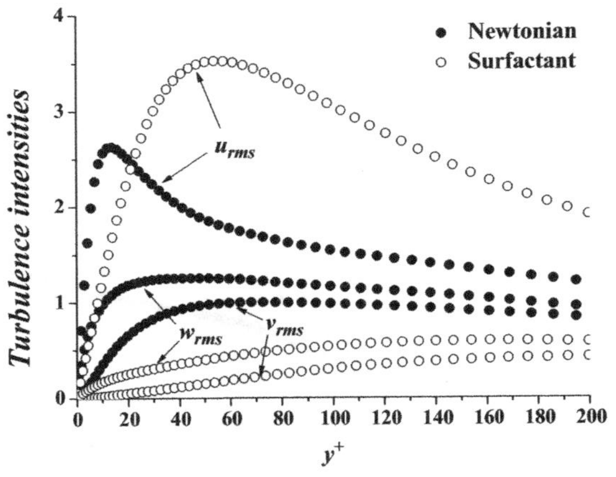

(a)

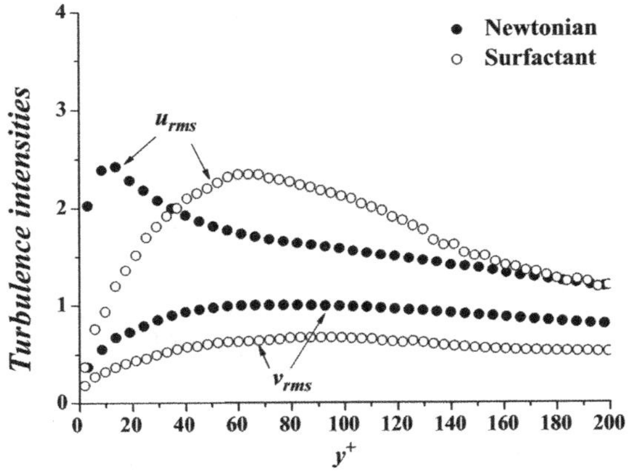

(b)

Figure 4.6 Root-mean-square velocity fluctuations: (a) numerical, and (b) experimental (Reprinted from *International Journal of Heat and Fluid Flow*, **25**, Yu, B., Li, F.-C. and Kawaguchi, Y., "Numerical and experimental investigation on turbulence characteristics in a drag-reducing flow with surfactant additives," 961–974, 2004, with permission from Elsevier.)

equation is satisfied:

$$\tau_{total} = -\overline{u'^+v'^+} + \frac{\partial U^+}{\partial y^+} + \beta \frac{C_{xy}^+}{We_\tau} \tag{4.37}$$

The $\tau_{total}\left(= 1 - \frac{y^+}{\mathrm{Re}_\tau}\right)$ is total stress. The last term on the right of Equation 4.37 is the viscoelastic stress, which is equal to zero for the Newtonian fluid flow. The mean velocity gradient in the viscous sublayer that becomes more moderate with the surfactant additives can be explained by Equation 4.37. In the viscous sublayer, the

Reynolds shear stress is negligible, due to the positive viscoelastic stress (as shown in Figure 4.5), where the mean velocity gradient becomes smaller to satisfy the balance equation.

Statistically, steady states were confirmed for both the Newtonian fluid and the CTAC surfactant solution (Figure 4.7a). Viscous shear stress is seen to increase appreciably, except for the decrease in the vicinity of the wall ($y^+ < 10$), with the addition of surfactant additives. The Reynolds shear stress decreases dramatically

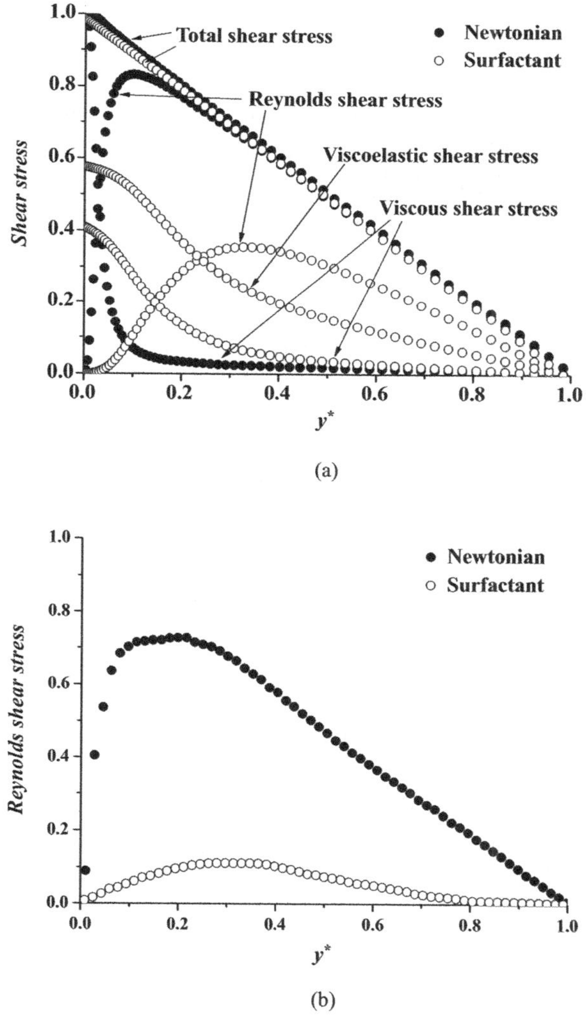

Figure 4.7 Budget of shear stress; (a) numerical, and (b) experimental (Reprinted from *International Journal of Heat and Fluid Flow*, **25**, Yu, B., Li, F.-C. and Kawaguchi, Y., "Numerical and experimental investigation on turbulence characteristics in a drag-reducing flow with surfactant additives," 961–974, 2004, with permission from Elsevier.)

with its maximum value around half that of the Newtonian fluid, and the location where Reynolds shear stress reaches its maximum value shifts from the wall region to the bulk flow. The decrease of the Reynolds shear stress is due to the increase in the viscous shear stress and the positive contribution made by the viscoelastic shear stress, with a greater effect from the latter. The viscoelastic stress is the largest component in the near-wall region, where the Reynolds shear stress decreases most appreciably. The viscoelastic stress is about 50% of the total shear stress in the whole region of the channel, indicating that a small amount of additives greatly changes the balance of the stress budget. The decrease in the Reynolds shear stress was observed (Figure 4.7b), and the measured Reynolds shear stresses were smaller than the predicted values.

To quantitatively identify the contributions made by each of these components to the friction factor, a twofold (double) integration is applied $\int_0^1 \int_0^y dy^* dy^*$ to Equation 4.37. Transforming the multiple integration to a single integration is performed by applying the integration by parts, to obtain an integration expression for the friction factor:

$$C_f = 12/\mathrm{Re}_b + 6\int_0^1 \frac{\left(-\overline{u'^+v'^+}\right)(1-y^*)}{U_b^{+2}} dy^* + 6\int_0^1 \beta \frac{C_{xy}^+}{We_\tau U_b^{+2}}(1-y^*)dy^* \quad (4.38)$$

The above equation shows that the friction factor is decomposed into a viscous contribution, which is identical to the laminar solution, the turbulence contribution, and the mean extra viscoelastic stress contribution. The turbulence contribution is proportional to the weighted average of the Reynolds shear stress, for which the weight linearly decreases with the distance from the wall. This fact quantitatively supports the observation for turbulent flow that turbulence structures that appear closer to the wall than the position of maximum Reynolds shear stress are more responsible for the frictional drag in wall turbulence. The viscoelastic contribution is also proportional to the weighted average viscoelastic shear stress with the weight decreasing linearly from 1 at the wall to zero at the center of the channel. For the Newtonian and viscoelastic flows, the fractional contribution made by each part is shown in Figure 4.8a. Apparently, at the same bulk Reynolds number, the absolute value of the viscous contribution is the same as that of the Newtonian fluid. Thus, the addition of surfactant additives dramatically decreases the turbulence contribution to friction drag, and introduces frictional drag by the viscoelastic contribution. Surfactant additives therefore have dual effects on frictional drag: (1) to introduce viscoelastic shear force, which has the function to increase frictional drag; and (2) to dampen the turbulent vortical structures, to decrease turbulent shear stress and then decrease frictional drag. The drag reduction occurs because the second effect exceeds that of the first. Figure 4.8b shows the cumulative contributions made by the Reynolds shear stress and viscoelastic shear stress. The decrease of the Reynolds shear stress contribution and the increase of the viscoelastic contributions occur primarily in the buffer layer.

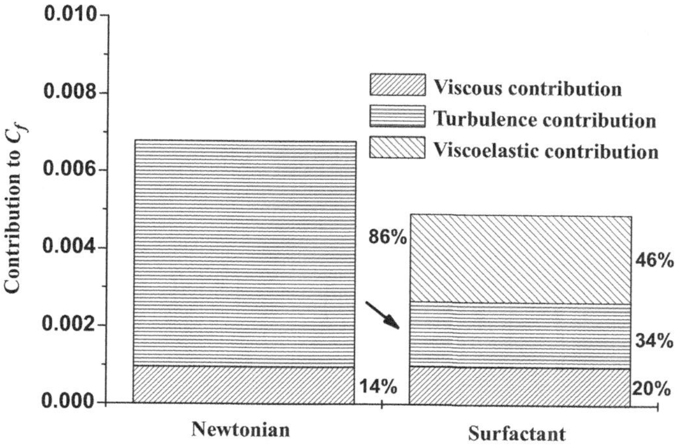

(a)

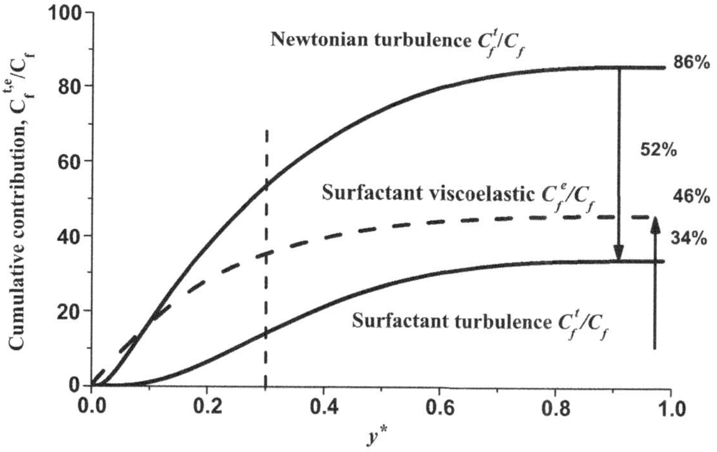

(b)

Figure 4.8 Fractional contribution and cumulative contribution to a friction factor in a calculation case; (a) fractional contribution, and (b) cumulative contribution (Reprinted from *International Journal of Heat and Fluid Flow*, **25**, Yu, B., Li, F.-C. and Kawaguchi, Y., "Numerical and experimental investigation on turbulence characteristics in a drag-reducing flow with surfactant additives," 961–974, 2004, with permission from Elsevier.)

The integrated balance equation for mean kinetic energy can be derived as follows:

$$\underbrace{\int_{-1}^{1} \frac{U^+}{Re_\tau} dy^*}_{\text{I}} = \underbrace{\int_{-1}^{1} -\overline{u'^+v'^+} \frac{\partial U^+}{\partial y^+} dy^*}_{\text{II}} + \underbrace{\int_{-1}^{1} \frac{dU^+}{dy^+} \frac{dU^+}{dy^+} dy^*}_{\text{III}} + \underbrace{\int_{-1}^{1} \frac{\beta}{We_\tau} C_{xy}^+ \frac{dU^+}{dy^+} dy^*}_{\text{IV}} \quad (4.39)$$

The work done by the average pressure gradient (I) is equal to the work of deformation by the Reynolds shear stress (II) and the dissipation (III) of mean flow energy by viscosity (we name it "viscous dissipation") and the work by mean viscoelastic stress (IV). For Newtonian flow, the last term on the right of Equation 4.30 is zero.

The integrated balance equation for turbulent kinetic energy can be derived as follows:

$$\underbrace{\int_{-1}^{1} -\overline{u'^{+}v'^{+}}\frac{\partial U^{+}}{\partial y^{+}}dy^{*}}_{\text{II}} = \underbrace{\int_{-1}^{1} \overline{\left(\frac{\partial u_i'^{+}}{\partial x_k^{+}}\right)\left(\frac{\partial u_i'^{+}}{\partial x_k^{+}}\right)}dy^{*}}_{\text{V}} + \underbrace{\int_{-1}^{1} \frac{\beta}{We_{\tau}}\left(\overline{c_{ik}'^{+}\frac{\partial u_i'^{+}}{\partial x_k^{+}}}\right)dy^{*}}_{\text{VI}} \quad (4.40)$$

For Newtonian flow, production of turbulence is lost by the dissipation due to the fluctuating velocity (V; we name it "turbulent dissipation"). With the addition of surfactant additives, turbulence is lost not only by turbulent dissipation but also through the effect of fluctuations in viscoelastic stress (VI).

The integrated balance equation for elastic energy can be derived as

$$\begin{aligned} 0 = &\underbrace{\int_{-1}^{1} \frac{\beta}{We_{\tau}}\frac{\partial U^{+}}{\partial y^{+}}C_{xy}^{+}dy^{*}}_{\text{IV}} + \underbrace{\int_{-1}^{1} \frac{\beta}{We_{\tau}}\overline{\frac{\partial u_i'^{+}}{\partial x_j^{+}}c_{ij}'^{+}}dy^{*}}_{\text{VI}} \\ &+ \underbrace{\int_{-1}^{1} \frac{\beta}{2We_{\tau}^{2}}\left[\left(3 - C_{ii}^{+}\right) - \alpha\left(C_{im}^{+} - \delta_{im}\right)^{2}\right]dy^{*}}_{\text{VII}} \end{aligned} \quad (4.41)$$

The last term (VII) of Equation 4.32 is elastic dissipation. From the above equations, the input energy by the mean pressure gradient is clearly seen to be dissipated by viscosity and elasticity. To identify the energy distribution quantitatively, each term is calculated for both a Newtonian fluid and the surfactant solution. Their fractional contributions are listed in Figure 4.9. Clearly, the energy transportation process for surfactant solution is quite different from that of the Newtonian fluid. The input energy by the mean pressure gradient is dissipated through Path 1 and Path 2. Path 1 is identical to the Newtonian case. One interaction is present between the two paths, namely, energy transport from turbulent production to a turbulence–elasticity interaction. Due to the addition of the surfactant additives, viscous dissipation becomes much smaller. Surfactant additives inhibit energy transfer to turbulent production and to turbulent dissipation. Most of the energy is transferred to the work by the mean elastic stress and to the work by fluctuating elastic stress, and finally dissipated by the relaxation of the deformed network structure to its equilibrium state. The following equations present details of the energy budgets.

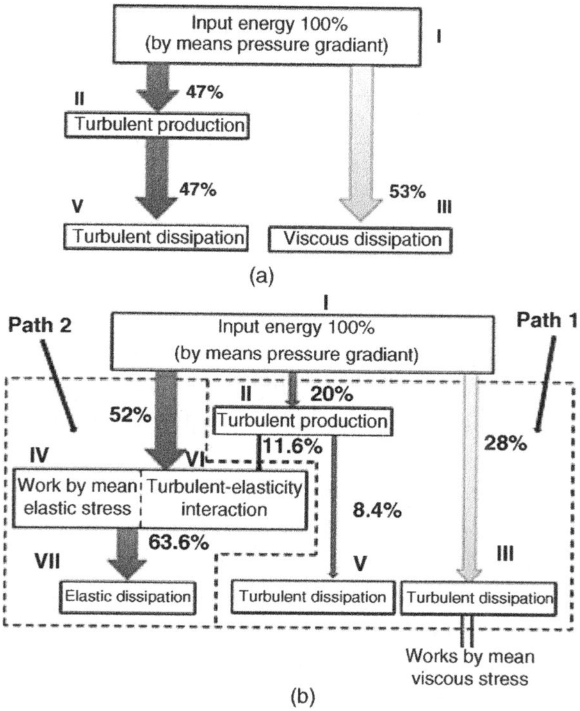

Figure 4.9 Comparison of the energy transportation process for: (a) Newtonian flow, and (b) surfactant solution (Reprinted from *International Journal of Heat and Fluid Flow*, **25**, Yu, B., Li, F.-C., Kawaguchi, Y., "Numerical and experimental investigation on turbulence characteristics in a drag-reducing flow with surfactant additives," 961–974, 2004, with permission from Elsevier.)

The budget terms of Reynolds stress $\overline{u'^+_i u^+_j}$, in the fully developed channel flow field, can be expressed as

$$\frac{D}{Dt}\left(\overline{u'^+_i u'^+_j}\right) = P_{ij} + T_{ij} + D_{ij} + \Pi_{ij} - \varepsilon_{ij} + E_{ij} \tag{4.42}$$

where the terms on the right-hand side of the above equation are identified as follows:

$$P_{ij} = -\left(\overline{u'^+_j u'^+_k}\frac{\partial U^+_i}{\partial x^+_k} + \overline{u'^+_i u'^+_k}\frac{\partial U^+_j}{\partial x^+_k}\right) \qquad \text{Production rate}$$

$$T_{ij} = -\frac{\partial}{\partial x^+_k}\left(\overline{u'^+_i u'^+_j u'^+_k}\right) \qquad \text{Turbulent transport rate}$$

$$\Pi_{ij} = -\left(\overline{u'^+_j \frac{\partial p'^+}{\partial x^+_i}} + \overline{u'^+_i \frac{\partial p'^+}{\partial x^+_j}}\right) \qquad \text{Vel:-press:-gradient term}$$

$$D_{ij} = \frac{\partial^2}{\partial x_k^{+2}} \left(\overline{u'^+_i u'^+_j} \right) \quad \text{Viscous diffusion rate}$$

$$\varepsilon_{ij} = 2 \overline{\left(\frac{\partial u'^+_i}{\partial x^+_k} \right) \left(\frac{\partial u'^+_j}{\partial x^+_k} \right)} \quad \text{Dissipation rate}$$

$$E_{ij} = \frac{\beta}{We_\tau} \left(\overline{u'^+_i \frac{\partial c'^+_{jk}}{\partial x^+_k}} + \overline{u'^+_j \frac{\partial c'^+_{ik}}{\partial x^+_k}} \right) \quad \text{Turbulence-elasticity interaction}$$

In Equation 4.33, as compared to Newtonian fluid, there is an extra term called the "turbulence–elasticity interaction." By setting $i = j$ in Equation 4.42, summing the index i, and dividing the sum by 2, the balance equation of mean turbulent kinetic energy is obtained.

The budget terms of the Reynolds normal stress $\overline{u'^+u'^+}$, $\overline{v'^+v'^+}$ and $\overline{w'^+w'^+}$ are shown in Figures 4.10–4.12. The magnitude of all budget terms becomes smaller and their peak-value positions are located closer to the bulk flow than those of the Newtonian flow, due to the addition of surfactant additives. The decrease in amplitude of the budget terms of $\overline{v'^+v'^+}$ and $\overline{w'^+w'^+}$ is larger than that of $\overline{u'^+u'^+}$. The production rate of $\overline{u'^+u'^+}$ is twice that of turbulent kinetic energy because production rates are zero for the normal component $\overline{v'^+v'^+}$ and the span-wise component $\overline{w'^+w'^+}$. For Newtonian flow, at $y^+ > 50$, one half of $\overline{u'^+u'^+}$ that is gained by production is lost by the dissipation rate, whereas the other half is redistributed to $\overline{v'^+v'^+}$ and $\overline{w'^+w'^+}$ through the pressure strain correlation. With the addition of surfactant additives, the production is primarily lost by the turbulence–elasticity interaction term and the dissipation term. A very small amount of energy in the production rate is lost by the pressure strain. Thus, compared to the Newtonian fluid, the energy transferred from the $\overline{u'^+u'^+}$ to $\overline{v'^+v'^+}$ and $\overline{w'^+w'^+}$ becomes much smaller. Since the cross-flow gained less energy from the bulk flow, the velocity fluctuation in the wall-normal and span-wise directions decreases, resulting in weakened stream-wise vortices. Figure 4.13 compares the stream-wise vortices in cross-section. With the addition of surfactant additives, the vortical structure becomes much larger and its strength becomes weaker. The violent stream-wise vortices in the near-wall region are significantly reduced, and the vortical structures are evenly distributed in cross-section. The weakening of the stream-wise vortices indicates that the additives significantly suppress the ejection and sweep motions occurring near the wall, which results in the suppressed turbulence production and Reynolds shear stress, indicating that the spacing between the high- and low-speed streaks becomes larger. Figures 4.11 and 4.12 show that normal and span-wise velocity–pressure gradient correlations (no direct production term is available for $\overline{v'^+v'^+}$ and $\overline{w'^+w'^+}$, as the velocity–pressure gradient correlations are usually considered to be pseudo-productions) of surfactant solutions are less than one tenth of the values of the Newtonian fluid. Thus, v^+_{rms} and w^+_{rms} for the surfactant

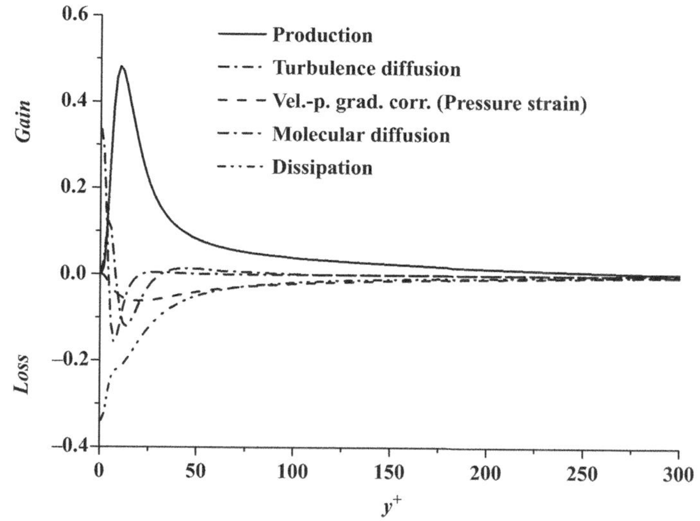

(a)

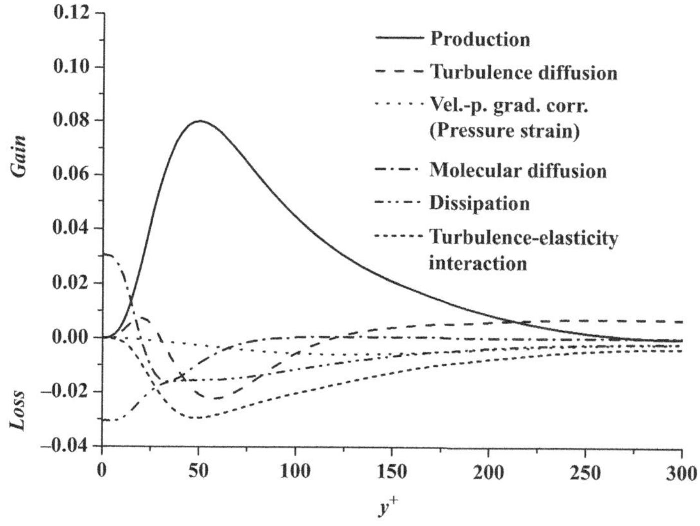

(b)

Figure 4.10 Budget of Reynolds stress $\overline{u'^{+}u'^{+}}$: (a) Newtonian fluid, and (b) surfactant solution (Reprinted from *International Journal of Heat and Fluid Flow*, **25**, Yu, B., Li, F.-C. and Kawaguchi, Y., "Numerical and experimental investigation on turbulence characteristics in a drag-reducing flow with surfactant additives," 961–974, 2004, with permission from Elsevier.)

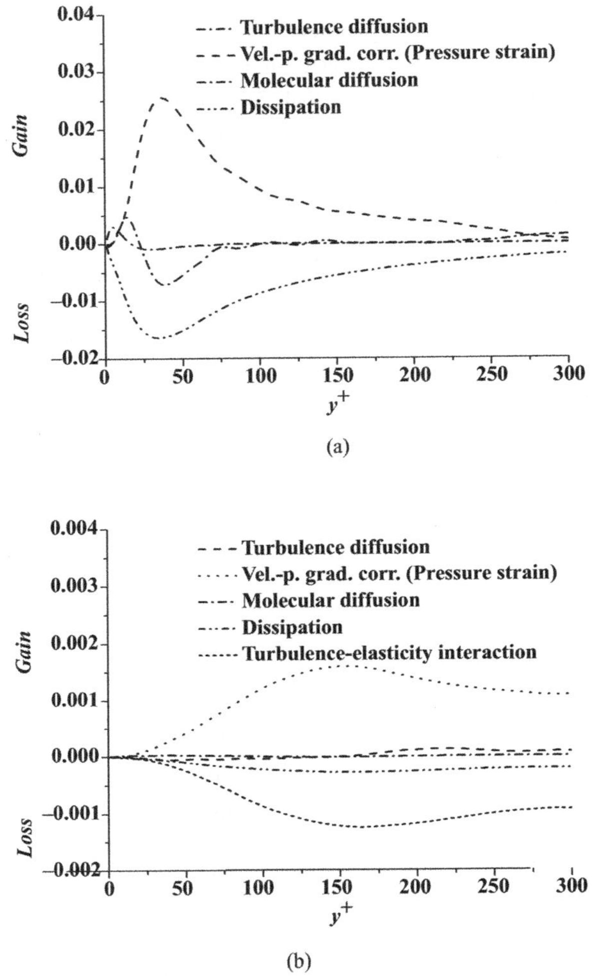

Figure 4.11 Budget of Reynolds stress $\overline{v'^+v'^+}$: (a) Newtonian fluid, and (b) surfactant solution (Reprinted from *International Journal of Heat and Fluid Flow*, **25**, Yu, B., Li, F.-C. and Kawaguchi, Y., "Numerical and experimental investigation on turbulence characteristics in a drag-reducing flow with surfactant additives," 961–974, 2004, with permission from Elsevier.)

solutions decrease significantly as compared to the Newtonian fluid. The changes of the budget terms of the Reynolds normal stress $\overline{v'^+v'^+}$ and $\overline{w'^+w'^+}$ are flat, which explains the flat distribution of the v^+_{rms} and w^+_{rms} along the channel height (Figure 4.6). The turbulent diffusion of $\overline{v'^+v'^+}$ for the surfactant solution becomes negligible, indicating that the turbulent transportation in the normal direction becomes very weak with the addition of the surfactant additives. In the Newtonian fluid, the dissipation terms are almost balanced with the velocity–pressure gradient terms in the budget of $\overline{v'^+v'^+}$ and

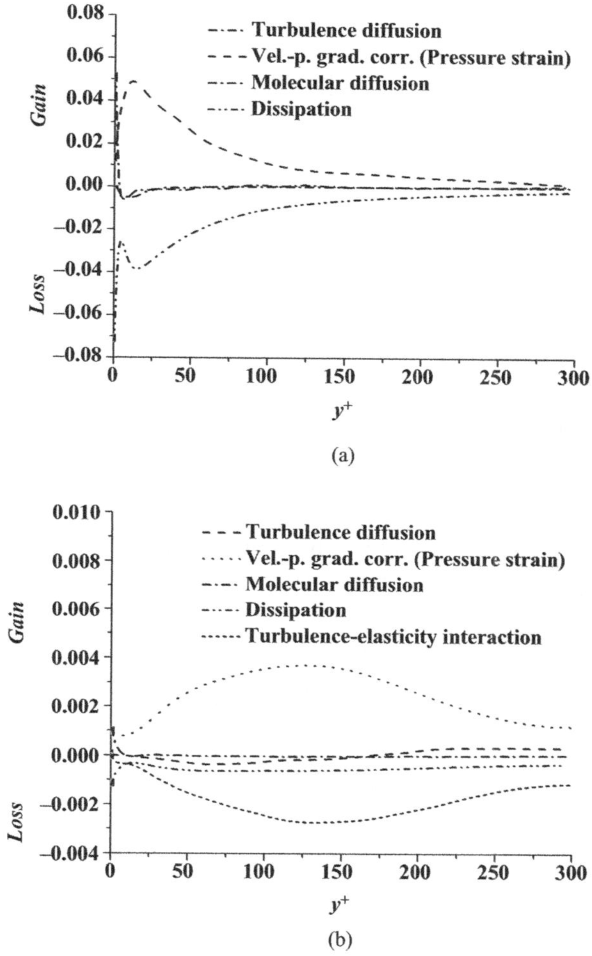

Figure 4.12 Budget of Reynolds stress $\overline{w'^+w'^+}$: (a) Newtonian fluid, and (b) surfactant solution (Reprinted from *International Journal of Heat and Fluid Flow*, **25**, Yu, B., Li, F.-C., Kawaguchi, Y., "Numerical and experimental investigation on turbulence characteristics in a drag-reducing flow with surfactant additives," 961–974, 2004, with permission from Elsevier.)

$\overline{w'^+w'^+}$. With the addition of surfactant additives, however, the dissipation term becomes very small and the turbulence–elasticity interaction becomes a dominant dissipation term, to balance the velocity–pressure gradient term. All of these observations show that viscoelasticity plays an important role in the redistribution of the budget of Reynolds stress.

The budget terms of turbulent kinetic energy are plotted in Figure 4.14 as a function of the dimensionless wall distance y^+ for both the Newtonian fluid and the CTAC surfactant solution. The production rates calculated by the measured instantaneous

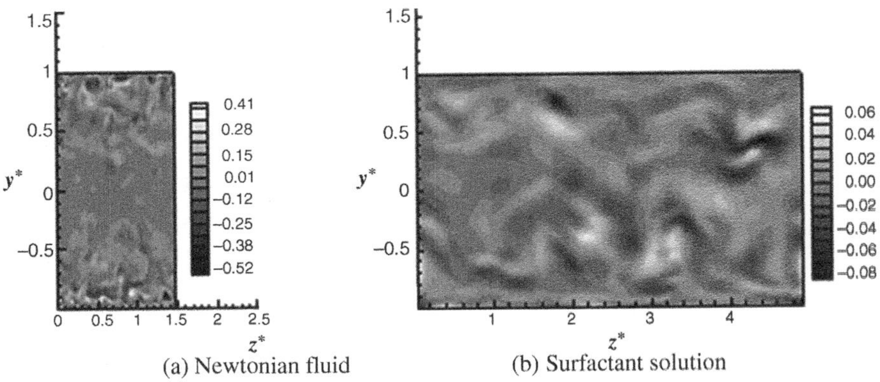

Figure 4.13 Comparison of instantaneous stream-wise vorticity for Newtonian flow and surfactant solution at a y–z cross-section: (a) Newtonian fluid, and (b) surfactant solution

velocity fluctuations are also included for comparison. It is clearly seen that the magnitudes of the budget terms significantly decrease with the addition of surfactant additives. The positions, where the production rate reaches its maximum value, and molecular diffusion and turbulence diffusion reach their minimum values, shift toward the bulk flow region. For example, the peak value position of the production rate shifts from $y^+ = 10$ to 50. These shifts are consistent with the expansion of the buffer layer. The maximum value of the production decreases by 83% as compared to Newtonian flow. This decrease must be due to the decrease of Reynolds shear stress. The wall friction factor is related to the turbulent production: as the turbulence level is decreased, the frictional drag also decreases. For Newtonian flow at $y^+ > 30$, the production term is balanced with the dissipation term. For the surfactant solution, however, the balancing is much more complicated. At $30 < y^+ < 100$, the production term becomes balanced with the sum of the turbulent diffusion, molecular diffusion, dissipation, and viscoelastic contribution terms. At $100 < y^+ < 150$, the production term is balanced with dissipation and turbulence–elasticity interaction terms. For $y^+ > 150$, the turbulent diffusion term becomes a large-gain term, and at the core region is the largest producing term. At $y^+ > 250$, the turbulent diffusion term is balanced with the dissipation term and turbulence–elasticity interaction term. The turbulence–elasticity interaction acts as a strong sink term along the channel height. Moreover, the negative peak position of the turbulence–elasticity interaction is almost the same as that of the production rate at $y^+ = 50$, which shows that the effect of the surfactant additives on turbulence flow is primarily in the buffer layer. Due to the strong turbulence–elasticity interaction, the percentage decrease of the dissipation term is larger than that of the production rate. For example, its peak value decreases by 93%. Although the surfactant additives significantly change the values and distribution of the budget terms, the production, turbulent diffusion, molecular diffusion, and other terms show identical trends in variation for both the Newtonian fluid and surfactant

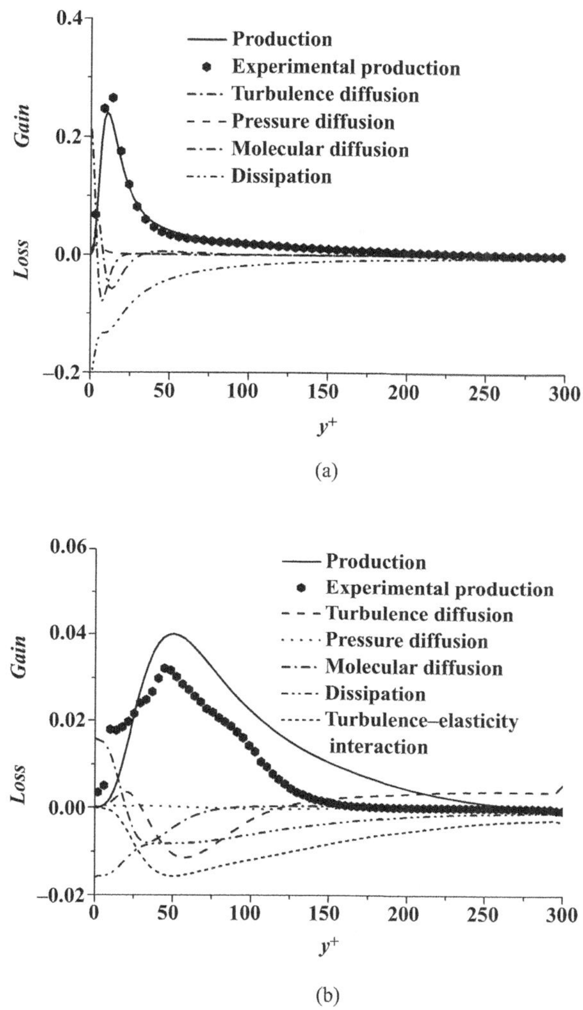

Figure 4.14 Budget of turbulent kinetic energy: (a) Newtonian fluid, and (b) surfactant solution

solution, with the terms in the latter varying flatly. The measured production rates of the Newtonian fluid agree well with the numerical predictions except in the near-wall region, where the experimental values are larger. The agreement of the production rates of the CTAC surfactant solution between experimental and numerical calculations is qualitatively good.

For our study, the viscoelastic contribution term is of the same order of magnitude as the production term. The dissipation term becomes a smaller term in the budget of Reynolds normal stress $\overline{v'^+v'^+}$ and $\overline{w'^+w'^+}$. The larger differences are the result of two factors, namely, the differences in the model and the computational parameters.

Table 4.3 Computational parameters (Reprinted from *International Journal of Heat and Mass Transfer*, **48**, Yu, B. and Kawaguchi, Y., "DNS of fully developed turbulent heat transfer of a viscoelastic drag-reducing flow," 4569–4578, 2005, with permission from Elsevier.)

	Re_τ	We_τ	β	Pr	α	$L_x \times L_y \times L_z$	$N_x \times N_y \times N_z$	Δx^+	Δy^+	Δz^+
Surfactant	150	40	0.2	0.71	0.001	10 h × 2 h × 4 h	96 × 128 × 96	15.6	0.2–4.5	6.3
Newtonian	150	×	0	0.71	×	7.5 h × 2 h × 2.5 h	128 × 128 × 128	8.8	0.2–4.5	2.9

Experiments showed that the addition of drag reducers dramatically depresses turbulence and decreases flow friction. However, the relating problem is reduction of heat transfer, which is harmful for this kind of system. Thus, it is very meaningful to explain the mechanism of drag-reducing heat transfer for scientific and engineering applications.

Table 4.3 shows the dimensionless computational parameters. The mean temperature distribution is given in Figure 4.15. The mean temperature profiles of the Newtonian fluid and drag-reducing flow collapse to the linear relationship $\Theta^+ = Pry^+$ in the viscous sublayer- and the profile of the drag-reducing flow upshifts in the logarithmic layer. Mean flow variables such as mean Reynolds number, friction factor, and Nusselt number are given in Table 4.4. In Table 4.4, C_f^* and Nu* were evaluated by Dean's correlation and Kays and Crawford's correlation, respectively, for comparison. As shown in Table 4.4, the friction factor and Nusselt number of the drag-reducing flow are much smaller than those of the Newtonian fluid, and a drag reduction rate of 46.8%

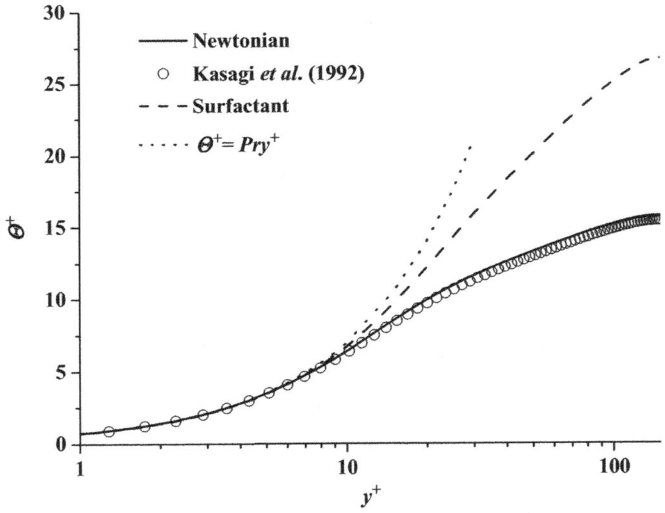

Figure 4.15 Mean temperature profiles (Reprinted from *International Journal of Heat and Mass Transfer*, **48**, Yu, B. and Kawaguchi, Y., "DNS of fully developed turbulent heat transfer of a viscoelastic drag-reducing flow," 4569–4578, 2005, with permission from Elsevier.)

Table 4.4 Mean flow variables (Reprinted from *International Journal of Heat and Mass Transfer*, **48**, Yu, B. and Kawaguchi, Y., "DNS of fully developed turbulent heat transfer of a viscoelastic drag-reducing flow," 4569–4578, 2005, with permission from Elsevier.)

	U_b	U_c	U_c/U_b	Re_b	C_f	C_f^*	Nu	Nu*	DR%	HTR%
Surfactant	21.5	26.6	1.24	6450	0.00433	0.00815	9.3	20.7	46.8	54.5
Newtonian	15.5	18.3	1.18	4640	0.00834	0.00884	15.3	15.9	×	×

and heat transfer reduction rate of 54.5% were obtained for the drag-reducing flow. The larger reduction rate than drag reduction rate has been confirmed by experiments, along with the upshift of mean temperature profile, increase of temperature fluctuations, decrease of wall-normal turbulent heat flux, and decrease of the cross-correlation coefficients between wall normal velocity and temperature.

The root-mean-square temperature fluctuations are compared in Figure 4.16. As compared to Newtonian flow, the temperature fluctuation intensity is appreciably enhanced. The maximum temperature fluctuation intensity of the surfactant solution is 4.2, much larger than the value of 2.4 of the Newtonian fluid. The peak value position of the temperature fluctuation is located at $y^+ = 35$, shifted to the bulk flow region as compared to the peak position $y^+ = 18$ of Newtonian fluid. Figure 4.17 compares the distribution of the stream-wise turbulent heat flux. The stream-wise heat flux of the surfactant solution is much larger, and the position where its maximum value attains shifts toward the bulk flow region. The significant increase of the stream-wise heat flux is primarily owing to the increase of the stream-wise velocity fluctuation intensity and the increase of the temperature fluctuation intensity.

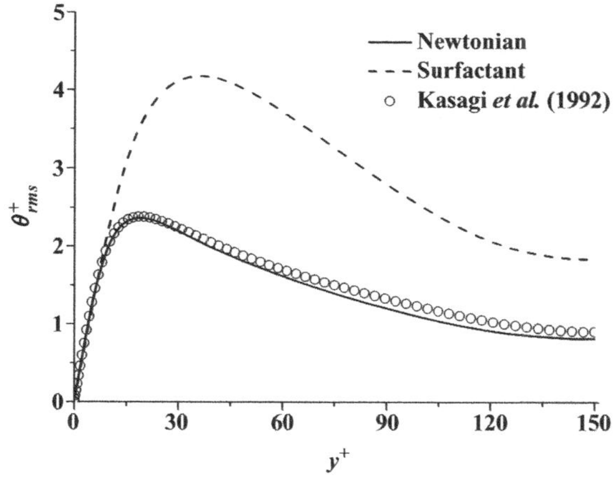

Figure 4.16 RMS of temperature fluctuation (Reprinted from *International Journal of Heat and Mass Transfer*, **48**, Yu, B. and Kawaguchi, Y., "DNS of fully developed turbulent heat transfer of a viscoelastic drag-reducing flow," 4569–4578, 2005, with permission from Elsevier.)

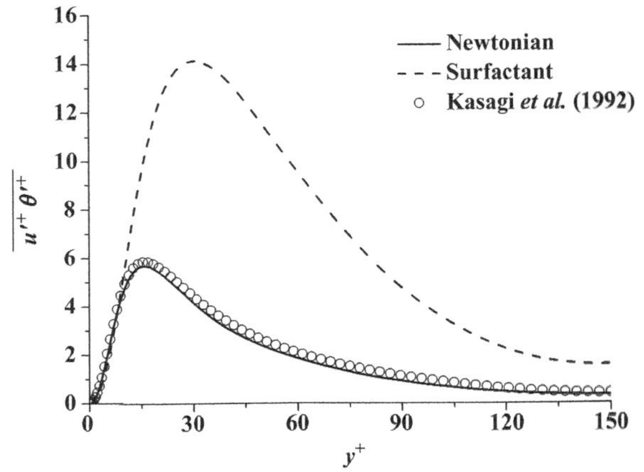

Figure 4.17 Stream-wise turbulent heat flux (Reprinted from *International Journal of Heat and Mass Transfer*, **48**, Yu, B. and Kawaguchi, Y., "DNS of fully developed turbulent heat transfer of a viscoelastic drag-reducing flow," 4569–4578, 2005, with permission from Elsevier.)

Figures 4.18a and 4.18b compare velocity streaks and thermal streaks at $y^+ = 15$ for the drag-reducing flow. It is seen that the thermal streaky structures closely resemble the velocity streaky structures, and high-temperature fluids are associated with high-speed streaks and low-temperature fluids are associated with low-speed steaks. The similarity indicates a close correlation between the stream-wise velocity component and temperature. In fact, as shown in Figure 4.19, the cross-correlation coefficients between u and θ have large values close to unity. As compared to the Newtonian fluid, $R_{u\theta}$ of the drag-reducing flow increases. The close correlation is another cause for the increase of the stream-wise turbulent heat flux. Figures 4.18b and 4.18c compare snapshots of the instantaneous temperature fluctuations in the x–z plane located at $y^+ = 15$ for drag-reducing flow and Newtonian flow. It is seen that the thermal streaks are more elongated and thermal streak spacing expands by additives.

For the fully developed channel flow between the isoflux walls, the total heat flux can be deduced from the averaged energy equation as follows:

$$q^+_{total} = \frac{1}{\Pr}\frac{\partial \Theta^+}{\partial y^+} - \overline{v'^+\theta'^+} \tag{4.43}$$

where, $q^+_{total}\left(= 1 - \dfrac{\int_0^{y^+} U^+ dy^+}{\int_0^{\mathrm{Re}_\tau} U^+ dy^+}\right)$ is total heat flux.

Unlike stream-wise turbulent heat flux, the wall-normal turbulent heat flux of drag-reducing flow by additives decreases as compared to Newtonian flow as shown in Figure 4.20. The position of its maximum value also shifts to the bulk flow. The strongly suppressed wall-normal velocity fluctuation is a cause of the reduction of

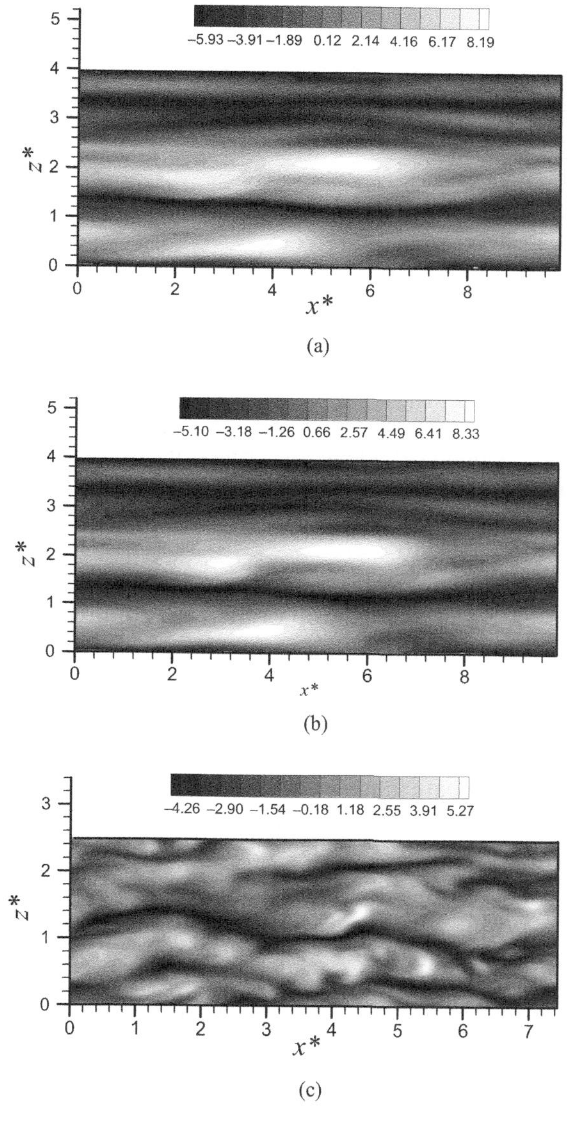

Figure 4.18 Instantaneous speed streaky and thermal streaky patterns of the *x–z* plane located at $y^+ = 15$. (a) Velocity streaks of viscoelastic flow, (b) thermal streaks of viscoelastic flow, and (c) thermal streaks of Newtonian flow (Reprinted from *International Journal of Heat and Mass Transfer*, **48**, Yu, B. and Kawaguchi, Y., "DNS of fully developed turbulent heat transfer of a viscoelastic drag-reducing flow," 4569–4578, 2005, with permission from Elsevier.)

wall-normal heat flux. For the drag-reducing flow, there is an increase of the conductive heat flux to compensate for the decrease of wall-normal heat flux as shown in Figure 4.20. This means conduction plays a more important role in heat transportation in the drag-reducing flow. Figure 4.19 shows that compared to the Newtonian case,

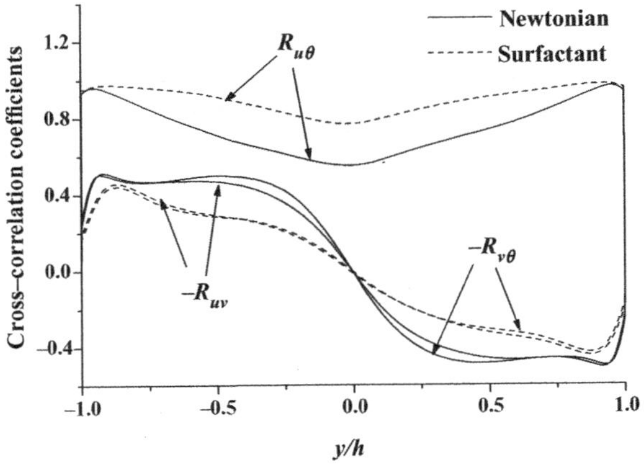

Figure 4.19 Cross-correlation coefficients (Reprinted from *International Journal of Heat and Mass Transfer*, **48**, Yu, B. and Kawaguchi, Y., "DNS of fully developed turbulent heat transfer of a viscoelastic drag-reducing flow," 4569–4578, 2005, with permission from Elsevier.)

both R_{uv} and $R_{v\theta}$ decrease and the position of their maximum value shifts to the bulk flow. The decreases of R_{uv} and $R_{v\theta}$ indicate that the correlation between the velocity components u and v, and that between v and θ, decrease. These decorrelations are believed to be the cause of drag reduction and heat transfer reduction. In addition, for both the Newtonian fluid and drag-reducing flow, the agreement between R_{uv} and $R_{v\theta}$ is good. These similarities indicate that the wall-normal turbulent heat flux and Reynolds shear stress are generated by a similar turbulent mechanism for both Newtonian flow and drag-reducing flow by additives.

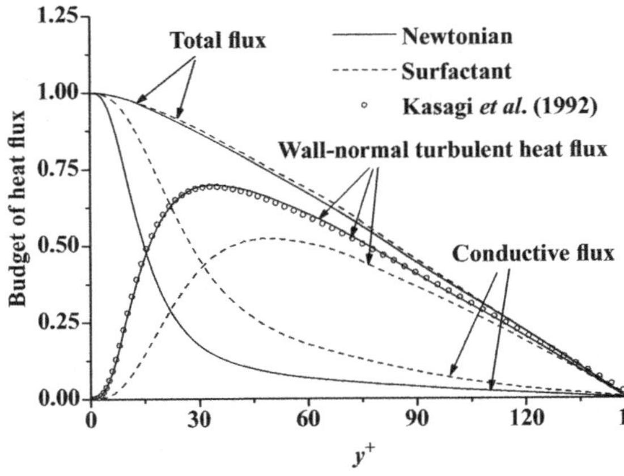

Figure 4.20 Budget of heat flux (Reprinted from *International Journal of Heat and Mass Transfer*, **48**, Yu, B. and Kawaguchi, Y., "DNS of fully developed turbulent heat transfer of a viscoelastic drag-reducing flow," 4569–4578, 2005, with permission from Elsevier.)

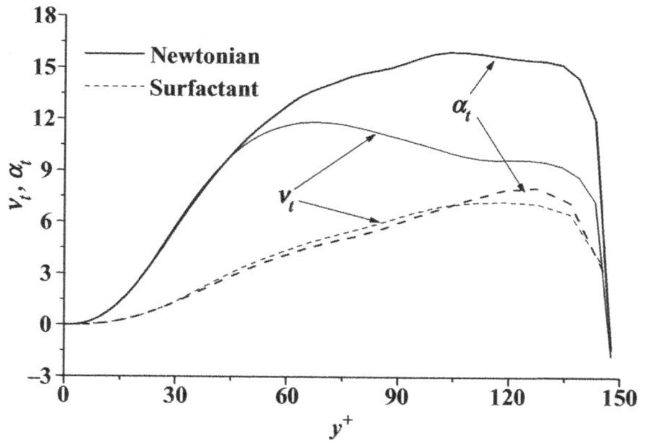

Figure 4.21 Eddy diffusivities (Reprinted from *International Journal of Heat and Mass Transfer*, **48**, Yu, B. and Kawaguchi, Y., "DNS of fully developed turbulent heat transfer of a viscoelastic drag-reducing flow," 4569–4578, 2005, with permission from Elsevier.)

The nondimensional eddy diffusivities of momentum and heat are shown in Figure 4.21. The decrease of both the eddy diffusivities of surfactant solution indicates that less frequent and weaker turbulent transportation occurs in the heat and fluid flow process. The turbulent Prandtl number, defined as the ratio between eddy diffusivities for momentum and heat, is shown in Figure 4.22. The Prandtl numbers for both the surfactant solution and Newtonian fluid at the vicinity of the wall approach to unity. In most logarithmic regions, the Prandtl numbers of Newtonian fluid are less than unity with those values of surfactant solution larger.

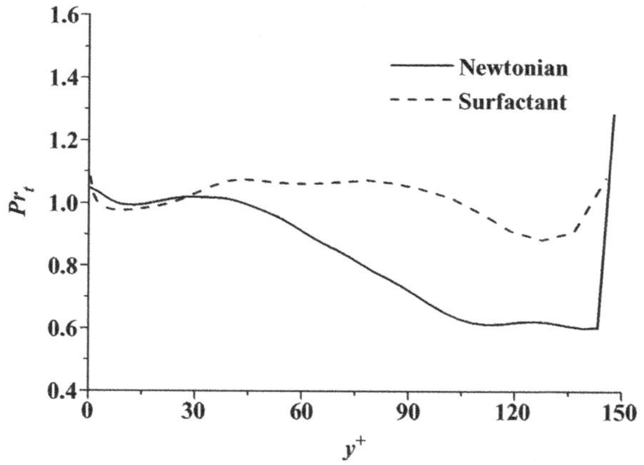

Figure 4.22 Turbulent Prandtl number (Reprinted from *International Journal of Heat and Mass Transfer*, **48**, Yu, B. and Kawaguchi, Y., "DNS of fully developed turbulent heat transfer of a viscoelastic drag-reducing flow," 4569–4578, 2005, with permission from Elsevier.)

Equations 4.35 and 4.36 are the budget equations for the stream-wise and wall-normal turbulent heat fluxes, respectively:

$$0 = -\overline{u'^+v'^+}\frac{\partial \Theta^+}{\partial y^+} + \overline{u'^+u'^+}\frac{\partial \langle T\rangle^+}{\partial x^+} - \overline{v'^+\theta'^+}\frac{\partial U^+}{\partial y^+}$$

Production

$$-\left(1+\frac{1}{Pr}\right)\overline{\frac{\partial u'^+}{\partial x_i^+}\frac{\partial \theta'^+}{\partial x_i^+}} - \overline{\theta'^+\frac{\partial p'^+}{\partial x^+}} \tag{4.44}$$

Dissipation Temperature pressure-gradient correlation

$$-\frac{\partial}{\partial y^+}\left(\overline{u'^+\theta'^+v'^+}\right) + \frac{\partial}{\partial x_i^+}\left(\overline{\theta'^+\frac{\partial u'^+}{\partial x_i^+}} + \frac{1}{Pr}\overline{u'^+\frac{\partial \theta'^+}{\partial x_i^+}}\right) + \frac{2\beta}{We_\tau}\overline{\theta'^+\frac{\partial c'^+_{xx_i}}{\partial x_i^+}}$$

Turbulent diffusion Molecular diffusion Viscoelastic contribution

$$0 = -\overline{v'^+v'^+}\frac{\partial \Theta^+}{\partial y^+} + \overline{u'^+v'^+}\frac{\partial \langle T\rangle^+}{\partial x^+} - \left(1+\frac{1}{Pr}\right)\overline{\frac{\partial v'^+}{\partial x_i^+}\frac{\partial \theta'^+}{\partial x_i^+}} - \overline{\theta'^+\frac{\partial p'^+}{\partial y^+}}$$

Production Dissipation Temperature pressure-gradient correlation (4.45)

$$-\frac{\partial}{\partial y^+}\left(\overline{v'^+\theta'^+v'^+}\right) + \left[\frac{\partial}{\partial x^+}\left(\overline{\theta'^+\frac{\partial v'^+}{\partial x_i^+}} + \frac{1}{Pr}\overline{v'^+\frac{\partial \theta'^+}{\partial x_i^+}}\right)\right] + \frac{2\beta}{We_\tau}\overline{\theta'^+\frac{\partial c'^+_{x_iy}}{\partial x_i^+}}$$

Turbulent diffusion Molecular diffusion Viscoelastic contribution

The budget equation of the fluctuating temperature variance $k_\theta\left(=\overline{\theta'^{+2}}/2\right)$, in the fully developed turbulent channel thermal field, can be written as

$$0 = \overline{u'^+\theta'^+}\frac{\partial \langle T\rangle^+}{\partial x^+} - \overline{v'^+\theta'^+}\frac{\partial \Theta^+}{\partial y^+} - \frac{1}{Pr}\left(\overline{\frac{\partial \theta'^+}{\partial x_i^+}\frac{\partial \theta'^+}{\partial x_i^+}}\right)$$

Production Dissipation

$$-\frac{\partial \overline{v'^+\theta'^{+2}}/2}{\partial y^+} + \frac{1}{Pr}\frac{\partial^2 \overline{\theta'^{+2}}/2}{\partial y^{+2}} \tag{4.46}$$

Turbulent diffusion Molecular diffusion

The various terms in the budget of the stream-wise turbulent heat flux and wall-normal turbulent heat flux are plotted in Figures 4.23 and 4.24 as a function of dimensionless

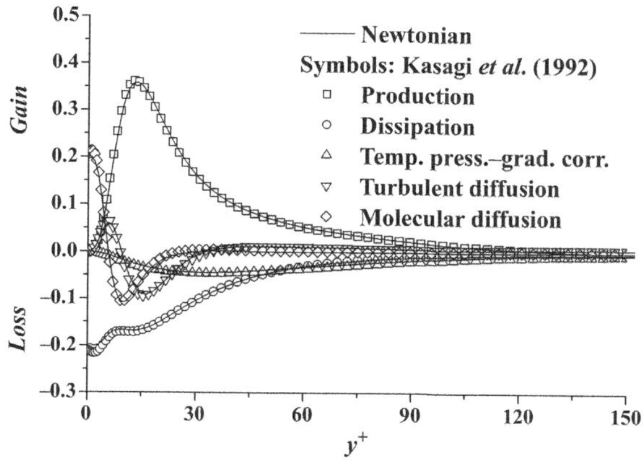

(a)

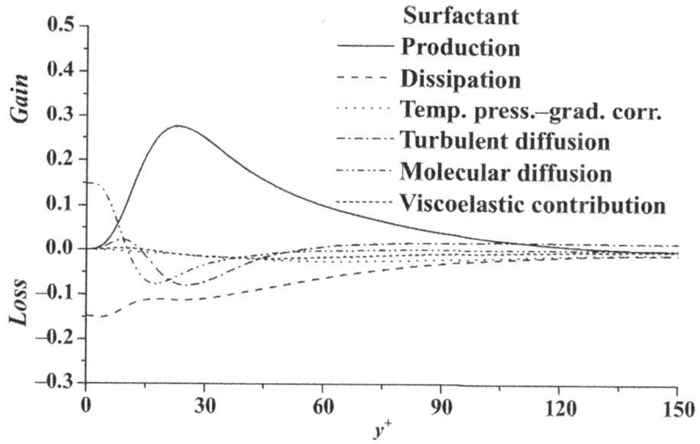

(b)

Figure 4.23 Budget terms of stream-wise turbulent heat flux: (a) Newtonian fluid, and (b) drag-reducing flow (Reprinted from *International Journal of Heat and Mass Transfer*, **48**, Yu, B. and Kawaguchi, Y., "DNS of fully developed turbulent heat transfer of a viscoelastic drag-reducing flow," 4569–4578, 2005, with permission from Elsevier.)

wall distance y^+. As compared to the Newtonian fluid, an additional viscoelastic contribution term is included in the budget of the turbulent heat fluxes. The maximum and minimum values associated with various turbulence quantities of the drag-reducing flow, such as production, display a dramatic decrease. The locations where production attains its maximum value, and where molecular diffusion and turbulent diffusion reach their minimum values, shift from the wall region to the bulk flow region. These shifts are consistent with the expansion of the buffer layer. Though the additives significantly change the values and distribution of the budget terms, the production, turbulent diffusion, molecular diffusion, and other terms show identical trends in variation between Newtonian solution and drag-reducing flow, with the latter

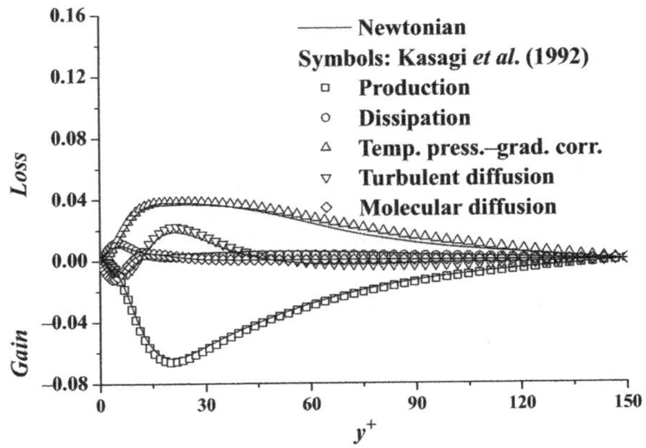

(a)

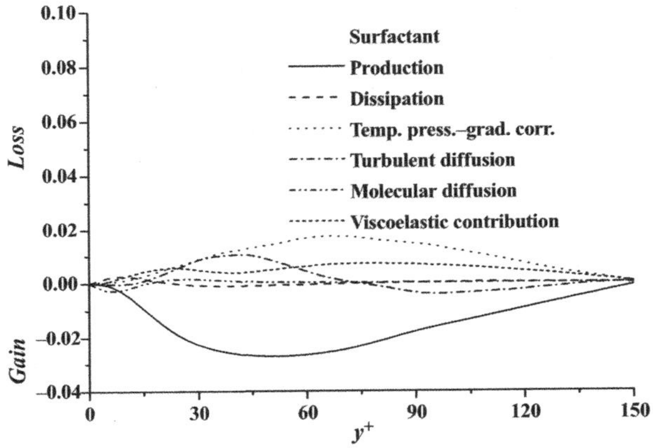

(b)

Figure 4.24 Budget terms of wall-normal turbulent heat flux: (a) Newtonian fluid, and (b) drag-reducing flow (Reprinted from *International Journal of Heat and Mass Transfer*, **48**, Yu, B. and Kawaguchi, Y., "DNS of fully developed turbulent heat transfer of a viscoelastic drag-reducing flow," 4569–4578, 2005, with permission from Elsevier.)

varying flatly. In the budget of stream-wise turbulent heat flux, the viscoelastic contribution acts as a weak sink term except in the vicinity of the wall, at which it acts as a minute production term. However, in the budget of wall-normal turbulent heat flux, the viscoelastic term is of the same order of magnitude as the other terms (e.g. production term) and it acts as a large sink term. There are three peaks (two positive and one negative) for the viscoelastic term in the budget of $\overline{v'^+\theta'^+}$ which are located near the peaks of the molecular diffusion, turbulent diffusion, and temperature pressure–gradient correlation.

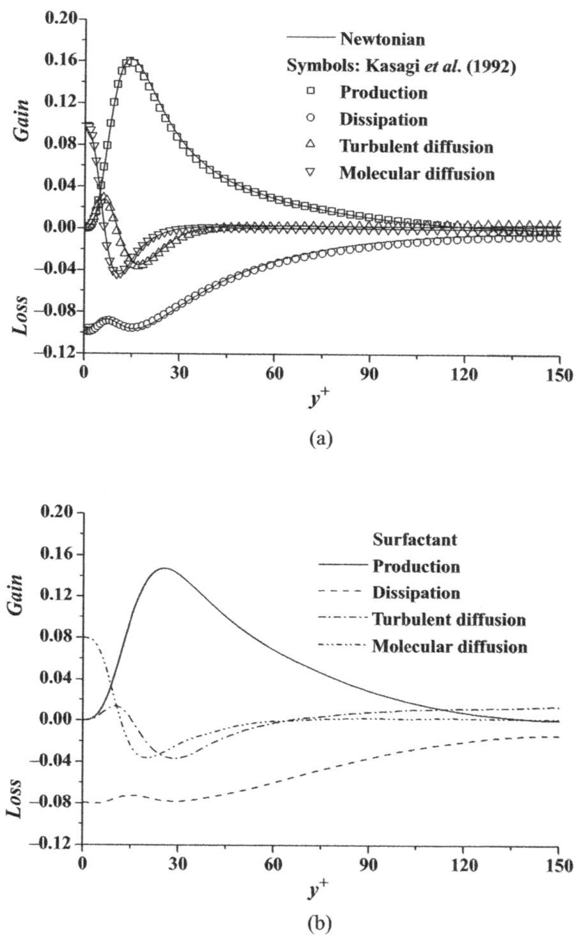

Figure 4.25 Budget terms of temperature variance: (a) Newtonian fluid, and (b) drag-reducing flow (Reprinted from *International Journal of Heat and Mass Transfer*, **48**, Yu, B. and Kawaguchi, Y., "DNS of fully developed turbulent heat transfer of a viscoelastic drag-reducing flow," 4569–4578, 2005, with permission from Elsevier.)

The budgets of the temperature variance are shown in Figure 4.25. Once again, the position of the peak value shifts toward the bulk flow due to the action of additives. Although the peak values decrease, the magnitudes of the quantities are not depressed in the whole region.

Figure 4.26 shows the one-dimensional energy (Fourier) spectrum of temperature fluctuation at $y^+ = 15$. E_{tt} is the one-dimensional energy spectrum of temperature fluctuation. Kz is the wavenumber for the span-wise direction. It is seen that the energy contained in low wavenumber modes increases and the peak of the energy spectrum shifts to a higher wavelength in the drag-reducing flow, indicating a larger thermal

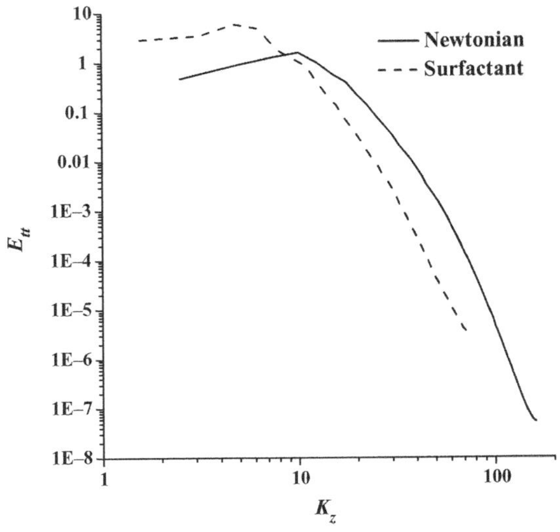

Figure 4.26 Energy spectrum of fluctuating temperature at $y^+ = 15$ (Reprinted from *International Journal of Heat and Mass Transfer*, **48**, Yu, B. and Kawaguchi, Y., "DNS of fully developed turbulent heat transfer of a viscoelastic drag-reducing flow," 4569–4578, 2005, with permission from Elsevier.)

streak spacing in the drag-reducing flow. This also indicates that large-scale temperature fluctuations are created but small scales of temperature fluctuations are dampened by additives.

To further study the characteristics of drag reduction and heat transfer reduction, the proper orthogonal decomposition (POD) method is introduced into DNS. Its main idea will be introduced below. Two hundred snapshots of the velocity and temperature field are obtained by DNS for POD analysis. First, the mean fields can be obtained

$$\overline{\tilde{F}(\vec{X})} = \frac{1}{N}\sum_{i=1}^{N} \tilde{F}(\vec{X}, t_i^*) \tag{4.47}$$

where F is a common variable for u^+,v^+,w^+, and θ^+; $\vec{X}$ represents x^*, y^*, and z^*; N is the number of snapshots; and "—" stands for mean values.

The fluctuating field can be extracted from the instantaneous field and the mean field:

$$\tilde{F}'(\vec{X}, t_i^*) = \tilde{F}(\vec{X}, t_i^*) - \overline{\tilde{F}(\vec{X})} \quad (i = 1, 2, \cdots, N) \tag{4.48}$$

The snapshot POD method is applied to the fluctuating field in the following steps:

1. Get the inner product in the Hilbert space:

$$C_{ij} = \frac{1}{N}\int_{\Omega} \tilde{F}'(\vec{X}, t_i^*)\tilde{F}'\left(\vec{X}, t_j^*\right)d\Omega \quad (i = 1, 2, \cdots, N, j = 1, 2, \cdots, N) \quad (4.49)$$

2. Get the eigenvalues and eigenvectors by singular value decomposition of the inner product C:

$$CA = \lambda A \quad (4.50)$$

whereλ is the eigenvalue, and A is the eigenvector corresponding to λ.

3. Get POD eigenmodes from the eigenvectors and the fluctuating field:

$$\phi_k(\vec{X}) = \sum_{i=1}^{N} A_{ki}\tilde{F}'(\vec{X}, t_i^*) \quad (4.51)$$

where k represents the kth POD eigenmode.

The total energy of the fluctuating field can be calculated by the sum of the eigenvalues:

$$E = \sum_{n=1}^{N} \lambda_n \quad (4.52)$$

Thus, the magnitude of the nth eigenvalue n represents the energy relative contribution to the total energy in the nth eigenmode:

$$\xi_n = \lambda_n / \sum_{n=1}^{N} \lambda_n \quad (4.53)$$

The cumulative energy relative contribution to the total energy, that is, the summation of the energy relative contributions for the first N_1 eigenmodes, is:

$$\zeta_n = \sum_{n=1}^{N_1} \lambda_n / \sum_{n=1}^{N} \lambda_n \quad (4.54)$$

The energy relative contributions to the total energy of each POD eigenmode obtained from the fluctuating field are compared in Figure 4.27. The first eigenmode, which has the highest energy contribution, contains the main energy of flow indicating that it may capture the largest scale vortex structure. The contributions of the following eigenmodes decrease, representing that the structure scales of flow and heat transfer become smaller. For drag-reducing flow, the energy contributions of the first amount of POD eigenmodes are higher than those of Newtonian fluid flow.

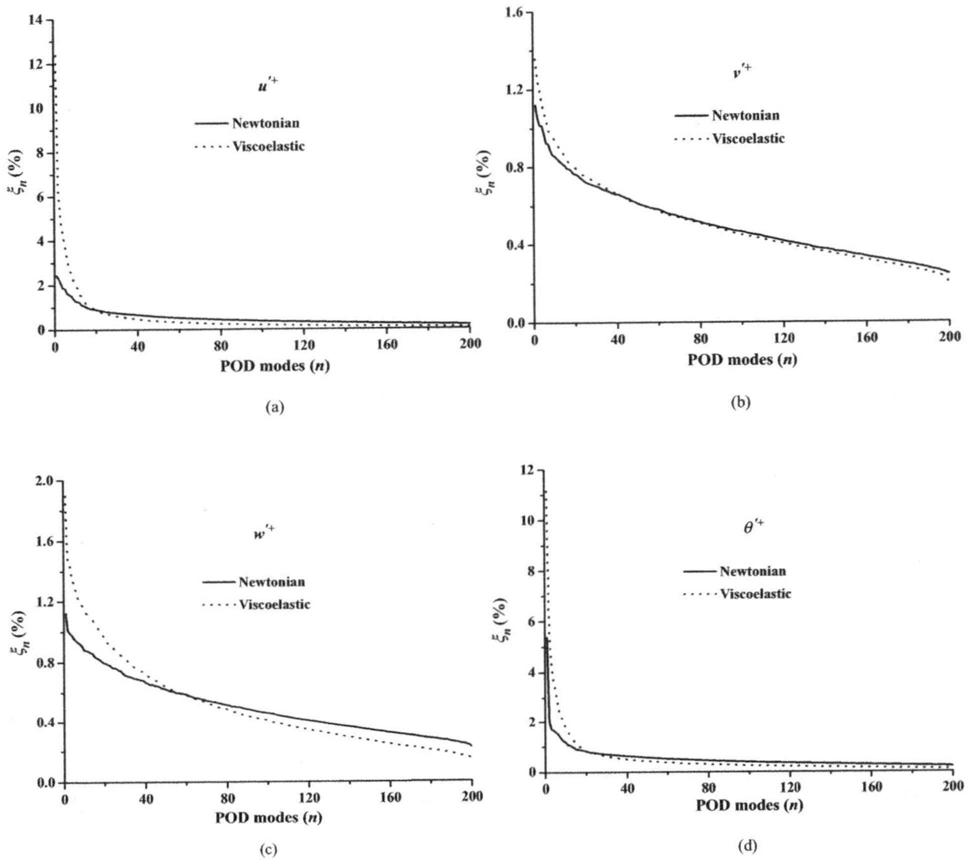

Figure 4.27 Energy contributions for different POD eigenmodes (Reproduced with permission from Wang, Y., Yu, B., Wu, X., Wei, J.J., Li, F.C. and Kawaguchi, Y. (2011), "POD study on the mechanism of turbulent drag reduction and heat transfer reduction based on Direct Numerical Simulation," *Progress in Computational Fluid Dynamics* **11**, 199–299, Inderscience Enterprises Limited.)

The increment decreases from the first eigenmode to the last one. The contributions of the latter eigenmodes are lower than those of Newtonian fluid flow. These phenomena show that drag-reducing additives mainly affect large-scale structures containing main turbulent kinetic energy.

The cumulative energy contributions of the POD eigenmodes are shown in Figure 4.28. For Newtonian fluid flow, the curve is flat, showing that all scale structures play important roles to the total energy and no dominant structure exists that contains the most energy. The convergence of the cumulative energy contribution of drag-reducing flow is much faster than that of Newtonian fluid flow. It is contacted with the increasing proportion of large-scale structures and decreasing proportion of small-scale structures in the energy spectrum.

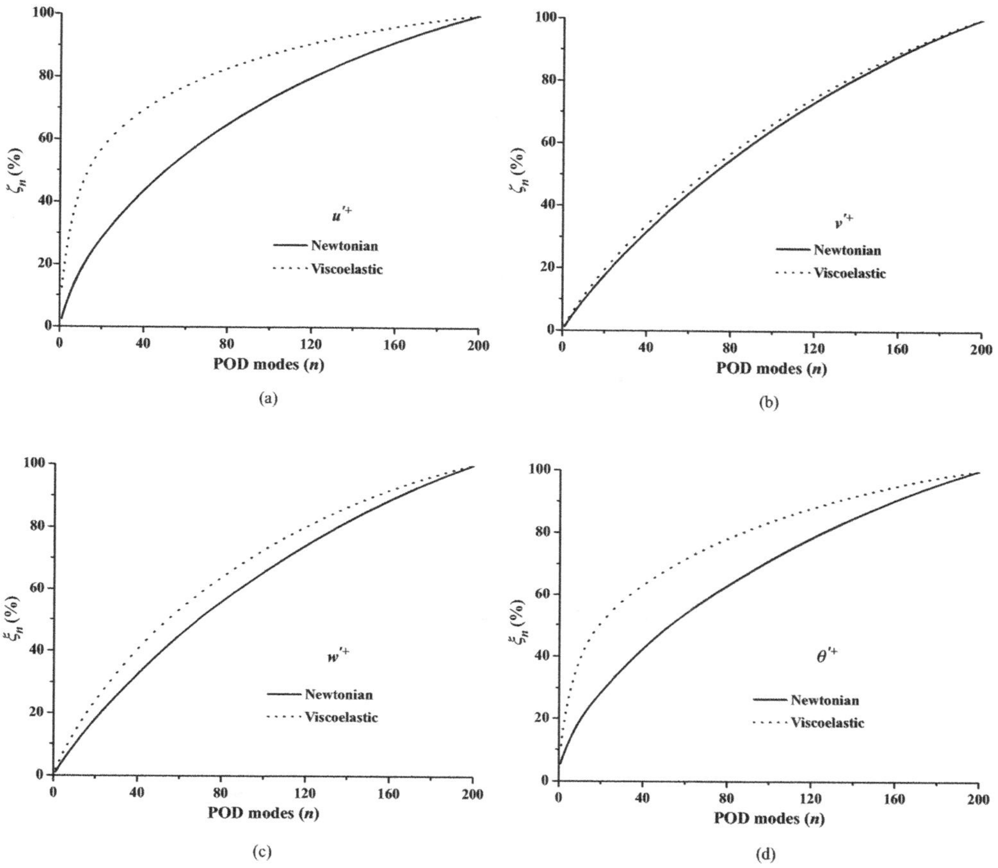

Figure 4.28 Cumulative energy contributions for different POD eigenmodes (Reproduced with permission from Wang, Y., Yu, B., Wu, X., Wei, J.J., Li, F.C. and Kawaguchi, Y. (2011), "POD study on the mechanism of turbulent drag reduction and heat transfer reduction based on Direct Numerical Simulation," *Progress in Computational Fluid Dynamics* **11**, 199–299, Inderscience Enterprises Limited.)

It can also be seen from Figures 4.27 and 4.28 that the energy distribution changes of u'^{+} and θ'^{+} are much greater than those of v'^{+} and w'^{+}. Taking the first POD eigenmode as an example, the energy contribution for u'^{+} increases from about 2.5% for Newtonian fluid flow to about 12.5% for drag-reducing flow. The energy contribution for θ'^{+} increases from about 5.5% to about 11%. However, the energy contribution of the first POD eigenmode only increases within the value of 1% for v'^{+} and w'^{+}. Therefore, u'^{+} and θ'^{+} are main fluctuations, and the flow and heat transfer become further anisotropic for drag0reducing flow. The POD eigenmodes of these main fluctuations will be analyzed to extract the proper characteristics of flow and heat transfer. Two typical POD eigenmodes are selected, that is, the eigenmodes with the highest energy contribution and the lowest energy contribution.

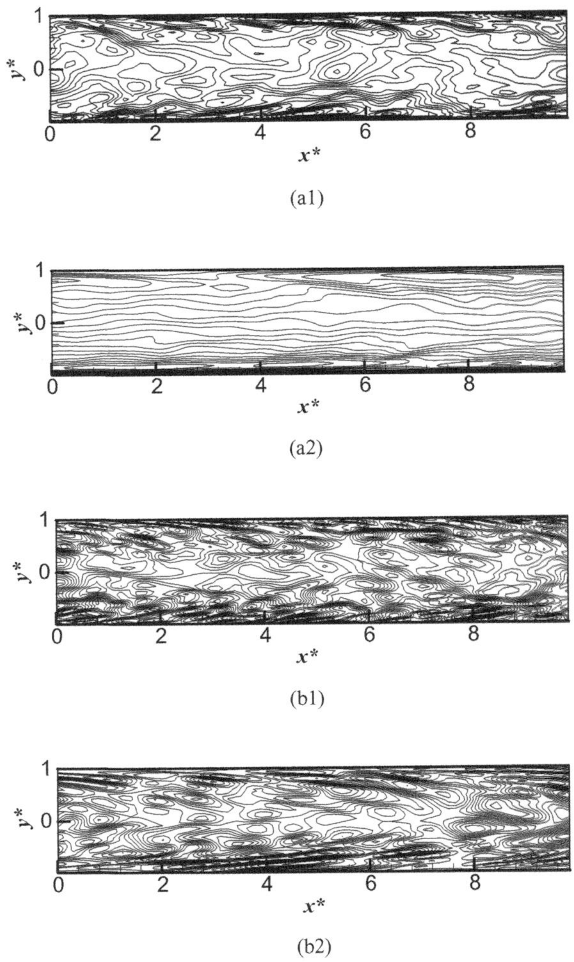

Figure 4.29 Typical POD eigenmodes of u'^+ in the x–y plane located at the middle z. a: the first POD eigenmode; b: the last POD eigenmode;1: Newtonian fluid flow; and 2: drag-reducing flow (Reproduced with permission from Wang, Y, Yu, B., Wu, X., Wei, J.J., Li, F.C. and Kawaguchi, Y. (2011), "POD study on the mechanism of turbulent drag reduction and heat transfer reduction based on Direct Numerical Simulation," *Progress in Computational Fluid Dynamics* **11**, 199–299, Inderscience Enterprises Limited.)

Figure 4.29 compares the typical POD eigenmodes of u'^+ in the x–y plane located at the middle z. "a" and "b" represent the first and the last POD eigenmode, respectively. "1" and "2" represent Newtonian fluid flow and drag-reducing flow, respectively. It can be seen that the POD eigenmodes with the highest and lowest energy contribution contain large-scale and small-scale fluctuation structures, verifying the analyses in Figures 4.27 and 4.28. The large-scale structure of drag-reducing flow enlarges apparently compared to that of Newtonian fluid flow, but the small-scale structures have no big differences. These indicate that energy contribution is corresponding to the scale of the fluctuating structure. The larger scale structure contains relatively higher energy, representing the main characteristic of flow. It is clear from Figures 4.29(a1) and 4.29(a2) that the fluctuating structures of drag-reducing flow are enlarged and the fluctuation gradient decreases in the wall-normal direction, reflecting the laminarization of the fluctuating velocity field.

Figure 4.30 compares the typical POD eigenmodes of u'^{+} in the x–z plane located at $y^{+}=5$. For both Newtonian fluid flow and drag-reducing flow, the POD eigenmodes with higher energy represent larger scale fluctuating structures (column a), showing proper characteristics of high-speed and low-speed streaks. It reflects the alternation of stream-wise fluctuation in different directions and at different intensities, that is, the strong positive fluctuations and the strong negative fluctuations appear in turn. The transitional area between these two strong fluctuation areas contains weak fluctuations, which are the superposition of the two strong fluctuations in the opposite directions. The POD eigenmodes with lower energy capture small and random structures (column b), showing the small-scale fluctuations caused by the collision and mix of different-speed fluids.

Figure 4.31 is the typical POD eigenmodes of u'^{+} in the y–z plane located at the middle x. The POD eigenmodes with higher energy mainly capture larger scale fluctuating structures near walls (column a), whereas the POD eigenmodes with lower energy mainly capture small-scale random fluctuating structures (column b). The large-scale structures of drag-reducing flow (Figure 4.31(a2)) are larger than those of Newtonian fluid flow (Figure 4.31(a1)), arranging in order along the span-wise direction. In this case, the whole y–z plane is occupied by these large-scale structures (Figure 4.31(a2)), without any medium- and small-scale structures in Newtonian fluid flow (Figure 4.31(a1)). Moreover, it is also found in Figure 4.31(a2) that the large-scale structures exist in pairs, that is, one positive large-scale structure always appears with one negative large-scale structure. All the phenomena above indicate that drag reduction is related to the alternative existence of a series of clockwise and counter-clockwise vortex structures. From Figures 4.31(b1) and 4.31(b2), it can be seen that the near-wall ($-1<y^{*}<-0.9$ and $0.9<y^{*}<1$) small-scale structures in drag-reducing flow become larger and their number becomes smaller than those in Newtonian fluid flow, showing that the energy contributions of the small-scale structures decrease compared to those of Newtonian fluid flow.

The typical POD eigenmodes of θ'^{+} are compared in the three typical planes shown in Figures 4.32–4.34. In the x–y plane (Figure 4.32), large-scale temperature structures mainly have y-gradient, showing main heat flux transfers to the inner fluid by the wall-normal temperature gradient. For drag-reducing flow (Figure 4.32(a2)), the large-scale structures containing main energy tend to be regular compared to those of Newtonian fluid flow. It reflects that the depression of heat convection and the characteristics of heat transfer are closer to heat conduction, which may be an important reason for the heat transfer reduction. The chaotic small structures in

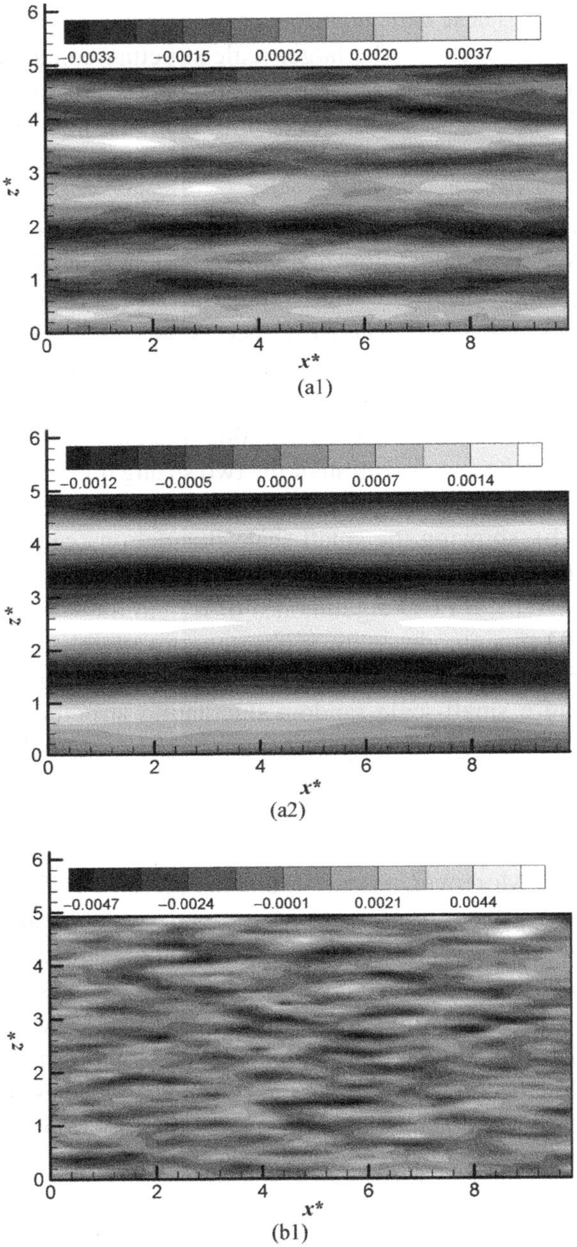

Figure 4.30 Typical POD eigenmodes of u'^{+} in the x–z plane near wall. a: the first POD eigenmode; b: the last POD eigenmode;1: Newtonian fluid flow; and 2: drag-reducing flow (Reproduced with permission from Wang, Y, Yu, B., Wu, X., Wei, J.J., Li, F.C. and Kawaguchi, Y. (2011), "POD study on the mechanism of turbulent drag reduction and heat transfer reduction based on Direct Numerical Simulation," *Progress in Computational Fluid Dynamics* **11**, 199–299, Inderscience Enterprises Limited.)

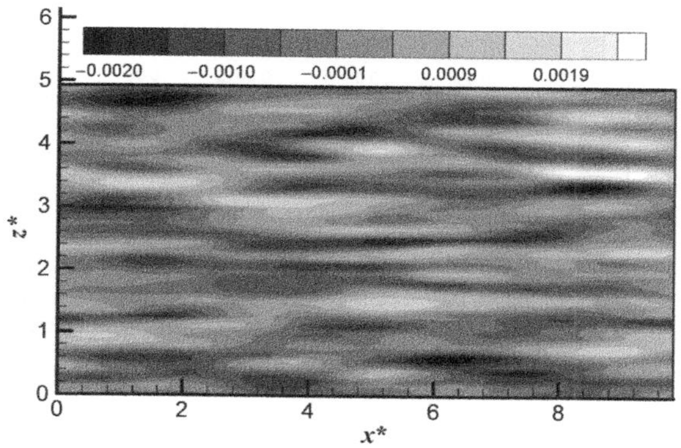

(b2)

Figure 4.30 (*Continued*)

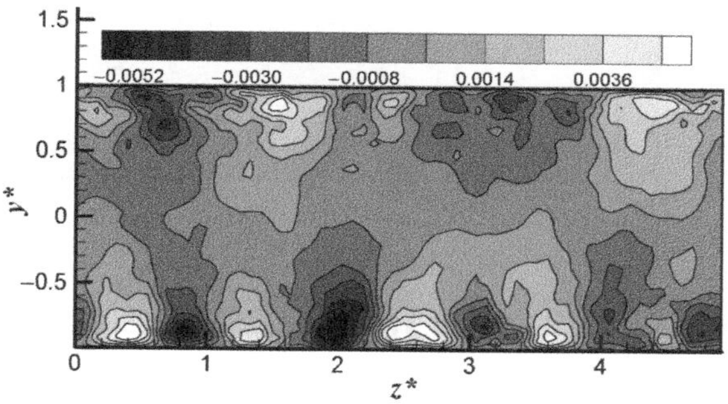

(a1)

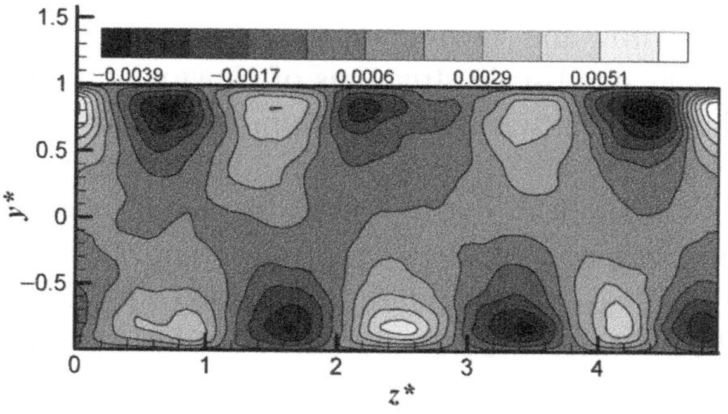

(a2)

Figure 4.31 Typical POD eigenmodes of u'^+ in the y–z plane located at the middle x. a: the first POD eigenmode; b: the last POD eigenmode; 1: Newtonian fluid flow; and 2: drag-reducing flow (Reproduced with permission from Wang, Y., Yu, B., Wu, X., Wei, J.J., Li, F.C. and Kawaguchi, Y. (2011), "POD study on the mechanism of turbulent drag reduction and heat transfer reduction based on Direct Numerical Simulation," *Progress in Computational Fluid Dynamics* **11**, 199–299, Inderscience Enterprises Limited.)

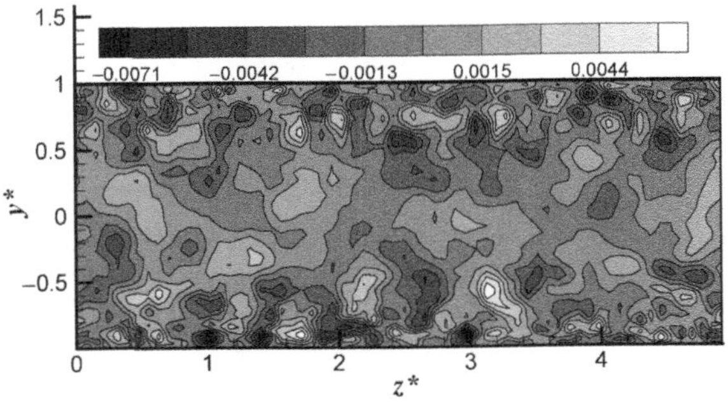

(b1)

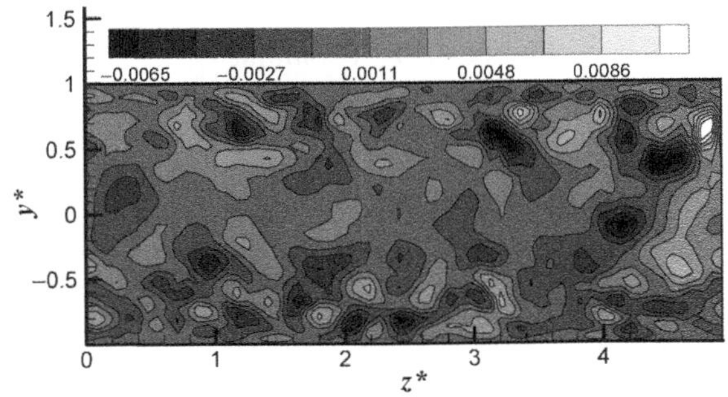

(b2)

Figure 4.31 (*Continued*)

Figures 4.32(b1) and 4.32(b2) exhibit small-scale temperature fluctuations, and irregular molecular motions have no big changes in the stream-wise direction between drag-reducing flow and Newtonian fluid flow. In the x–z plane (Figure 4.33), the captured large- and small-scale structures of Newtonian fluid flow are tiny and short, indicating that the heat transfer between high- and low-temperature fluids is good (Figures 4.33(a1) and 4.33(b1)). When the drag-reducing additives exist, the temperature streaks captured by the first POD eigenmode become larger and the distinction between the high- and low-temperature streaks is much clearer (Figure 4.33(a2)). The small-scale temperature fluctuations represented by the last POD eigenmode are in better order (Figure 4.33(b2)) than those of Newtonian fluid flow (Figure 4.33(b1)). The variation shows that the temperature fluctuations are suppressed in all scales and arranged regularly. The suppression of the temperature fluctuations near the wall causes heat transfer reduction. In the y–z plane (Figure 4.34), all scale temperature structures of Newtonian fluid flow occupy the whole plane, reflecting the active temperature fluctuation and heat transfer in all scales (Figures 4.34(a1) and 4.34(b1)). For drag-reducing flow, the small-scale temperature structures are still active compared to those of Newtonian fluid flow (Figure 4.34(b2)), but the large-scale temperature structures centralize near the walls (Figure 4.34(a2))

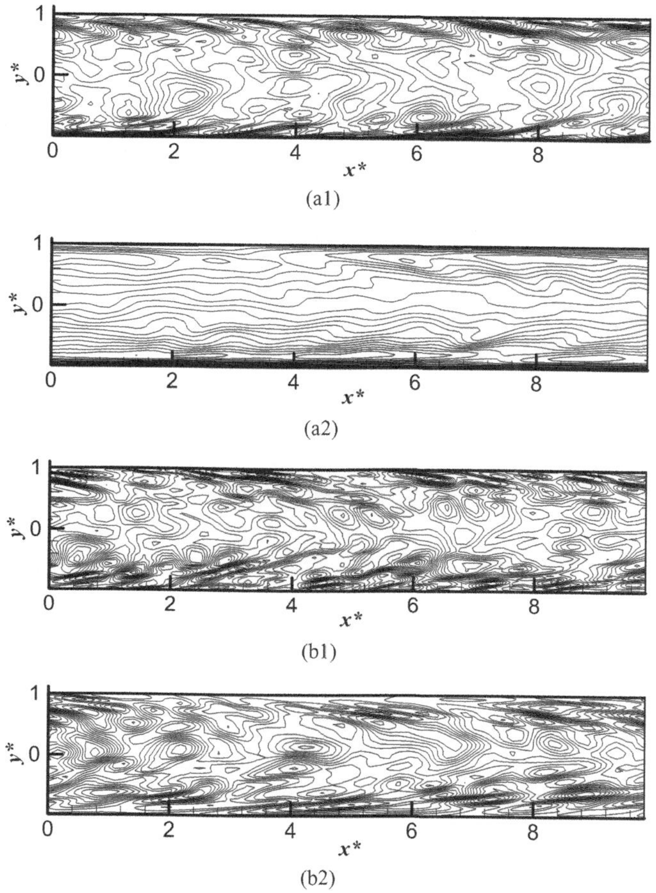

Figure 4.32 Typical POD eigenmodes of θ'^{+} in the x–y plane located at the middle z. a: the first POD eigenmode; b: the last POD eigenmode;1: Newtonian fluid flow; and 2: drag-reducing flow (Reproduced with permission from Wang, Y., Yu, B., Wu, X., Wei, J.J., Li, F.C. and Kawaguchi, Y. (2011), "POD study on the mechanism of turbulent drag reduction and heat transfer reduction based on Direct Numerical Simulation," *Progress in Computational Fluid Dynamics* **11**, 199–299, Inderscience Enterprises Limited.)

and appear as clockwise and counter-clockwise stream-wise fluctuating structures alternately along the span-wise direction, causing the heat transfer ability of fluid in central area reduced.

It can be seen from the analyses above that the large- and small-scale coherent structures extracted from the POD eigenmodes can represent main characteristics of drag reduction and heat transfer reduction. More detailed conclusions can be made as follows:

1. Drag-reducing additives mainly affect large-scale structures, which contain main proportion of turbulent kinetic energy. The energy relative contribution to the total

energy of the large-scale structures becomes larger, and the contribution of the small-scale structures, which represents energy dissipation, becomes smaller in drag-reducing fluid flow. These changes cause stronger anisotropy in the flow and heat transfer characteristics of drag-reducing flow than those of Newtonian fluid flow.

2. The energy distributions of stream-wise velocity fluctuation and temperature fluctuations are affected by drag-reducing additives more deeply than those of

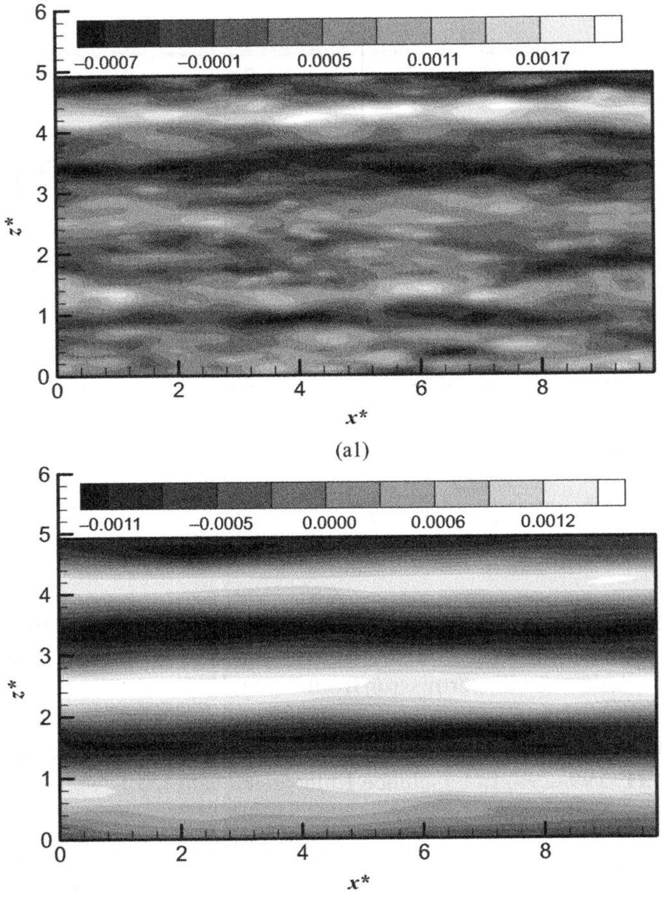

Figure 4.33 Typical POD eigenmodes of θ'^{+} in the x–z plane near wall. a: the first POD eigenmode; b: the last POD eigenmode; 1: Newtonian fluid flow; and 2: drag-reducing flow (Reproduced with permission from Wang, Y., Yu, B., Wu, X., Wei, J.J., Li, F.C. and Kawaguchi, Y. (2011), "POD study on the mechanism of turbulent drag reduction and heat transfer reduction based on Direct Numerical Simulation," *Progress in Computational Fluid Dynamics* **11**, 199–299, Inderscience Enterprises Limited.)

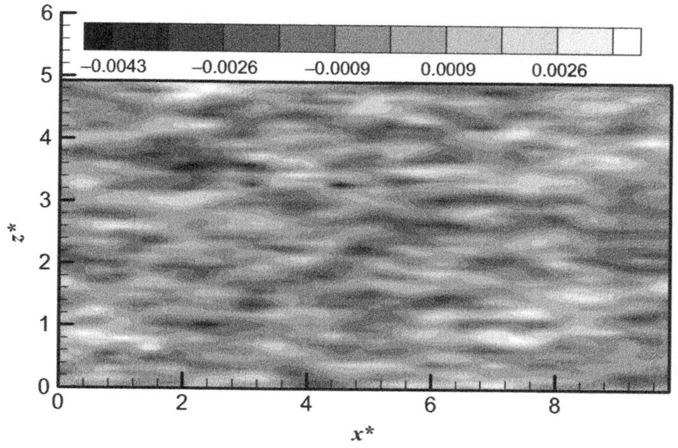

(b1)

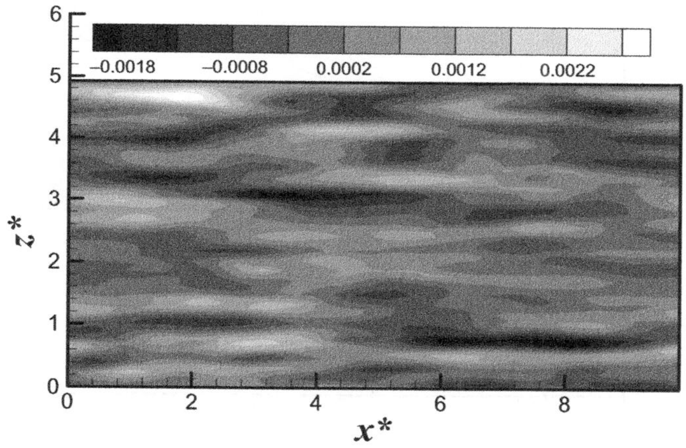

(b2)

Figure 4.33 (*Continued*)

velocity fluctuations in the wall-normal and the span-wise directions. For drag-reducing flow, the large-scale structures are enlarged in the stream-wise direction and align in the span-wise direction alternately as well-organized clockwise and counter-clockwise stream-wise swirling motions. The gradient of these large-scale structures becomes smaller in the wall-normal direction. The large-scale temperature fluctuations concentrate near the walls of the channel, while the main characteristic of heat transfer is closer to heat conduction. This may be an important reason for heat transfer reduction. The small-scale structures are weakened, so the flow develops to the aspect of lower dissipation.

4.4.2 The Rheological Parameter Effect of DNS on Surfactant Drag Reduction

The above overall DNS analyzed the characteristics of drag reduction and heat transfer and found that viscoelasticity is the main reason for drag reduction. Therefore, the influence of Weissenberg number, representing viscoelasticity, on drag reduction is studied here. Key parameters are $We_\tau = 2$, 12.5, 30, 45, $Re_\tau = 150$, $\alpha = 0.001$, and $\beta = 0.9$ (the definition of β is Equation 4.16). Table 4.5 listed the drag reduction rate, mean Reynolds number, and friction factor on different Weissenberg numbers. Drag reduction is 12.1% at $We_\tau = 12.5$, but no drag reduction occurs at $We_\tau = 8$. Thus, the onset Weissenberg number is about 10.

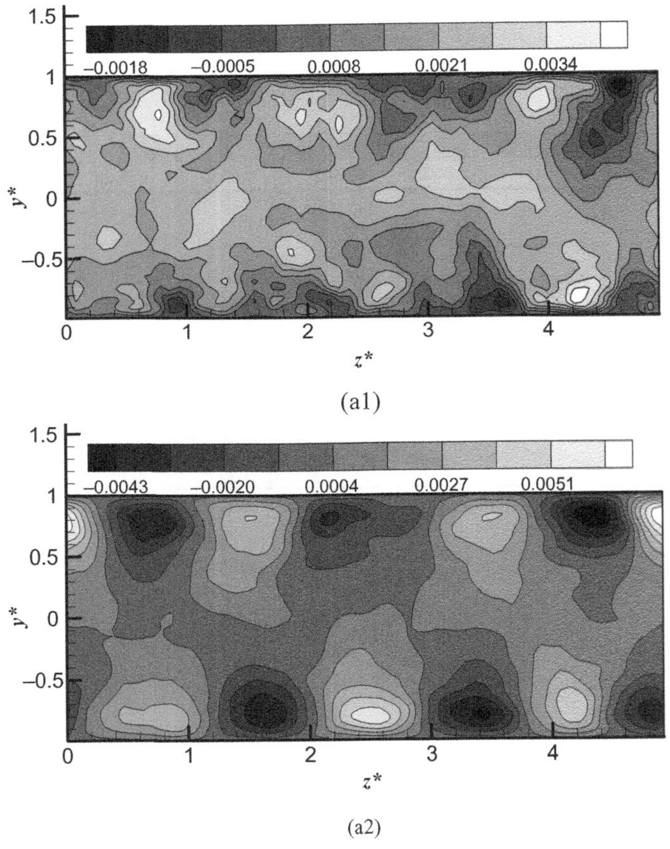

Figure 4.34 Typical POD eigenmodes of θ'^+ in the y–z plane located at the middle x. a: the first POD eigenmode; b: the last POD eigenmode; 1: Newtonian fluid flow; and 2: drag-reducing flow (Reproduced with permission from Wang, Y., Yu, B., Wu, X., Wei, J.J., Li, F.C. and Kawaguchi, Y. (2011), "POD study on the mechanism of turbulent drag reduction and heat transfer reduction based on Direct Numerical Simulation," *Progress in Computational Fluid Dynamics* **11**, 199–299, Inderscience Enterprises Limited.)

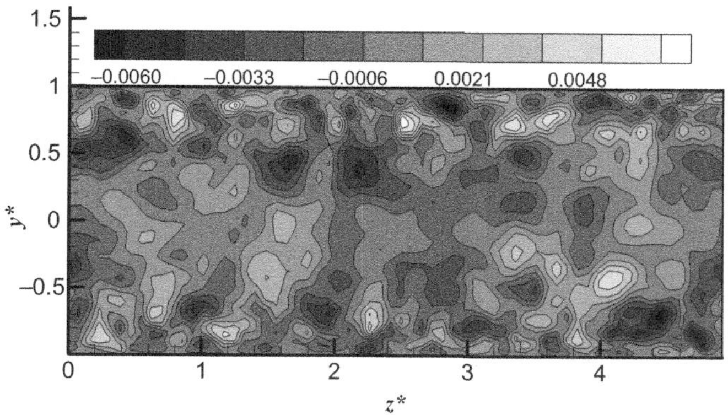

(b1)

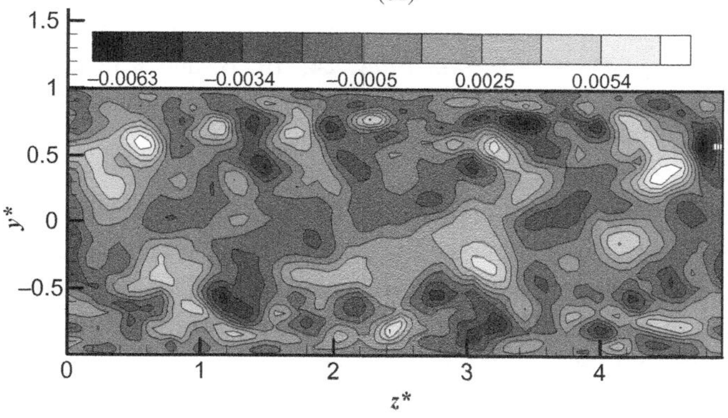

(b2)

Figure 4.34 *(Continued)*

Figure 4.35 compares the root mean square of c_{xx} fluctuations. It is seen that the conformation fluctuations become much stronger as the Weissenberg number increases. We believe that the strong fluctuations at high Weissenberg numbers may be another cause of the numerical instability, and our calculation broke down in the cases of an even higher Weissenberg number, $We_\tau = 60$.

Table 4.5 Mean Reynolds numbers, friction factors, and drag reduction rate (Reprinted from *International Journal of Heat and Fluid Flow*, **24**, Yu, B. and Kawaguchi, Y., "Effect of Weissenberg number on the flow structure DNS study of drag-reducing flow with surfactant additives," 491–499, 2003, with permission from Elsevier.)

We_τ	12.5	30	45
Re_b	4838	6180	6936
C_f	0.00769	0.00471	0.00374
DR%	12.1%	42.8%	53.2%

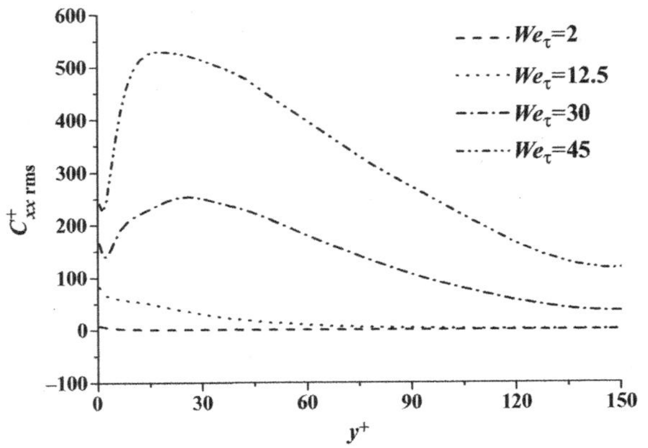

Figure 4.35 Root-mean-square fluctuations for a conformation component (Reprinted from *International Journal of Heat and Fluid Flow,* **24**, Yu, B. and Kawaguchi, Y., "Effect of Weissenberg number on the flow structure DNS study of drag-reducing flow with surfactant additives," 491–499, 2003, with permission from Elsevier.)

Figure 4.36 shows the instantaneous contour maps of the conformation component c^+_{xx} in the middle vertical x–y plane of the channel. It is seen that as the Weissenberg number increases, the value of c^+_{xx} increases greatly. The gradients near the wall become larger as the Weissenberg number increases. From this figure, we can partly explain why the calculation easily breaks down as the Weissenberg number increases. The steep conformation gradient is difficult to be captured accurately.

Figure 4.37 shows instantaneous snapshots of the velocity fluctuation fields in the middle vertical x–y plane of the channel at different Weissenberg numbers (the stream-wise velocity components are subtracted by a local mean velocity $u^+(y)$). For comparison, an instantaneous velocity field for the Newtonian case is also presented. It is clearly seen that the flow structure of the no drag reduction case ($We_\tau = 2$) is similar to that of the Newtonian case. As the Weissenberg number increases, the vortex structure becomes elongated in the stream-wise direction, especially in the region near the walls. In Figure 4.37e, at the region near the bottom wall, the fluid flows from right to left, which means that elongated low-speed stream-wise streaks exist. To solve the larger flow structure, a larger computational domain may be required. The effect of computational domain size on the solutions is now being studied by our research group.

Figure 4.38 shows instantaneous snapshots of the stream-wise fluctuating velocity in the x–z plane at $y^+ = 15$. It is seen that as the Weissenberg number increases, the low-speed streaks become more elongated and the average spacing of the streaks becomes wider. The larger spacing is connected with the larger flow structure such as that shown in Figure 4.38. Figure 4.39 shows the two-point correlations of stream-wise velocity R_{uu} in the span-wise direction. The separation at which the minimum R_{uu} occurs can be used to estimate the mean spacing between high- and low-speed streaks,

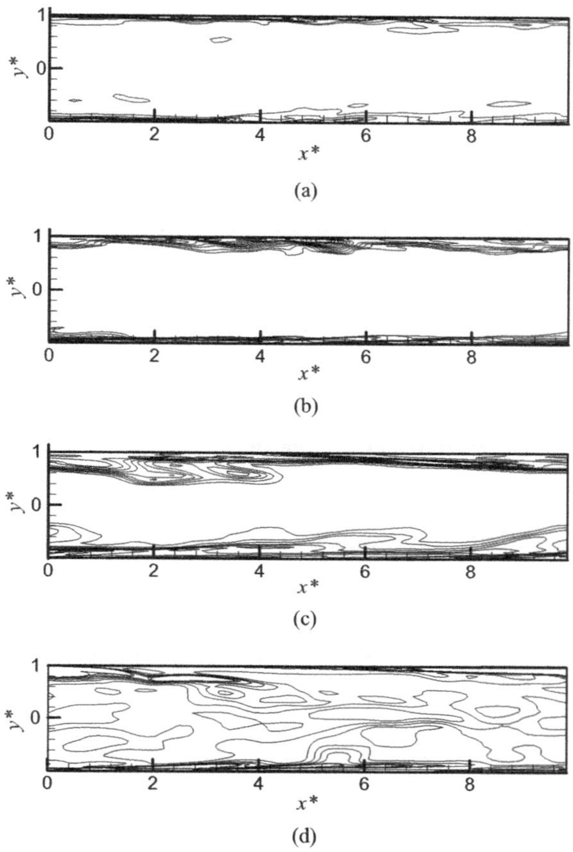

Figure 4.36 Contours of instantaneous c_{xx}^{+} in the middle vertical x–y plane of the channel. (a) $We_\tau = 2, c_{xx}^{+} = 1.5 - 11$; (b) $We_\tau = 12.5, c_{xx}^{+} = 37 - 545$; (c) $We_\tau = 30, c_{xx}^{+} = 68 - 857$; and (d) $We_\tau = 45, c_{xx}^{+} = 107 - 1520$ (Reprinted from *International Journal of Heat and Fluid Flow*, **24**, Yu, B. and Kawaguchi, Y., "Effect of Weissenberg number on the flow structure DNS study of drag-reducing flow with surfactant additives," 491–499, 2003, with permission from Elsevier.)

that is, the mean streak spacing is roughly twice the distance to the negative peak. It is seen more clearly from this picture that as the Weissenberg number increases, the streak spacing becomes larger.

Other rheological parameters in the Giesekus model are also important for drag reduction. They will be analyzed as follows. Computational parameters and results under different rheological parameters are listed in Table 4.6 and Table 4.7, where the superscript "*" represents the definition based on wall effective viscosity and β still uses the form of Equation 4.16.

From the numerical simulations, the following conclusions can be drawn. As the Weissenberg number increases, the flow structures become larger. The larger the drag

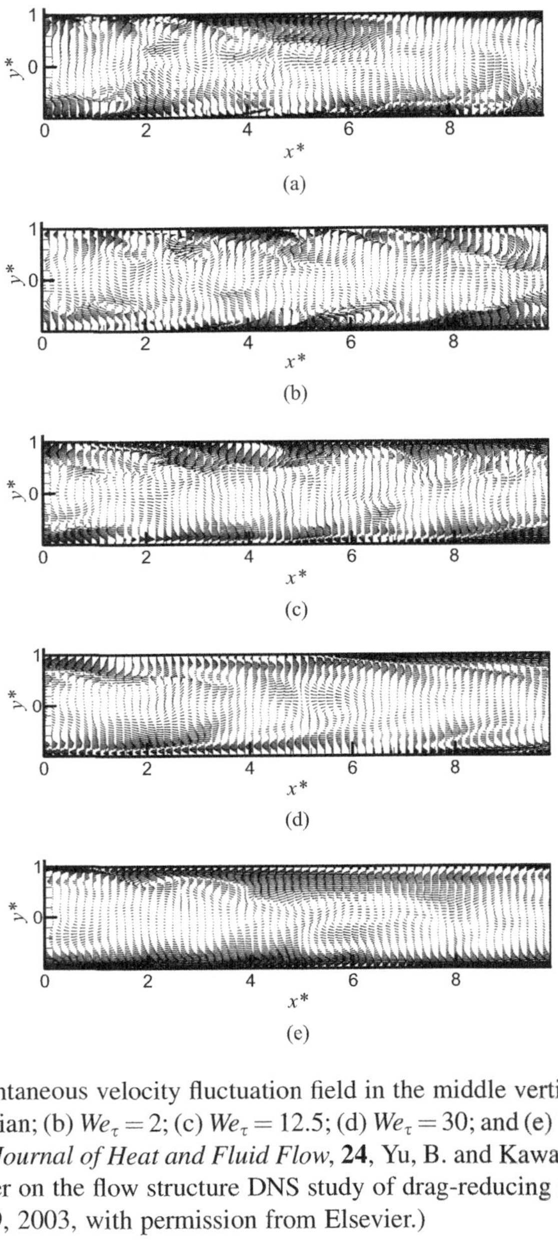

Figure 4.37 Instantaneous velocity fluctuation field in the middle vertical *x–y* plane of the channel. (a) Newtonian; (b) $We_\tau = 2$; (c) $We_\tau = 12.5$; (d) $We_\tau = 30$; and (e) $We_\tau = 45$ (Reprinted from *International Journal of Heat and Fluid Flow*, **24**, Yu, B. and Kawaguchi, Y., "Effect of Weissenberg number on the flow structure DNS study of drag-reducing flow with surfactant additives," 491–499, 2003, with permission from Elsevier.)

reduction rate, the larger u^+_{rms} increases and the smaller v^+_{rms} and w^+_{rms} decrease. The Reynolds shear stress becomes smaller as the Weissenberg number increases. As the Weissenberg number increases, both the streak spacing and the drag reduction become larger. The onset Weissenberg number obtained in the present study was around 10, and the maximum drag reduction obtained was 53%. We also partly

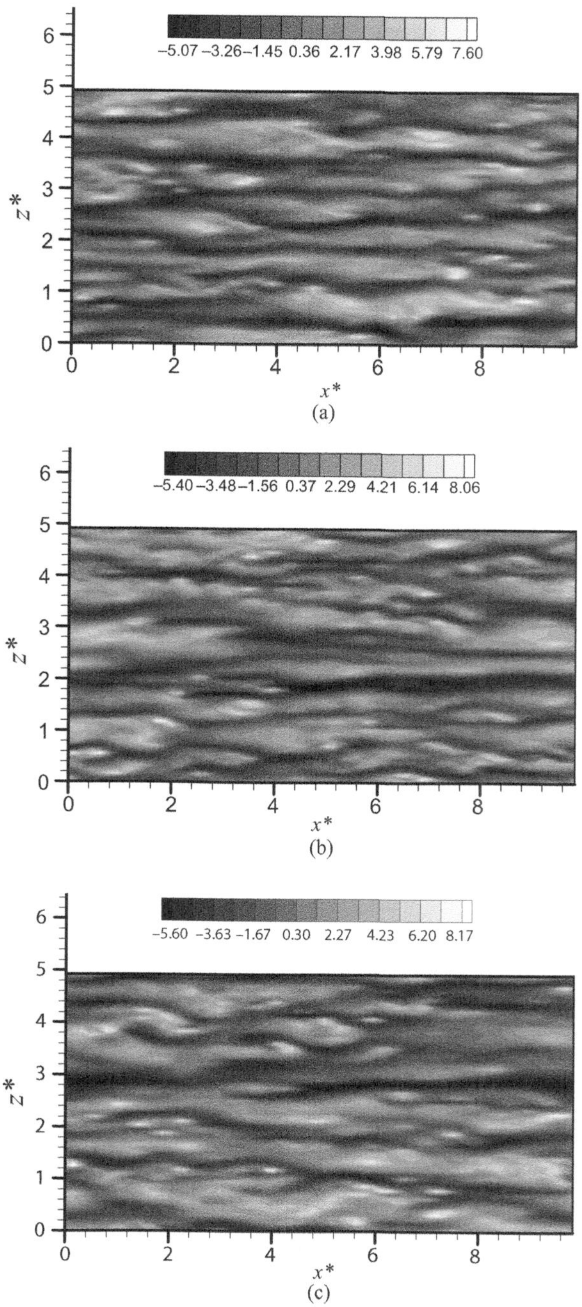

Figure 4.38 Instantaneous snapshot of stream-wise velocity fluctuation in the x–z plane at $y^+ = 15$. (a) Newtonian; (b) $We_\tau = 2$; (c) $We_\tau = 12.5$; (d) $We_\tau = 30$; and (e) $We_\tau = 45$ (Reprinted from *International Journal of Heat and Fluid Flow*, **24**, Yu, B. and Kawaguchi, Y., "Effect of Weissenberg number on the flow structure DNS study of drag-reducing flow with surfactant additives," 491–499, 2003, with permission from Elsevier.)

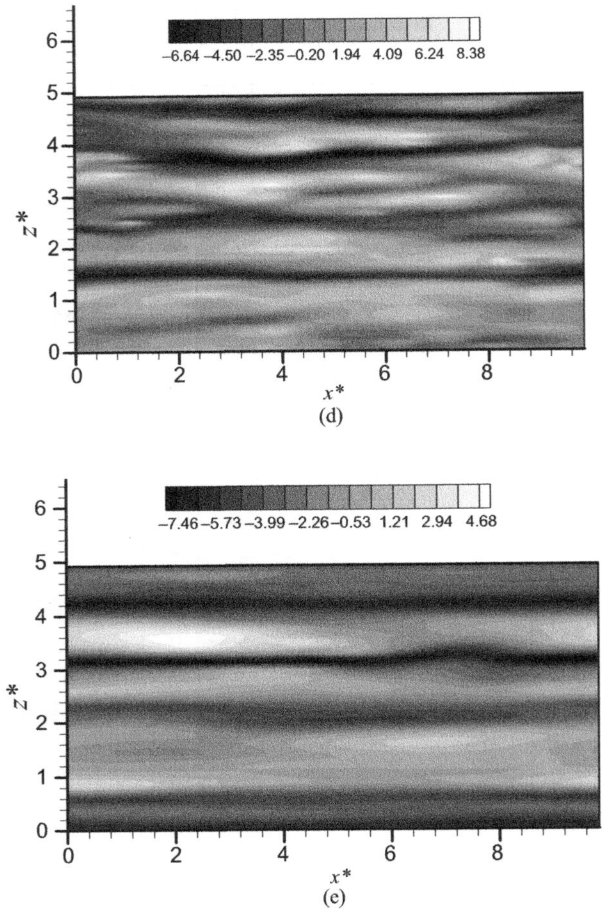

Figure 4.38 (*Continued*)

explained why the calculations tend to break down in the case of a high Weissenberg number for viscoelastic flow simulation.

Some important results are listed in Table 4.6 and Table 4.7. Table 4.6 shows that with the increase of We_τ, with the decrease of α, and with the decrease of β, the drag reduction rate increases. Table 4.7 shows that, generally speaking, the larger DR rates are associated with larger $\overline{u^+_{rms}}$, smaller $\overline{v^+_{rms}}$, and smaller $\overline{w^+_{rms}}$. With the increase of DR rate, the peak value position δ^{+*}_{urms} of the RMS of the stream-wise velocity fluctuation shifts further to the bulk flow region. δ^{+*}_{uu} and δ^{+*}_{vv} are mean spacing between low-speed streaks and the mean diameter of stream-wise vortex at $y^{+*} = 15$, which are estimated by the separation of minimum span-wise two-point correlations R_{uu} and R_{vv}. Generally the large DR rates are related to large δ^{+*}_{uu} and δ^{+*}_{vv}. $\overline{v_t}$ and $\overline{k_e}$ are the mean turbulent viscosity of the flow and mean elastic energy of the networks structures, respectively.

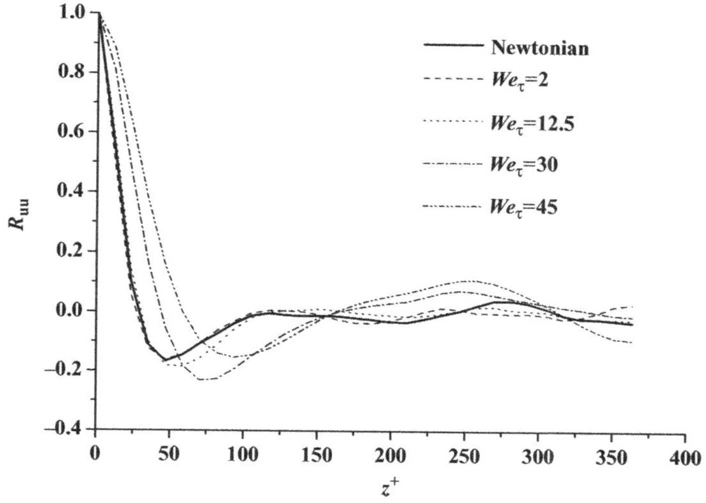

Figure 4.39 Two-point spanwise correlation of the velocity component in the stream-wise direction at $y^+ = 15$ (Reprinted from *International Journal of Heat and Fluid Flow*, **24**, Yu, B. and Kawaguchi, Y., "Effect of Weissenberg number on the flow structure DNS study of drag-reducing flow with surfactant additives," 491–499, 2003, with permission from Elsevier.)

The mean turbulent viscosities of all the viscoelastic fluids are smaller than that of Newtonian fluid and decrease with the increase of DR rate. From Table 4.7, it is seen that drag reduction is related to not only the magnitude of elastic energy but also its spatial distribution.

Figure 4.40 shows that the elastic energy of Fluid A is large only in the viscous sublayer; the elastic energy of Fluid B is larger than Fluid G and Fluid I in the viscous

Table 4.6 Computational parameters and some important results (Reprinted from *International Journal of Heat and Fluid Flow*, **27**, Yu, B. and Kawaguchi, Y., "Parametric study of surfactant-induced drag-reduction by DNS," 887–894, 2006, with permission from Elsevier.)

Fluid	β	We_τ	α	η_0/η_w	U_m^+	Re_τ^*	Re_m^*	DR*%
A	0.5	8	0.001	1.107	15.08	166	5000	0
B	0.5	12.5	0.001	1.184	18.14	178	6440	25.4
C	0.5	20.0	0.001	1.311	22.14	197	8710	46.0
D	0.5	30.0	0.001	1.455	25.92	218	11 300	57.9
E	0.5	40.0	0.001	1.533	30.36	230	13 970	67.7
F	0.5	50.0	0.01	1.598	33.32	240	15 980	72.0
G	0.5	20.0	0.01	1.668	22.33	250	11 180	43.5
H	0.3	30.0	0.001	1.907	29.15	286	16 670	63.4
I	0.8	30.0	0.001	1.126	22.06	169	7450	46.3
N	×	×	×	1.000	14.78	150	4440	×

Table 4.7 Some important results (Reprinted from *International Journal of Heat and Fluid Flow*, **27**, Yu, B. and Kawaguchi, Y., "Parametric study of surfactant-induced drag-reduction by DNS," 887–894, 2006, with permission from Elsevier.)

$DR^{*}\%$	Fluid	$\overline{u_{rms}^{+}}$	$\overline{v_{rms}^{+}}$	$\overline{w_{rms}^{+}}$	δ_{urms}^{+*}	δ_{uu}^{+*}	δ_{vv}^{+*}	$\overline{\nu_t}$	$\overline{k_e}$
0	A	1.46	0.570	0.721	11.2	130	26	9.1	0.40
25.4	B	1.65	0.528	0.679	17.1	166	42	8.6	0.62
43.5	G	1.85	0.526	0.684	20.3	234	58	8.4	0.47
46.0	C	1.98	0.474	0.634	22.2	215	46	8.4	0.98
46.3	I	2.18	0.479	0.679	22.2	211	52	8.1	0.65
57.9	D	2.48	0.433	0.588	28.6	307	68	6.5	1.39
63.4	H	2.68	0.402	0.564	32.3	358	89	5.0	2.02
67.7	E	2.75	0.349	0.515	30.2	323	72	4.1	1.79
72.0	F	3.19	0.308	0.474	36.3	486	75	3.2	2.01
×	N	1.43	0.593	0.747	12.2	94	35	10.9	×

sublayer but smaller in the buffer layer, and it has a smaller DR rate than Fluid G and Fluid I; the elastic energy of Fluid F is the same as those of Fluid D and Fluid E in the viscous sublayer but larger in the buffer layer, and it has a larger DR rate. All these show that generally the occurrence of a large DR rate requires that the elastic energy has a large value in a wide buffer layer.

Figure 4.41 shows the Reynolds shear stress. Generally, the DR rate is associated with the decrease of Reynolds shear stress, but there is not a certain relationship between the DR rate and Reynolds shear stress. A maximum DR rate up to 72% of Fluid F was obtained in the present study, but we did not find a diminishment of Reynolds shear stress, which was observed in the experiments (Li *et al.*, 1998).

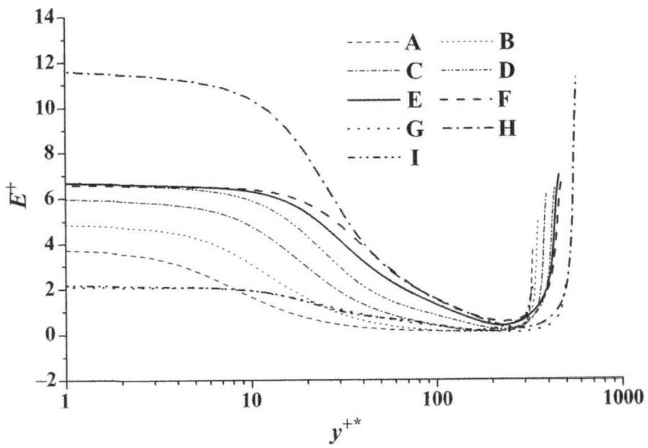

Figure 4.40 Mean elastic energy (Reprinted from *International Journal of Heat and Fluid Flow*, **27**, Yu, B. and Kawaguchi, Y., "Parametric study of surfactant-induced drag-reduction by DNS," 887–894, 2006, with permission from Elsevier.)

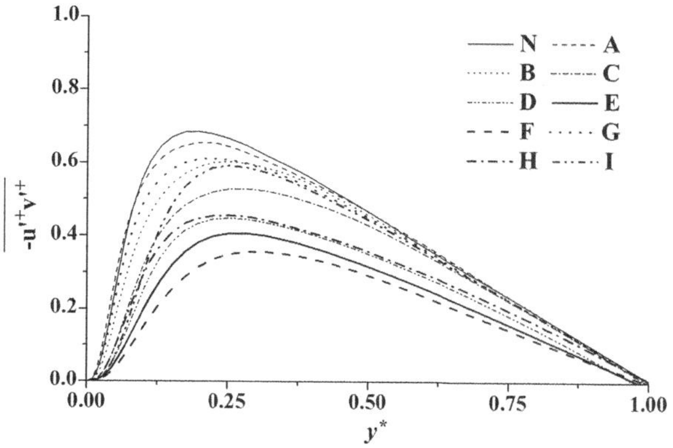

Figure 4.41 Reynolds shear stress (Reprinted from *International Journal of Heat and Fluid Flow*, **27**, Yu, B. and Kawaguchi, Y., "Parametric study of surfactant-induced drag-reduction by DNS," 887–894, 2006, with permission from Elsevier.)

This indicates that the diminishment of Reynolds shear stress is not a must for a large DR rate. The Reynolds shear stress of Fluid A is smaller than that of Newtonian fluid, but no drag reduction occurs. Fluid C, Fluid G, and Fluid I have similar DR rates, but the Reynolds shear stresses differ appreciably. Fluid B has a smaller Reynolds shear stress than Fluid G but a larger DR rate. All the above indicates that a large decrease of Reynolds shear stress does not certainly mean a large DR rate.

The quadrant analysis of the Reynolds shear stress provides detailed information on the contributions from various events occurring in the flow. Here we investigate how the Reynolds shear stress is changed for different fluids by the quadrant analysis. Fluids A, B, C, F, G, and I are compared with Newtonian fluid, as shown in Figure 4.42. It is seen that for all the viscoelastic fluids, the four quadrant events (Q1–Q4) at the bulk flow region are either almost the same as those of Newtonian fluid or stronger than those of Newtonian fluid. These stronger events are probably due to the effect of Re_{τ}^*, that is, the Re_{τ}^* of Newtonian fluid is smaller than that of the viscoelastic fluid. We can expect the large values in each quadrant for the Newtonian fluid for larger corresponding Re_{τ}^*. At the near-wall region, the ejection and sweep events of Fluids B, C, F, G, and I are reduced but the outward motion of high-speed fluids (Q1) and the inward motion of low-speed fluids (Q3) are either reduced or enhanced. For Fluid A, the ejection and sweep events do not change as compared to with Newtonian fluid. The decrease of Reynolds shear stress of Fluids B, C, F, and I at the near-wall region is due to the reduction of the ejection evens (Q2) and sweep events (Q4). For Fluid I, the decrease of sweep events is the largest contribution to the reduction of Reynolds shear stress. However, the decrease of Reynolds shear stress for Fluid G is primarily due to the increase of the outward motion of high-speed fluids (Q1) and the inward motion of low-speed fluids (Q3), and the decrease of sweep events (Q4). In a word, generally drag reduction is associated with the suppression of the ejection and sweep events.

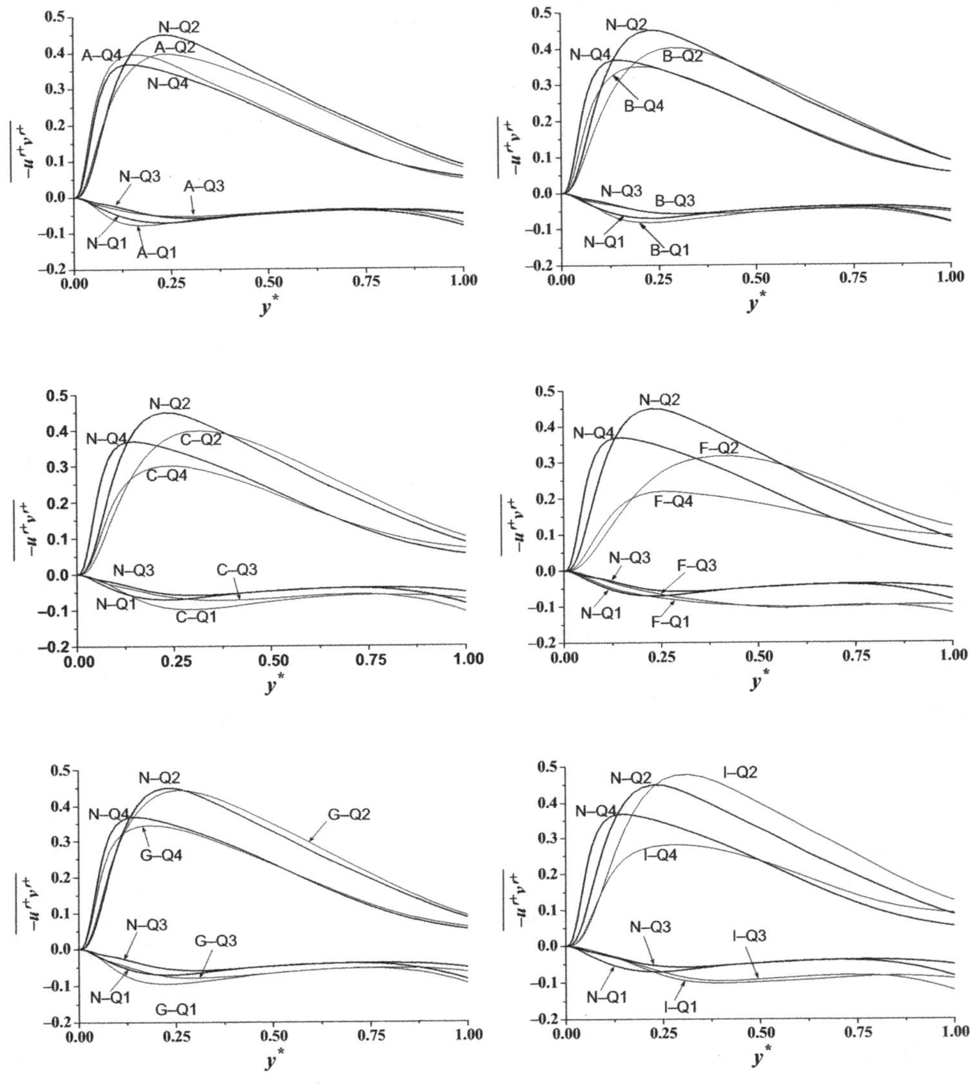

Figure 4.42 Reynolds shear stress from each quadrant (Reprinted from *International Journal of Heat and Fluid Flow*, **27**, Yu, B. and Kawaguchi, Y., "Parametric study of surfactant-induced drag-reduction by DNS," 887–894, 2006, with permission from Elsevier.)

The typical instantaneous stream-wise velocity contours, secondary velocity vectors, and elastic energy at a cross-section in the *y*–*z* plane are shown in Figure 4.43. It is seen that the size of the stream-wise vortex becomes larger with the increase of drag reduction rate. The occurrence of ejection and sweep events is greatly reduced with the increase of DR% rates. The high elastic energy is released during the ejection events. The isosurfaces of instantaneous stream-wise vorticities are shown in Figure 4.44, in

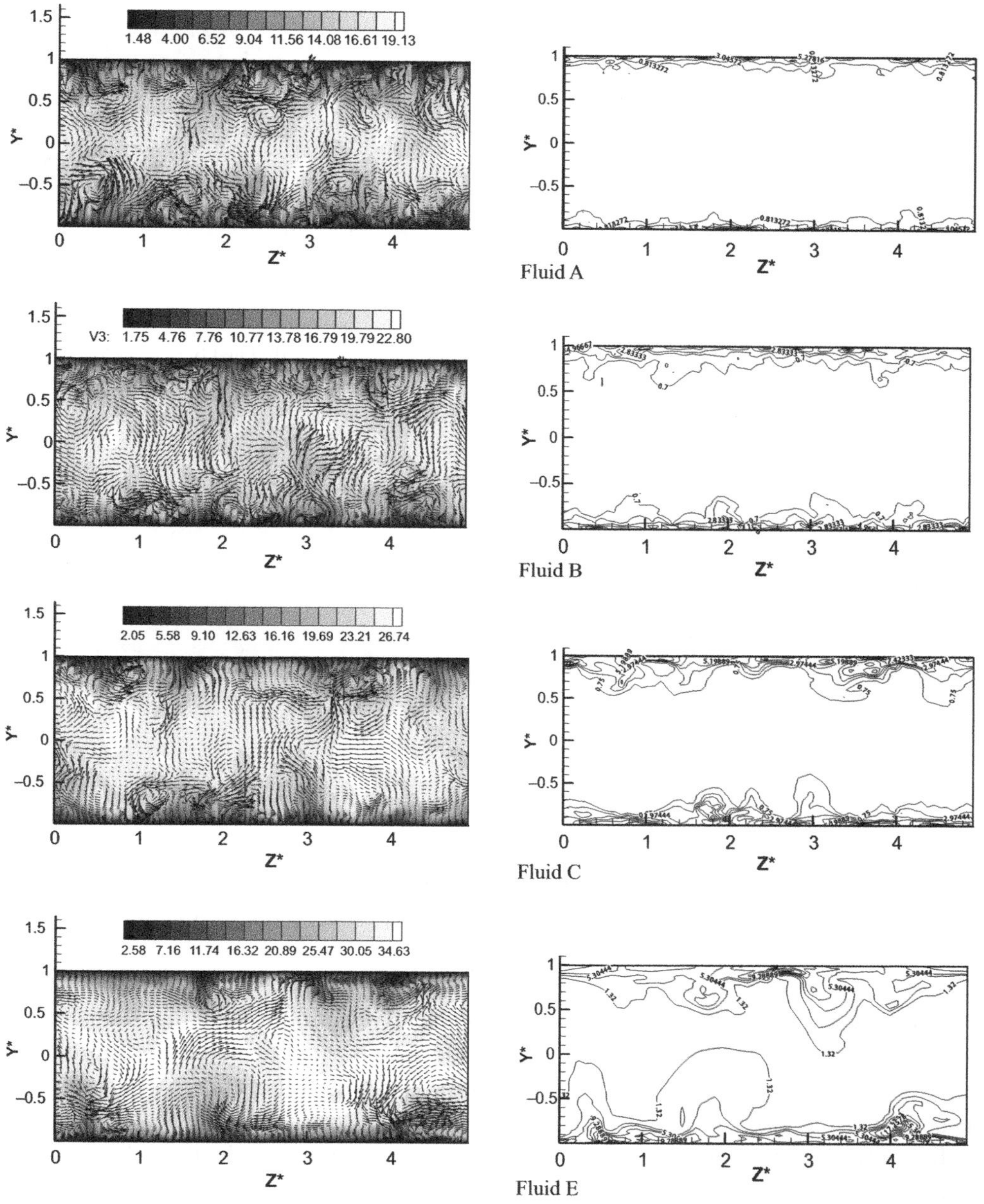

(a) (b)

Figure 4.43 Instantaneous velocities and elastic energy at a y–z plane. (a) Contour of streamwise velocity, and v and w velocity vectors; and (b) elastic energy (Reprinted from *International Journal of Heat and Fluid Flow*, **27**, Yu, B. and Kawaguchi, Y., "Parametric study of surfactant-induced drag-reduction by DNS," 887–894, 2006, with permission from Elsevier.)

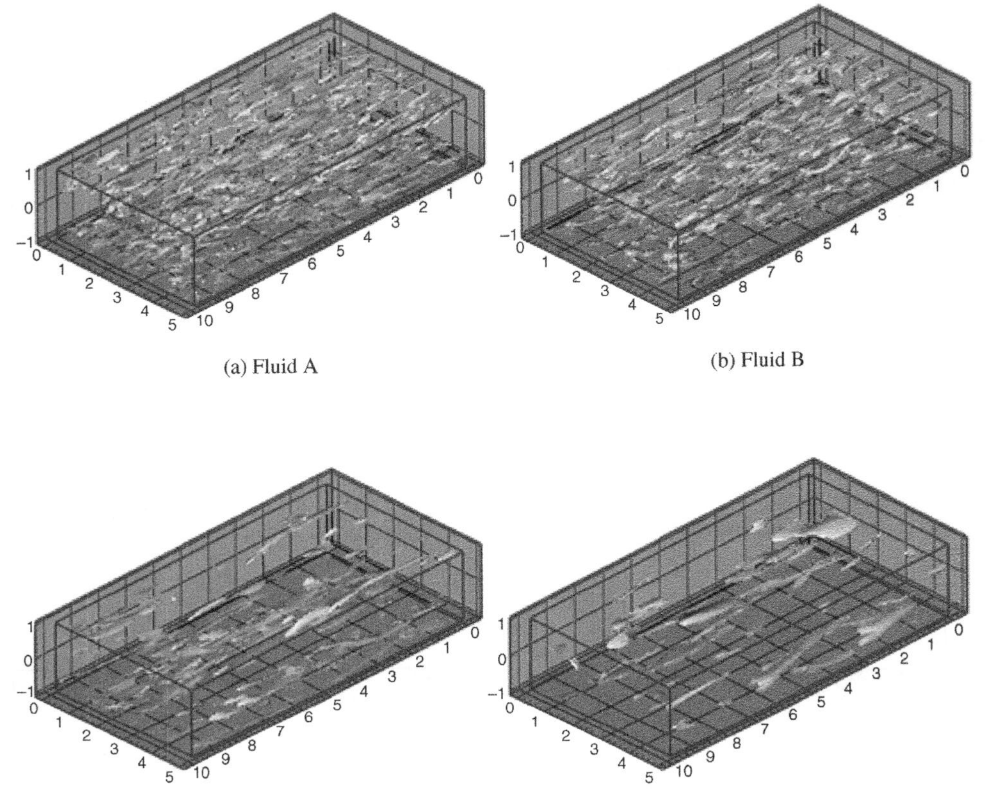

(a) Fluid A (b) Fluid B

(c) Fluid D (d) Fluid F

Figure 4.44 Isosurfaces of an instantaneous stream-wise vorticity ($\omega_x = 0.1$) (Reprinted from *International Journal of Heat and Fluid Flow*, **27**, Yu, B. and Kawaguchi, Y., "Parametric study of surfactant-induced drag-reduction by DNS," 887–894, 2006, with permission from Elsevier.)

which the change of vortical structures can be clearly seen. It is seen that the stream-wise vortices become much weaker and more elongated with the increase of large drag reduction rates. The significant weakening of the strength of the stream-wise vorticity near the wall leads to the drag reduction.

To study the effect of elastic energy and vorticities on drag reduction, the balance equations of elastic energy and vorticity are derived as follows. Similar to Equation 4.32, the balance equation of elastic energy has another form based on the definition of Equation 4.16:

$$\underbrace{\int_{-1}^{1} \frac{1-\beta}{We_\tau} \frac{dU^+}{dy^+} C_{xy}^+ dy^*}_{\text{work by viscoelastic stress}} + \underbrace{\int_{-1}^{1} \frac{1-\beta}{We_\tau} \left(\overline{-c'^+_{ik} \frac{\partial u'^+_i}{\partial x_k^+}} \right) dy^*}_{\text{turbulence-elasticity interaction}} + \underbrace{\int_{-1}^{1} \frac{1-\beta}{2We_\tau^2} \left[(3 - C_{ii}^+) - \alpha (C_{im}^+ - \delta_{im})^2 \right] dy^*}_{\text{elastic dissipation}} = 0 \tag{4.55}$$

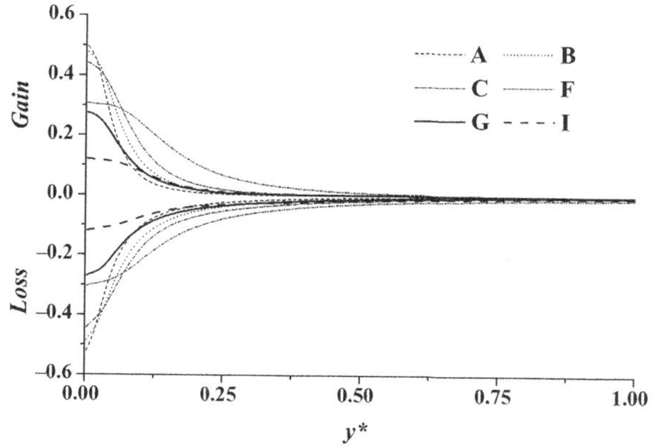

Figure 4.45 The elastic energy budget terms IV and VII (Reprinted from *International Journal of Heat and Fluid Flow*, **27**, Yu, B. and Kawaguchi, Y., "Parametric study of surfactant-induced drag-reduction by DNS," 887–894, 2006, with permission from Elsevier.)

Figure 4.45 compares all terms in the above equation. The relative contributions of the turbulent production of the Newtonian fluid and Fluid G are 41 and 34%, respectively. This means that for Fluid G, turbulence has also been effectively suppressed.

The derived vorticity balance equation for drag-reducing flow is as follows:

$$\frac{\partial \varpi_i^+}{\partial t^+} + u_j^+ \frac{\partial \varpi_i^+}{\partial x_j^+} = \varpi_j^+ \frac{\partial u_i^+}{\partial x_j^+} + \beta \frac{\partial^2 \varpi_i^+}{\partial x_j^+ \partial x_j^+} + \frac{\varepsilon_{ijk}(1-\beta)\mathrm{Re}_\tau}{We_\tau} \frac{\partial}{\partial x_j^+} \left(\frac{\partial c_{kl}^+}{\partial x_l^*} \right) \tag{4.56}$$

Figure 4.46 shows the stretching mode ($\partial u/\partial x > 0$) and squeezing mode ($\partial u/\partial x < 0$) of the stream-wise vortex. It is seen that both the stretching and squeezing modes of the stream-wise vortex have been appreciably suppressed in drag reducing flow. For Fluid A, the stretching of the stream-wise vortex is enhanced as compared to that of Newtonian fluid, which is consistent with a smaller vortex size.

Numerical simulations show that with the increase of We_τ, with the decrease of α, and with the decrease of β, the drag reduction rate increases. Large drag reduction is associated with large elastic energy in the buffer layer. The vortical structures become weaker and more elongated in the stream-wise direction, and the stretching of the stream-wise vortex is greatly reduced for a large drag reduction rate.

4.4.3 DNS with the Bilayer Model of Flows with Newtonian and Non-Newtonian Fluid Coexistence

This section will introduce a bilayer model of surfactant drag reduction, in which the non-Newtonian fluid part uses the Giesekus model to study the effect of drag-reducing additives on local areas. The flow to be studied was a fully developed

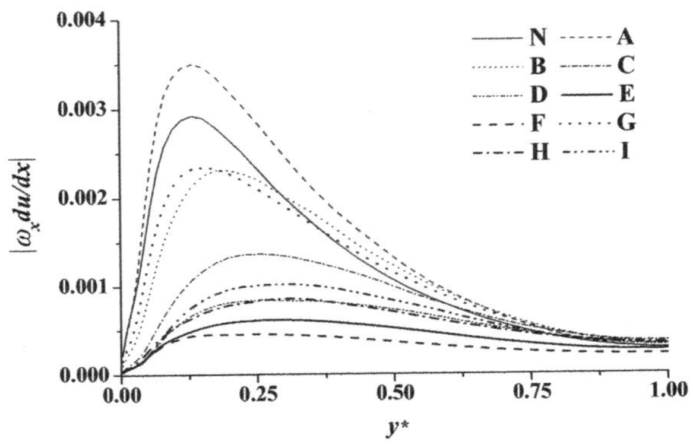

(a)

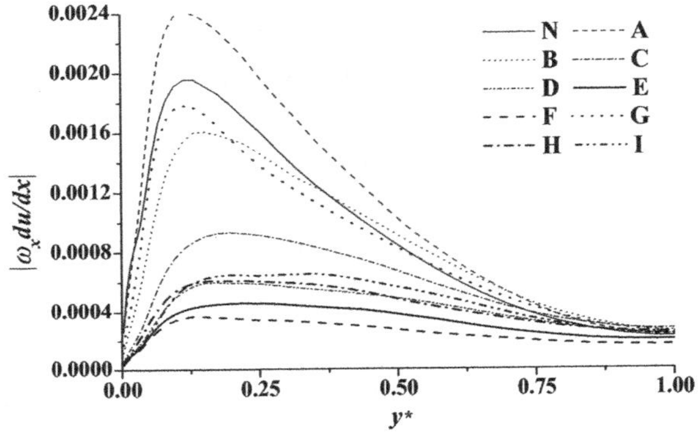

(b)

Figure 4.46 Intensity of the stream-wise vortex stretching: (a) stretching mode, and (b) squeezing mode (Reprinted from *International Journal of Heat and Fluid Flow*, **27**, Yu, B. and Kawaguchi, Y., "Parametric study of surfactant-induced drag-reduction by DNS," 887–894, 2006, with permission from Elsevier.)

channel flow. Two types of fluid motions, Flow A and Flow B (shown in Figure 4.47), were studied, where Newtonian and non-Newtonian fluid separately flow at different layers with the interface of different fluids parallel to the walls. In Flow A, the network structures exist at the center region of the channel. In Flow B, they exist at the near-wall region.

By moving the interface position, how the network structures reduce frictional drag at different flow layers can be studied. For Flow A, we did four calculations with the thickness of the non-Newtonian fluid: $2 \times 0h$, $2 \times 0.4h$, $2 \times 0.6h$, and $2 \times 0.9h$, respectively. For flow B, we carried out three computations with the thickness of the non-Newtonian fluid $2 \times 0.2h$, $2 \times 0.4h$, and $2 \times h$, respectively. The first case of flow A is Newtonian fluid, and the last case of Flow B is non-Newtonian fluid. For all the other cases, the Newtonian fluid and non-Newtonian fluid coexist.

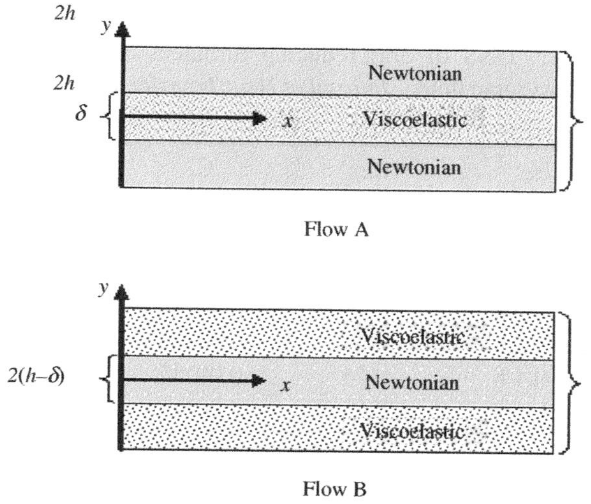

Figure 4.47 Bilayer model of flows with Newtonian and non-Newtonian fluid coexistence (Reprinted with permission from Yu, B. and Kawaguchi, Y., "DNS of drag-reducing turbulent channel flow with coexisting Newtonian and non-Newtonian fluid," *Journal of Heat Transfer* **127**, 929–935, © 2005 ASME International.)

At the interface of the Newtonian and non-Newtonian fluid region, the following shear stress and normal stress balance equations were satisfied:

$$\left(\frac{\partial u^+}{\partial y^*}\right)_N = \left(\frac{\partial u^+}{\partial y^*}\right)_{NN} + \beta\frac{\mathrm{Re}_\tau c^+_{xy}}{We_\tau} \tag{4.57}$$

$$\left(\frac{\partial v^+}{\partial y^*}\right)_N = \left(\frac{\partial v^+}{\partial y^*}\right)_{NN} + \beta\frac{\mathrm{Re}_\tau (c^+_{yy}-1)}{We_\tau} \tag{4.58}$$

$$\left(\frac{\partial w^+}{\partial y^*}\right)_N = \left(\frac{\partial w^+}{\partial y^*}\right)_{NN} + \beta\frac{\mathrm{Re}_\tau c^+_{yz}}{We_\tau} \tag{4.59}$$

Calculations were performed with parameters $Re_\tau = 125$, $We_\tau = 25$, $\alpha = 0.001$, and $\beta = 0.1$ in the non-Newtonian fluid region and $Re_\tau = 125$ in the Newtonian fluid region. Table 4.8 shows the mean Reynolds number, the calculated frictional factor C_f, the evaluated frictional factor C_f^D, and the drag reduction rate $DR\%$. It is seen that there is no drag reduction for Flow A(2) and the largest drag reduction rate for Flow B(3). For Flow A(3), a small drag reduction rate around 4% is obtained. Flow A(2) and Flow A(3) show that the network structures are not effective in reducing

Table 4.8 Reynolds number, friction factor, and DR rate (Reprinted with permission from Yu, B. and Kawaguchi, Y., "DNS of drag-reducing turbulent channel flow with coexisting Newtonian and non-Newtonian fluid," *Journal of Heat Transfer* **127**, 929–935, © 2005 ASME International.)

	δ	Re_b	C_f	C_f^D	DR%
Flow A(1)	2×0 h	3653	0.00936	0.00939	0%
Flow A(2)	2×0.4 h	3651	0.00937	0.00939	0%
Flow A(3)	2×0.6 h	3726	0.00900	0.00934	3.63%
Flow A(4)	2×0.9 h	4071	0.00754	0.00914	17.5%
Flow B(1)	2×0.2 h	3848	0.00844	0.00927	8.94%
Flow B(2)	2×0.4 h	4175	0.00717	0.00908	21.0%
Flow B(3)	$2 \times$ h	4263	0.00688	0.00903	23.9%

frictional drag in the bulk flow region, especially in the center of the channel. In Flow B(1), the non-Newtonian fluid region is 20% of the entire flow region; about 9% drag reduction rate is obtained. In Flow B(2), the non-Newtonian fluid region is 40% of the entire flow region and the drag reduction rate approaches that in Flow B(3). Flow B(1) and Flow B(2) indicate that the network structures are most effective at the region near the wall. By comparing Flow A(4) and Flow B(3), it is seen in Flow A(4) that the network structures cover most of the fluid region except the viscous sublayer ($0 \leq y^+ \leq 5$) and a small part of the buffer layer ($5 < y^+ \leq 12.5$), but its drag reduction ability is 25% less than that of Flow B(3). In addition, the non-Newtonian fluid region of A(4) is 2.25 times as large as Flow B(2), but it has a smaller drag reduction rate. All these further verify the hypothesis of Lumley that the primary action of additives is in the buffer layer.

The local fractional contribution to DR in Figure 4.48 shows more clearly that additives are important in the near-wall region. There are some discrepancies in the contributions in flow A and flow B, which must come from the nonlinear characteristics of the governing equations. This indicates that fluid dynamics with Newtonian and non-Newtonian fluid is complex and the contribution cannot be simply superposed.

The Reynolds shear stresses are compared in Figure 4.49. Drag reduction is often explained by the decrease of Reynolds shear stress. It is seen that in our calculations, the Reynolds shear stress decreases for all the DR cases. However, flow A(4) has a larger Reynolds shear stress than flow B(1), but it also has a larger DR rate. This is because in flow B(1), although the Reynolds shear stress is smaller, the viscoelastic shear stress is large in the near-wall region, which increases frictional drag. The comparison of Reynolds shear stress indicates that the DR rate is associated with, but not proportional to, the decrease of Reynolds shear stress. The decrease of Reynolds shear stress is only one important factor for achieving a large DR rate.

Table 4.9 lists elastic dissipations. It is seen that generally the larger the elastic dissipation, the larger the drag reduction rate. But flow B(1) is an exception; it has a

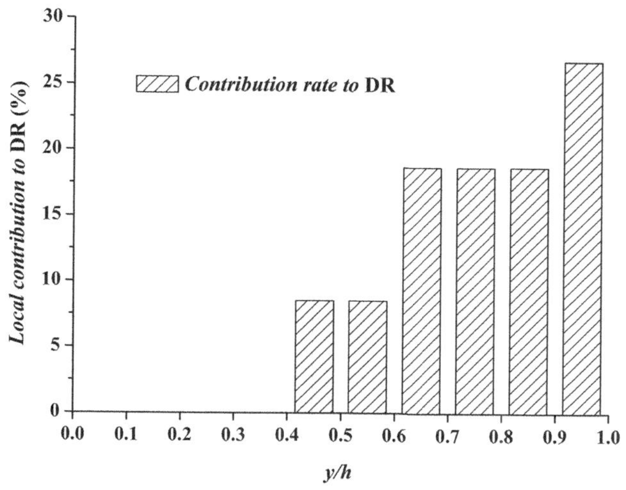

(a)

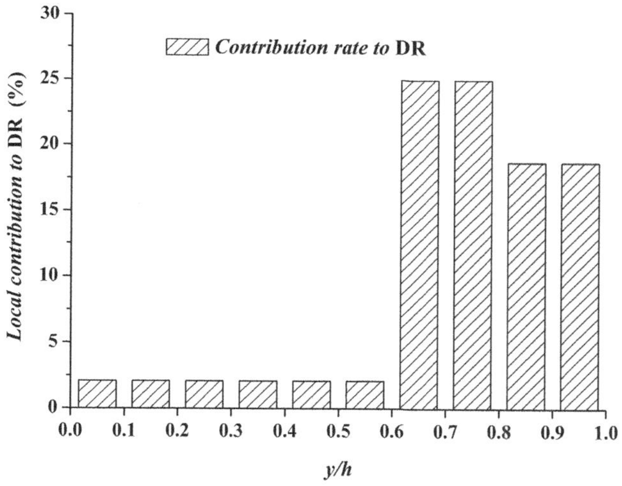

(b)

Figure 4.48 Local fractional contribution to the drag reduction rate (Reprinted with permission from Yu, B. and Kawaguchi, Y., "DNS of drag-reducing turbulent channel flow with coexisting Newtonian and non-Newtonian fluid," *Journal of Heat Transfer* **127**, 929–935, © 2005 ASME International.)

larger elastic dissipation than flow A(4) and flow B(2), but a smaller drag reduction rate. This is because in flow B(1), most of the energy of the elastic dissipation comes from mean kinetic energy, and only a small amount of energy comes from turbulent kinetic energy. Thus the turbulence has not been effectively suppressed.

Table 4.9 Elastic dissipation (Reprinted with permission from Yu, B. and Kawaguchi, Y., "DNS of drag-reducing turbulent channel flow with coexisting Newtonian and non-Newtonian fluid," *Journal of Heat Transfer* **127**, 929–935, © 2005 ASME International.)

	DR%	Elastic dissipation
Flow A(1)	0%	0
Flow A(2)	0%	0.00544
Flow A(3)	3.63%	0.0110
Flow A(4)	17.5%	0.0180
Flow B(1)	8.94%	0.0210
Flow B(2)	21.0%	0.0203
Flow B(3)	23.9%	0.0285

The above analyses show that drag-reducing additives mainly affect the buffer layer. The drag reduction rate is related to not only the decrease of Reynolds stress but also the viscoelastic stress.

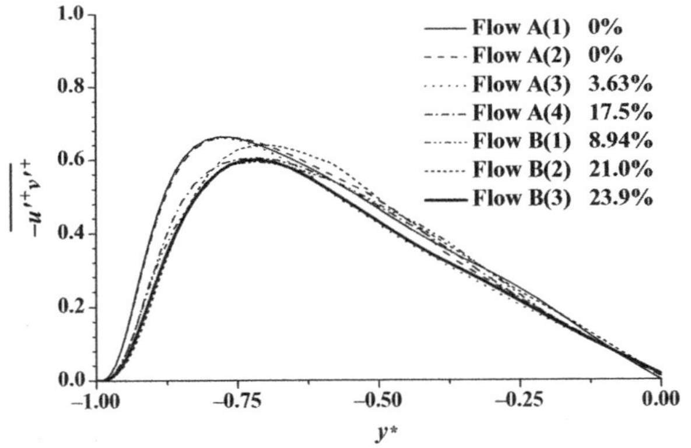

Figure 4.49 A comparison of Reynolds shear stresses (Reprinted with permission from Yu, B. and Kawaguchi, Y., "DNS of drag-reducing turbulent channel flow with coexisting Newtonian and non-Newtonian fluid," *Journal of Heat Transfer* **127**, 929–935, © 2005 ASME International.)

4.5 Conclusion and Future Work

Until now, many researchers have studied the mechanisms of drag reduction and reduction by DNS and achieved fruitful results. However, higher Reynolds number DNS is still hard to achieve restricted to the computational condition. Therefore, numerical simulation of turbulent drag reduction and reduction still needs advancement. The following points can be further studied mainly:

1. Numerical treatment of constitutive equations still needs to be made. Although high-resolution schemes such as MINMOD resolved the stability and precision and convective terms in constitutive equation, some artificial diffusion still exists. Thus, higher resolution schemes, and even essential non-artificial diffusion and non-oscillatory schemes, should be developed.
2. Enhance the comparability between numerical results and experimental data. Parameters in constitutive equations need to be determined by rheological experiments.
3. Develop viscoelastic turbulent LES for engineering applications.

References

1. Gennes, D. (1990) *Introduction to Polymer Dynamics*, Cambridge University Press, Cambridge.
2. Lu, B., Li, X., Zakin, J.L. and Talmon, Y. (1997) A non-viscoelastic drag reducing cationic surfactant system. *Journal of Non-Newtonian Fluid Mechanics*, **71** (1–2), 59–72.
3. Zhang, Z.S., Cui, G.X. and Xu, C.X. (2005) *Turbulent Theory and Simluation*, Qinghua University Press, Beijing, p. 163.
4. Moin, P. and Mahesh, K. (1998) Direct numerical simulation: a tool in turbulence research. *Annular Review of Fluid Mechanics*, **30**, 553–573.
5. Zhang, Z.S., Cui, G.X. and Xu, C.X. (2005) *Turbulent Theory and Simluation*, Qinghua University Press, Beijing.
6. Tao, W.Q. (2002) *Numerical Heat Transfer*, Xi'an University Press, Xi'an.
7. Lee, J., Dünweg, B. and Schumacher, J. (2010) Multiscale modelling strategy using the lattice Boltzmann method for polymer dynamics in a turbulent flow. *Computers and Mathematics with Applications*, **59**, 2374–2379.
8. Orlandi, P. (1995) A tentative approach to the direct simulation of drag reduction by polymers. *Journal of Non-Newtonian Fluid Mechanics*, **60**, 277–301.
9. Toonder, J.M.J., Hulsen, M.A., Kuiken, G.D.C. and Nieuwstadt, F.T.M. (1997) Drag reduction by polymer additives in a turbulent pipe flow: numerical and laboratory experiments. *Journal of Fluid Mechanics*, **337**, 193–231.
10. Sureshkumar, R., Beris, A.N. and Handler, R.A. (1997) Direct numerical simulation of turbulent channel flow of a polymer solution. *Physics of Fluids*, **9**, 743–755.
11. Dimitropoulos, C.D., Sureshkumar, R. and Beris, A.N. (1998) Direct numerical simulation of viscoelastic turbulent channel flow exhibiting drag reduction: effect of the variation of rheological parameters. *Journal of Non-Newtonian Fluid Mechanics*, **79**, 433–468.
12. Angelis, E.D., Casciola, C.M. and Piva, R. (2002) DNS of wall turbulence: dilute polymers and self-sustaining mechanisms. *Computers in Fluids*, **31**, 495–507.
13. Ilg, P., Angelis, E.D., Karlin, I.V. *et al.* (2002) Polymer dynamics in wall turbulent flow. *Europhysics Letters*, **58** (4), 616–622.
14. Ptasinski, P.K., Boersma, B.J., Nieuwstadt, F.T.M. *et al.* (2003) Turbulent channel flow near maximum drag-reduction: simulations, experiments and mechanisms. *Journal of Fluid Mechanics*, **490**, 251–291.

15. Housiadas, K.D. and Beris, A.N. (2003) Polymer-induced drag reduction: Effects of the variations in elasticity and inertia in turbulent viscoelastic channel flow. *Physics of Fluids*, **15** (8), 2369–2384.
16. Vaithianathan, T. and Collins, L.R. (2003) Numerical approach to simulating turbulent flow of a viscoelastic polymer solution. *Journal of Computational Physics*, **187**, 1–21.
17. Dimitropoulos, C.D., Dubief, Y., Shaqfeh, E.S.G. *et al.* (2005) Direct numerical simulation of polymer-induced drag reduction in turbulent boundary layer flow. *Physics of Fluids*, **17**, 011705∼1–4.
18. Sibilla, S. and Beretta, C.P. (2005) Near-wall coherent structures in the turbulent channel flow of a dilute polymer solution. *Fluid Dynamics Research*, **37**, 183–202.
19. Li, C.F., Gupta, V.K., Sureshkumar, R. and Khomami, B. (2006) Turbulent channel flow of dilute polymeric solutions: drag reduction scaling and an eddy viscosity model. *Journal of Non-Newtonian Fluid Mechanics*, **139**, 177–189.
20. Sureshkumar, R. (2001) Local linear stability characteristics of viscoelastic periodic channel flow. *Journal of Non-Newtonian Fluid Mechanics*, **97**, 125–148.
21. Min, T., Jung, Y.Y., Choi, H. and Joseph, D.D. (2003) Drag-reduction by polymer additives in a turbulent channel flow. *Journal of Fluid Mechanics*, **486**, 213–238.
22. Min, T., Choi, H. and Jung, Y.Y. (2003) Maximum drag-reduction in a turbulent channel flow by polymer additives. *Journal of Fluid Mechanics*, **492**, 91–100.
23. Eckhardt, B., Kronjäger, J. and Schumacher, J. (2002) Stretching of polymers in a turbulent environment. *Computer Physics Communications*, **147**, 538–543.
24. Min, T. and Choi, H. (2005) Combined effects of polymers and active blowing suction on drag reduction. *Journal of Non-Newtonian Fluid Mechanics*, **131**, 53–58.
25. Li, C.F., Sureshkumar, R. and Khomami, B. (2006) Influence of rheological parameters on polymer induced turbulent drag reduction. *Journal of Non-Newtonian Fluid Mechanics*, **140**, 23–40.
26. Tamano, S., Itoh, M., Hoshizaki, K. and Yokota, K. (2007) Direct numerical simulation of the drag-reducing turbulent boundary layer of viscoelastic fluid. *Physics of Fluids*, **19**, 075106∼1– 17.
27. Yu, B. and Kawaguchi, Y. (2003) Effect of Weissenberg number on the flow structure DNS study of drag-reducing flow with surfactant additives. *International Journal of Heat and Fluid Flow*, **24**, 491–499.
28. Yu, B. and Kawaguchi, Y. (2004) Direct numerical simulation of viscoelastic drag-reducing flow a faithful finite-difference method. *Journal of Non-Newtonian Fluid Mechanics*, **116**, 431–466.
29. Yu, B., Li, F.C. and Kawaguchi, Y. (2004) Numerical and experimental investigation on turbulence characteristics in a drag-reducing flow with surfactant additives. *International Journal of Heat and Fluid Flow*, **25**, 961–974.
30. Yu, B. and Kawaguchi, Y. (2005) DNS of drag-Reducing turbulent channel flow with coexisting Newtonian and non-Newtonian fluid. *Journal of Fluids Engineering*, **127**, 929–935.
31. Yu, B. and Kawaguchi, Y. (2005) DNS of fully developed turbulent heat transfer of a viscoelastic drag-reducing flow. *International Journal of Heat and Mass Transfer*, **48**, 4569–4578.
32. Yu, B. and Kawaguchi, Y. (2006) Parametric study of surfactant-induced drag-reduction by DNS. *International Journal of Heat and Fluid Flow*, **27**, 887–894.
33. Kagawa, Y., Yu, B., Kawaguchi, Y. *et al.* (2008) Turbulent heat transfer of viscoelastic fluid flow accompanied by drag reduction with DNS analysis. *Progress in Computational Fluid Dynamics*, **8** (7/8), 477–485.
34. Yu, B., Wu, X., Wei, J.J. *et al.* (2011) DNS study by a bilayer model on the mechanism of heat transfer reduction in drag-reduced flow induced by surfactant. *International Communications in Heat and Mass Transfer*, **38**, 160–167.
35. Tsukahara, T., Ishigami, T., Yu, B. and Kawaguchi, Y. (2011) DNS study on viscoelastic effect in drag-reduced turbulent channel flow. *Journal of Turbulence*, **12** (13), 1–25.

36. Tsukahara T. and Kawaguchi, Y. (2011) "Comparison of heat-transfer reduction in drag-reduced turbulent channel flows with different fluid and thermal boundary conditions," *Progress in Computational Fluid Dynamics*, **11**(3/4), 216–226.
37. Tsukahara, T. and Kawaguchi, Y. (2011) "Turbulent heat transfer in drag-reducing channel flow of viscoelastic fluid," In: Heat Transfer / Book 3, ISBN 978-953-307-583-9, InTech, in press.
38. Tsukahara, T., Kawase, T. and Kawaguchi, Y. (2011), "DNS of viscoelastic turbulent channel flow with rectangular orifice at low Reynolds number," *International Journal of Heat and Fluid Flow*, **32**, 529–538.
39. Gu, W.G. (2010) Experimentaland DNS study on drag-reducing mechanism of surfactant drag-reducing fluid, PhD dissertation, Shanghai Jiaotong University.
40. White, C.M., Somandepalli, V.S.R. and Mungal, M.G. (2004) The turbulence structure of drag-reduced boundary layer flow. *Experiments in Fluids*, **36**, 62–69.
41. Dimitropoulos, C.D., Dubief, Y., Shaqfeh, E.S.G. and Moin, P. (2006) Direct numerical simulation of polymer-induced drag reduction in turbulent boundary layer flow of inhomogeneous polymer solution. *Journal of Fluid Mechanics*, **566**, 153–162.
42. Kajishima, T. and Miyake, Y. (1998) Drag reduction by polymer additives in turbulent channel flow simulated by discrete-element models. *Transactions of JSME (B)*, **64**, 3636–3643 (in Japanese).
43. Wang, Y., Da, X., Yu, B. *et al.* (2008) Direct numerical simulation on drag reducing flow by polymer additives. *Journal of Engineering Thermophysics*, **29** (4), 617–620.
44. Wang, Y., Yu, B., Wei, J.J. *et al.* (2009) Direct Numerical Simulation on drag-reducing flow by polymer additives using a spring-dumbbell model. *Progress in Computational Fluid Dynamics*, **9** (3/4/5), 217–224.
45. Hagiwara, Y., Hana, H., Tanaka, M. and Murai, S. (2000) Numerical simulation of the interactions of highly entangled polymers with coherent structure in a turbulent channel flow. *International Journal of Heat and Fluid Flow*, **21**, 589–598.
46. Sureshkumar, R. and Beris, A.N. (1995) Effect of artificial stress diffusivity on the stability of numerical calculations and the flow dynamics of time-dependent viscoelastic flows. *Journal of Non-Newtonian Fluid Mechanics*, **60**, 53–80.
47. Min, T., Yoo, J.Y. and Choi, H. (2001) Effect of spatial discretization schemes on numerical solutions of viscoelastic fluid flows. *Journal of Non-Newtonian Fluid Mechanics*, **100**, 27–47.
48. Poreh, M. and Dimant, Y. (1972) Velocity distribution in flows with drag reduction. Proceedings of the 9th International Symposium of Naval Hydrodynamics, 2, pp. 1350–1323.
49. Edwards, M.F. and Smith, R. (1980) The turbulent flow of non-Newtonian fluids in the absence of anomalous wall. *Journal of Non-Newtonian Fluid Mechanics*, **7**, 153–169.
50. Azouz, I. and Shirazi, S.A. (1997) Numerical simulation of drag-reducing turbulent flow in annular conduits. *Transactions of ASME Journal of Fluids Engineering*, **119**, 838–846.
51. Durst, F. and Rastogi, A.K. (1977) Calculations of turbulent boundary layer flows with drag reducing polymer additives. *Physics of Fluids*, **20**, 1975–1985.
52. Hassid, S. and Poreh, M. (1975) A turbulent energy model for flows with drag reduction. *Transactions of ASME Journal of Fluids Engineering*, **97**, 234–241.
53. Hassid, S. and Poreh, M. (1978) A turbulent energy model for flows with drag reduction. *Transactions of ASME Journal of Fluids Engineering*, **100**, 107–112.
54. Poreh, M. and Hassid, S. (1977) Mean velocity and turbulent energy closures for flows with drag reduction. *Physics of Fluids*, **20**, S193–196.
55. Jones, W.P. and Launder, B.E. (1973) The calculation of low-Reynolds-number phenomena with a two-equation model of turbulence. *International Journal of Heat and Mass Transfer*, **16**, 1119–1130.

56. Cruz, D.O.A. and Pinho, F.T. (2003) Turbulent pipe flow predictions with a low Reynolds number k-ε model for drag reducing fluids. *Journal of Non-Newtonian Fluid Mechanics*, **114** (2–3), 109–148.
57. Pinho, F.T. (2003) A GNF framework for turbulent flow models of drag reducing fluids and proposal for a k-ε type closure. *Journal of Non-Newtonian Fluid Mechanics*, **114** (2–3), 149–184.
58. Cruz, D.O.A., Pinho, F.T. and Resende, P.R. (2004) Modelling the new stress for improved drag reduction predictions of viscoelastic pipe flow. *Journal of Non-Newtonian Fluid Mechanics*, **121** (2–3), 127–141.
59. Resende, P.R., Escudier, M.P., Presti, F. *et al.* (2006) Numerical predictions and measurements of Reynolds normal stresses in turbulent pipe flow of polymers. *International Journal of Heat and Fluid Flow*, **27**, 204–219.
60. Park, T.S., Sung, H.J. and Suzuki, K. (2003) Development of a nonlinear near-wall turbulence model for turbulent flow and heat transfer. *International Journal of Heat and Fluid Flow*, **24**, 29–40.
61. Pinho, F.T., Li, C.F., Younis, B.A. and Sureshkumar, R. (2008) A low Reynolds number turbulence closure for viscoelastic fluids. *Journal of Non-Newtonian Fluid Mechanics*, **154**, 89–108.
62. Nagano, Y. and Hishida, M. (1987) Improved form of the k-ε model for wall turbulent shear flows. *Transactions of ASME Journal of Fluids Engineering*, **109**, 156–160.
63. Resende, P.R., Kim, K., Younis, B.A. *et al.* (2011) A FENE-P k-ε turbulence model for low and intermediate regimes of polymer-induced drag reduction. *Journal of Non-Newtonian Fluid Mechanics*, **166**, 639–660.
64. Leighton, R. (2003) Proceedings of ASME FEDSM'03. Honolulu, Hawaii, USA, July 6–10.
65. Zhang, Z.S., Cui, G.X. and Xu, C.X. (2005) *Turbulent Theory and Simluation*, Qinghua University Press, Beijing, pp. 25–30.
66. Harlow, F.H. and Welch, J.E. (1965) Numerical calculation of time-dependent viscous incompressible flow of fluid with free surface. *Physics of Fluids*, **8**, 2182–2189.
67. Virk, P., Mickley, H. and Smith, K. (1970) The ultimate asymptote and mean flow structure in Toms' phenomenon. *ASME Journal of Applied Mechanics*, **37**, 480–493.

5

Microstructures and Rheological Properties of Surfactant Solution

5.1 Microstructures in Surfactant Solution and Its Visualization Methods

5.1.1 Microstructures in Surfactant Solution

5.1.1.1 Formation of Micellar Structure

A monomer molecule of surfactant contains a hydrophobic head and a hydrophilic tail. When surfactants are dispersed in a polar solvent, the hydrophilic tail tends to contact with the solvent, while the hydrophobic head prefers to leave the polar environment. By the repulsion of water, the hydrophobic groups gets together to form nonpolar interfaces. There exists a critical micelle concentration (CMC), above which spherical micelles are formed with the hydrophobic groups at the center and the hydrophobic groups at the surface of the spherical structures. When the concentration of surfactant solution is increased to second critical micelle concentration (CMC_{II}), the spherical micelles will be transformed into rod-like micelles. A micellar network structure will be formed by the connection of rod-like micelles at their ends when the surfactant concentration is further increased.

Under certain conditions, rod-like micelles may be transformed into a vesicle or lamellar structure. Compared to the concentration, temperature has an opposite effect on surfactant micelle formation. As the temperature is increased, the thermal motion of solvent molecules is enhanced, reducing the interaction between surfactant molecules and thus preventing the formation of micellar structures. The change of surfactant microstructures at different concentrations and temperatures has been shown in Figure 1.2 in Chapter 1 [1]. We can see that the value of CMC decreases to the

Turbulent Drag Reduction by Surfactant Additives, First Edition. Feng-Chen Li, Bo Yu, Jin-Jia Wei and Yasuo Kawaguchi.

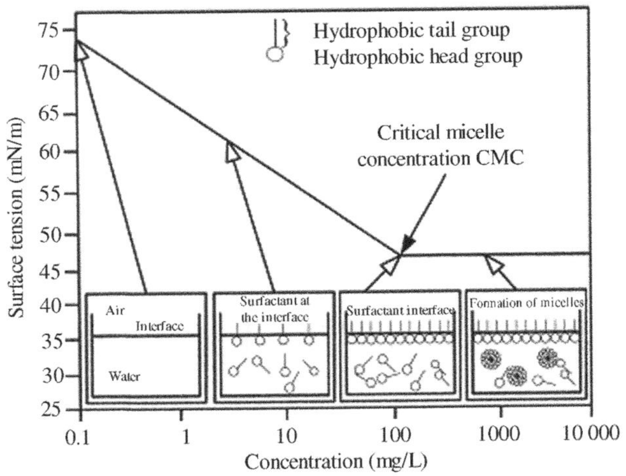

Figure 5.1 Schematics of surfactant solution near CMC

minimum and then increases gradually as the temperature is increased. The effect of temperature on CMC is small.

5.1.1.2 Critical Micellar Concentration (CMC)

The critical micellar concentration depends on many factors, such as the surfactant hydrocarbon chain length, surfactant head group, counterion, added salts, and temperature.

A usual way to obtain CMC is to measure the surface tension of the surfactant solution. On the plot of surface tension–logarithmic concentration (γ-ln$c$), an intersection point obtained by extending the linear decreasing part and constant part of the surface tension curve can be regarded as the value of CMC, as shown in Figure 5.1 [2].

The CMC is determined by surfactant structure. In aqueous solution, the CMC decreases as the length of the surfactant hydrocarbon chain increases up to 16 carbon atoms, and the decrease is slowed down when the chain length exceeds 16; the CMC remains unchanged for the chain length larger than 18 even if the chain length is further increased. The existence of a branch on the hydrocarbon chain may decrease the CMC. The CMC of the surfactant with double bonds in the hydrocarbon chain is higher than the corresponding saturated surfactants [3].

For pure surfactant solution, ionic surfactants have much higher CMCs than nonionic surfactants with equivalent hydrophobic groups due to the electrical repulsion of head groups with the same charge. The addition of counterion into surfactant solution can neutralize electric charge, so it can reduce the CMC of ionic surfactant solution greatly, but it has a small effect on nonionic surfactants [2]. The CMC is higher when the charge on the head group is closer to the α-carbon atom of the alkyl chain.

The surfactant concentration at which spherical micelles are transformed into rod-like micelles is defined as CMC_{II}. The effects of the surfactant hydrocarbon chain length, head groups, counterion, and added salts on CMC_{II} are similar to those on CMC. The effect of temperature on the transformation from spherical to rod-like is larger than that on CMC. Although CMC_{II} is defined, there is a broad range for the sphere–rod transformation. Spherical micelles may still exist even if the rod-like or worm-like micellar network structure has been formed.

5.1.1.3 Micelle Shape

There exist four major micelle types: spherical micelle, rod-like, lamellar, and vesicle. When the surfactant concentration reaches CMC, micelles are considered as spherical or nearly spherical. When the concentration is further increased up to CMC_{II}, rod-like micelles are formed. When the rod-like micelles become long, the micelles will bend [4]. We call these threadlike micelles (TLM) or wormlike micelles.

The formation of surfactant micelles is the equilibrium result of the hydrophobic action of the alkyl tail group and the electric action of the head group. The shape of surfactant molecules and counterion play an important role in the micelle formation. Long alkyl chains ($n > 16$) with counterion can enhance the axial growth of the micelles and the formation of threadlike structures. A packing parameter, $p = V/(al)$ (V represents the volume of hydrophobic tail groups per molecule, a is the cross-sectional area of the head group per molecule, and l denotes the alkyl chain length), is used to determine the micelle shape. Different micelle shapes correspond to different p values. For $p < 1/3$, spherical micelles are formed; for $1/3 < p < 1/2$, threadlike micelles are formed (TLM); for $1/2 < p < 1$, lamellar micelles are formed; and for $p > 1$, reversed micelles of vesicles are formed [5]. Larger head groups tend to reduce the packing parameter p. Therefore, a smaller aggregation size of micelles can be obtained by increasing the size of the head group.

5.1.1.4 Micelle Size

The micelle size depends on many factors, such as the surfactant hydrocarbon chain structure, the head group, counterions, added salts, and temperature. From geometric consideration, the aggregation number of spherical micelle increases with increasing surfactant hydrocarbon chain length, and decreases with increasing the head group size. Geer *et al.* suggested that larger head groups tend to keep the counterions from approaching the charge center and thus are not favorable for micellar growth [6]. Zhang *et al.* suggested that the existence of benzene rings in the head group may be helpful to the micelle growth due to the interaction of benzene rings in the head group and in the counterion [7]. Rehage and Haffmann explained the head group effects on micelle shape based on geometrical considerations [8]. If the hydrophilic head group requires an area larger than the corresponding cross-section of the hydrocarbon chain, the system tends to aggregate with convex curvatures. If both sizes are the same, planar structures are favored. If the hydrophilic head requires an area less than that of the hydrocarbon chain, inverse micelles are formed.

Tanford developed two equations to calculate volume occupied by an alkyl chain of N carbon atoms and the maximum possible extension of a hydrocarbon chain [9]:

$$V = 27.4 + 26.9N(\mathring{A}^3) \tag{5.1}$$

$$l_{\max} = 1.5 + 1.265N(\mathring{A}) \tag{5.2}$$

To ensure an aqueous environment for the polar groups, no matter what shape the micelles are, at least one dimensional size of the micelles in the three dimensions must not exceed $2l$. Therefore, the radius of spherical and rod-like micelles should be about $l_{\max}$.

The length of rod-like micelles increases with increasing surfactant concentration and decreasing temperature. Elson and Garside suggested that rod-like micelles are longest and strongest around the surfactant Kraftt point [10]. Hoffman found that the rod length increases linearly with increasing concentration to a maximum value and then decreases. The concentration corresponding to the longest rod is noted as c^*. It is considered that the length approaches the average distance between the rods at c^*. c^* increases with decreasing the alkyl chain length, and the aggregation number may be very large at some moderate concentration [11].

Added salts play an important role in the micellar growth since the micellar formation is strongly controlled by the positive charges on cationic surfactant head groups. Salts have a great ability to neutralize these charges, and will favor both micellar formation and growth. For cationic surfacant solution, micelle length increases to a peak and then begins to decrease with increasing salt concentration. The higher the chemical valence of anion, the more effective the salts.

Counterions can reduce the area of head groups by reducing the electrostatic repulsive force between head groups, enhancing the sphere–rod transformation. The average micelle size is proportional to the square root of micelle concentration. The intrinsic viscosity of micellar solution (say, zero shear viscosity) should depend on micelle concentration. The flexibility of rod-like micelles is determined by the following factors: particle strength, charge density of micelles, surface tension, counterion concentration, and hydrocarbon chain length of surfactant [12]. Different flexibility may exist even for the same surfactant solution under different conditions. Added counterion generates the electrostatic repulsive force between head groups, thus reducing the distance between head groups and the effective area of micelle interface.

Micellar structure is also affected obviously by the flow of surfactant solution. Under the action of shear stress, a shear-induced network structure can be formed by the connection between micelles. When the shear stress becomes too strong, the network can be destroyed. In the solution containing vesicles, flow can destroy the vesicles to form thread-like micelles. After the solution flow is stopped, the inner structure recovers gradually to that before flow due to strong electrostatic interaction among surfactant and solvent molecules [13].

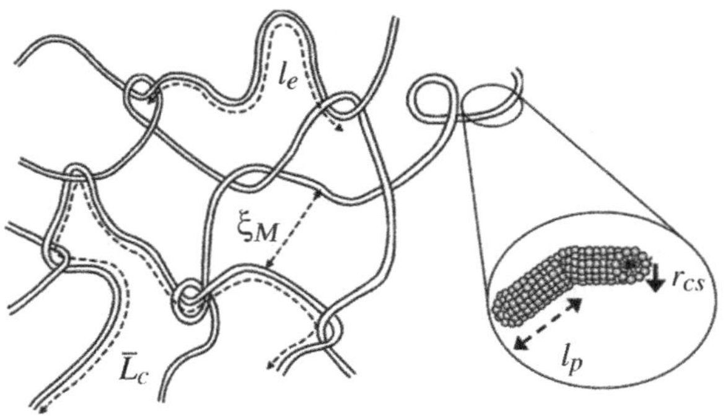

Figure 5.2 Micellar network structure

All the above surfactant micellar structures are those of aqueous solution. If polar organic solvents such as the mixture of ethylene glycol and water are used, the aggregation and growth of micelles will be greatly affected. Polar solvents go against the aggregation of surfactant molecules because the polar organic–water mixture solvents can dissolve surfactant molecules more easily than water [14,15].

For the micellar network structure, the following characteristic lengths are defined (Figure 5.2 and Table 5.1): contour length L_c, entangle length l_e, mesh size ξ_M, persistence length l_p, and cross-sectional radius r_{CS}. The determination of characteristic lengths can provide computational parameters for numerical simulation. There are also two characteristic times defined for the micelle motion: λ_{br} and λ_{rep}. λ_{br} is the breakage time of worm-like micelles, λ_{rep} is the reptation time, and λ_{br} is the reciprocal of G''_{min}; when the solution relaxation time λ_t is determined, λ_{rep} can be determined as $\lambda_{rep} = \lambda_t^2/\lambda_{br}$.

5.1.2 *Visualization Methods for Microstructures in Surfactant Solution*

There are several visualization methods of solution microstructure such as cryo-transmission electron microscopy (cryo-TEM), flow birefringence, nuclear magnetic

Table 5.1 Characteristic length and time of network structures [16]

Characteristic length or time	Unit	Value
ξ_M	nm	55 ± 1.7
l_p	nm	10.3 ± 0.2
l_e	nm	170 ± 11
L_c	nm	940 ± 70
r_{CS}	nm	2.3 ± 0.2
λ_{br}	s	0.013 ± 0.001
λ_{rep}	s	10.0 ± 0.4

resonance (NMR), small-angle scattering (SAS), light scattering, and fluorescence quenching. The most commonly used NMR method is proton NMR or ^{1}H NMR. The small-angle scattering method includes small-angle neutron scattering (SANS) and small-angle X-ray scattering (SAXS). Light-scattering methods include static light scattering (SLS) and dynamic light scattering (DLS). In addition, according to the viewpoint that flow drag reduction is related to the length and number of thread-like micelles [1], drag reduction can also be used to speculate microstructures of surfactant solution.

5.1.2.1 Cryo-Transmission Electron Microscopy (cryo-TEM)

Cryo-TEM can get direct images of micellar structures without altering the structures. The sample is prepared in a controlled environmental vitrification system which achieves steady state at 30° and 100% relative humidity. A thin film (about 100~300 nm thick) of sample is formed by placing a tiny drop of the liquid on a holey polymer support film which has been coated with carbon, and the assembly is then vitrified by rapidly plunging it into liquid ethane at its freezing point (about 90 K). The vitrified sample can be examined by TEM. It is noted that the surfactant micellar structures will be affected by the flow state, and large shear stress may exist in the sample preparation, which may get different images for the same sample [17].

5.1.2.2 Flow Birefringence [1]

For a fluid system, no matter what shape the particles have, when the solution is static, particles will have random orientations. The solution is isotropic from the statistical consideration. When external forces are applied, the isotropic particles will align with flow direction, and the solution will be anisotropic. The alignment of particles in a shear or elongational flow field gives birefringence.

Flow birefringence is an effective method for studying micellar shape and size. For spherical micelles, there is no birefringence due to isotropy of the solution; while for rod- or thread-like micelles, flow birefringence can be observed due to the anisotropy of solution caused by the alignment of micelles. Therefore, flow birefringence can distinguish between spherical and rod-like micellar shapes.

There are two key parameters for flow birefringence: the extinction angle χ and birefringence Δn. The extinction angle χ is the angle between the principle optical axis and the flow direction. Flow birefringence can be used to calculate the stress tensor by $\Delta n \cdot \sin(2\chi) = 2C\tau_{12}$ and $\Delta n \cdot \cos(2\chi) = C(\tau_{11} - \tau_{22})$, where C is the optical–stress coefficient, τ_{12} is shear stress, and $(\tau_{11}-\tau_{22})$ is first normal stress difference. This method is called the "stress optical rule," and it is suitable for flexible chain solutions. The advantage of the optical technique is that it does not destroy the flow state.

5.1.2.3 Nuclear Magnetic Resonance (NMR)

NMR can be used to determine the molecular structures. Proton NMR uses ^{1}H (proton) as an emission source (some use ^{13}C). D_2O is usually used instead of H_2O because solvents affect the measured results.

5.1.2.4 Small-Angle Scattering (SAS)

Small-angle X-ray scattering (SAXS) is one kind of SAS. The X-ray passes through a non-uniform sample with the magnitude of nanometers within a very small angle range (0.1~10°), and the reflective angle range consists of the shape and size of large molecules, the characteristic distance of part ordered material, the hole size, and other information. SAXS is used to determine microscale or nanoscale structures of a particle system. The size, shape, distribution, and surface-to-volume ratio of aggregated particles can be measured. SAXS is accurate and does not destroy the sample, and it only needs a small amount of sample.

Small-angle neutron scattering (SANS) has the same principle as SAXS. However, the energy scattered by a neutron is much less than that scattered by an X-ray, and thus SANS has much smaller disturbance. Moreover, the wavelength of a neutron bean is shorter than that of an X-ray, so SANS can get more accurate results. For investigating the structures of surfactant solutions, D_2O is usually used instead of deionized water to get maximum contrast.

5.1.2.5 Light Scattering

When light passes through a particle less than its wavelength, it will be scattered to different directions (Rayleigh scattering). If the light source is laser (a continuous single color light), the fluctuation of scattering intensity with time can be observed. The fluctuation is caused by the Brownian motion of microparticles in the solution, which leads to the change of distance between scatterers with time. In the fluctuation range of scattering intensity, scattering lights experienced adjacent particles that have constructive or destructive interferences, and thus the time scale of the scatterer's motion can be obtained. The light-scattering method includes static light scattering (SLS) and dynamic light scattering (DLS).

5.2 Rheology and Measurement Methods of Surfactant Solution

When the surfactant molecules are in a dispersed state in the solution, the solution shows the characteristics of Newtonian fluid due to isotropy. When the micelles are formed by the monomers, the solution may change greatly, showing non-Newtonian rheological characteristics. The parameters related to rheology are viscosity, linear viscoelasticity, normal stress, elongational viscosity, and others.

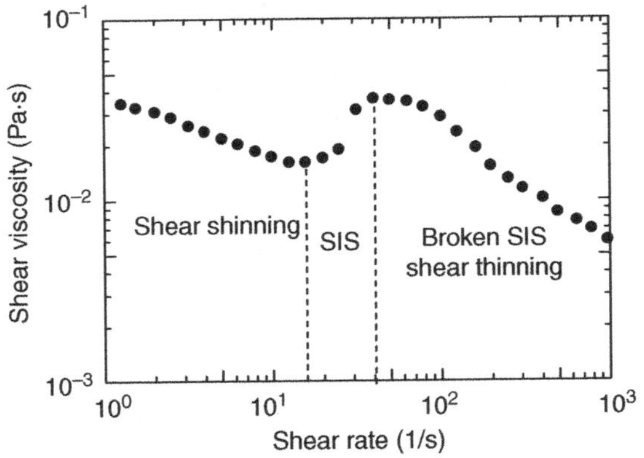

Figure 5.3 Shear viscosity versus shear rate

5.2.1 Rheological Parameters

5.2.1.1 Viscosity

The viscosity of surfactant solutions changes with the flow state. Usually, the variation of shear viscosity with shear rate can be divided into four phases: the Newtonian region, shear-thinning region, shear-thickening region, and shear-rethinning region [17], as shown in Figure 5.3. In some cases, the shear-thickening region is not obvious. A remarkable change of viscosity is considered to be caused by the change of micellar structures. For example, in the surfactant solution with linear micelles, the micelles are distributed randomly at the initial stage, the solution shows a homogeneous state, and thus the viscosity change is not obvious, showing Newtonian features (zero-shear viscosity can be deduced as a characteristic viscosity at this stage); as the shear rate is increased, the micelles are affected by flow and align with flow direction, and thus the viscosity decreases; the micelles can form a network by the action of flow shear, the elasticity of the network plays a dominant role, and thus the viscosity increases with increasing shear rate in the shear-thickening region; as the shear rate is further increased, the network is destroyed by shear, and the viscosity decreases with increasing shear rate; and when the shear rate is increased up to a critical value, the micelles align with flow direction completely, and the viscosity does not change obviously.

It is noted that the four phases of the variation of shear viscosity with shear rate are not always met.

Shear thickening is a special phenomenon of surfactant solution, which is an important feature distinguished from polymer solution.

Usually, zero-shear viscosity is used to define the intrinsic viscosity of surfactant solution. Zero-shear viscosity is the viscosity when the shear rate approaches zero. Doi and Edwards developed a zero-shear viscosity prediction formula for dilute solution with rod-like micelles [18]:

$$\eta_0 = \eta_s\left(1 + \hat{c}L^3\right) \tag{5.3}$$

where η_0 is zero-shear viscosity, η_s is the viscosity of solvent, $\hat{c}$ is the rod number per unit volume, and L is the rod length. $\hat{c}$ can be computed by

$$\hat{c} = \frac{c_M M}{\pi\rho r^2 L} \tag{5.4}$$

where c_M is the surfactant solution concentration, M is the molecular weight of surfactant molecules, r is the radius of rod-like micelles, and ρ is the solution density. Substituting Equation 5.3 into Equation 5.4, we can get

$$\eta_0 = \eta_s\left(1 + KL^2\right) \tag{5.5}$$

where $K = c_M M/\pi\rho r^2$. The above formula is valid only for dilute solution. From the formula, we can see that the viscosity increases with the square of rod length. Adding salt can promote the growth of rod-like micelles, so the viscosity can increase greatly.

When the rotational volume of a micelle contacts one or more other micelles, the micelles start to overlap, and have an interference with each other. The condition for overlap is $\hat{c} >> 1/L^3$. For dense solution meeting this condition, Doi and Edwards recommend a formula as follows [18]:

$$\eta_0 = \eta_s\left(1 + \left(\hat{c}L^3\right)^3\right) \tag{5.6}$$

Substituting Equation 5.4 into the above equation, one can get

$$\eta_0 = \eta_s\left(1 + K^3L^6\right) \tag{5.7}$$

Therefore, for a dense solution or a solution with long micelles, the viscosity increases rapidly with increasing micelle length.

Shear thickening is considered to be caused by shear-induced structure (SIS) or shear-induced phase transition, which is related to molecular structures, concentration, solvent, counterion, and temperature.

5.2.1.2 Viscoelasticity

When the elastic material is imposed with a strain, the stress relaxes with an exponential form. The stress relaxation module can be expressed as a relaxation module by dividing strain:

$$G(t) = \frac{\tau(t)}{\gamma} \tag{5.8}$$

For polymer fluid, at $\gamma < \gamma_c \approx 0.5$, all $G(t)$- t curves decrease with the same tendency. The stress relaxation is linearly dependent on strain, and is called linear viscoelasticity. In a sinusoidal oscillation experiment (usually called a "dynamic experiment"), the material is imposed with a sinusoidal strain, and the stress also has a sinusoidal oscillation but with a drift of phase angle δ:

$$\gamma = \gamma_0 \sin \omega t \tag{5.9}$$

$$\tau = \tau_0 \sin(\omega t + \delta) = \tau_0' \sin \omega t + \tau_0'' \cos \omega t \tag{5.10}$$

The stress can be decomposed into two waves with the same frequency, and generates two dynamic modules:

$$G' = \frac{\tau_0'}{\gamma_0} = \frac{\tau_0}{\gamma_0} \cos \delta \tag{5.11}$$

$$G'' = \frac{\tau_0''}{\gamma_0} = \frac{\tau_0}{\gamma_0} \sin \delta \tag{5.12}$$

where G' is the elastic (or storage) module, and G'' is the viscous (or dissipation) module. For an ideal solid, the phase angel $\delta = 0°$; while for an ideal liquid, the phase angle $\delta = 90°$. The value of the phase angle is dependent on a time scale. For water, if the time can be accurate to within 1 ns, it may show solid characteristics, while the ice may show liquid characteristics at high shear stresses if the year is selected as the time scale. For actual oscillation measurements, the sample is usually imposed with a sinusoidal stress with a large frequency range for investigating the relationship of the two modules with frequency.

We define

$$\tan \delta = \frac{G''(\omega)}{G'(\omega)} \tag{5.13}$$

If we use a complex form to express the strain, $\gamma(t') = \gamma_0 \exp(i\omega t')$, the corresponding strain rate becomes

$$\dot{\gamma}(t') = i\omega\gamma_0 \exp(i\omega t') \tag{5.14}$$

We define shear module G^* in a complex form:

$$\sigma(t) = G^*(\omega)\gamma(t) \tag{5.15}$$

$$G^* = G' + iG'' \tag{5.16}$$

The shear viscosity η^* in a complex form is defined as

$$\sigma(t) = \eta^* \dot{\gamma}(t) \tag{5.17}$$

5.2.1.3 Normal Stress Difference

Surfactant solutions belong to viscoelastic fluids, and their normal stress differences are not zero.

$$N_1(\dot{\gamma}) = \sigma_{xx} - \sigma_{yy} \tag{5.18}$$

$$N_2(\dot{\gamma}) = \sigma_{yy} - \sigma_{zz} \tag{5.19}$$

N_1 and N_2 are first and second normal stress differences, respectively.

Two normal stress difference coefficients Ψ_1 and Ψ_2, corresponding to N_1 and N_2, are defined as

$$N_1 = \dot{\gamma}^2 \Psi_1, \quad N_2 = \dot{\gamma}^2 \Psi_2 \tag{5.20}$$

Normal stress differences can be measured by parallel plates or cone–plate systems. Generally N_2 is negative or zero. The absolute value of N_2 is about 1/10 of N_1, and N_2 has a minor effect on fluid characteristics. Therefore, only N_1 is studied.

It is generally considered that the elastic force originates from the rod-like micelles. When the rod length exceeds the average distance between rods, viscosity and elastic viscosity increase greatly due to the formation of three-dimensional dynamic network structures.

5.2.1.4 Elongational Viscosity

For Newtonian fluids, the elongational viscosity in one-dimensional elongational flow is three times the shear viscosity. The ratio of elongational viscosity to shear viscosity is defined as the Trontou number $\mathrm{Tr} = \eta_E/\eta_s$. For surfactant solutions, the Trontou number is not a fixed value, and it is usually larger than 3. Similar to shear viscosity, elongational viscosity changes with elongation rate.

Generally, the shear viscosity and normal stress difference of surfactant solutions change with the shear rate, and the surfactant solutions show high elongational viscosity. However, for some surfactant solutions, for example Arquad S-50 5 mM/NaSal 12.5 mM [1], the shear viscosity and normal stress difference show little change with increasing shear rate but show high elongational viscosity.

Besides the above parameters, swirl decay time (SDT) and delayed time can also be used to describe rheological properties.

The Giesekus constitutive equation is considered as a relatively accurate equation for describing the rheology of surfactant solutions. For low frequency of shear rate, the Giesekus model can be simplified to the Maxwell model.

5.2.2 *Measurement Method of Rheological Parameters*

5.2.2.1 Shear Viscosity

Parallel Plate System

A schematic of a parallel plate rheometer is shown in Figure 5.4. For the rotational flow between two parallel plates, the shear rate at the plate edge ($r=R$) is

$$\dot{\gamma}_R = R\Omega/h \tag{5.21}$$

where Ω is angular velocity, and M is torque.

The viscosity can be deduced as

$$\eta = \frac{3Mh}{2\pi R^4\Omega\left(1+\frac{1}{3}\frac{\mathrm{d}\ln M}{\mathrm{d}\ln\Omega}\right)} \tag{5.22}$$

Cone–Plate System

The cone–plate system can also be used to measure rheology, and the schematic is shown in Figure 5.5. To obtain a steady shear flow field in the gap between the cone and plate, the following conditions or assumptions should be met:

1. Steady, laminar, and isothermal flow.

$$v_r = v_\theta = 0$$

$$\beta < 0.10\,\mathrm{rad}(\sim 6^\circ)$$

2. Body force is neglected.
3. Spherical liquid boundary.

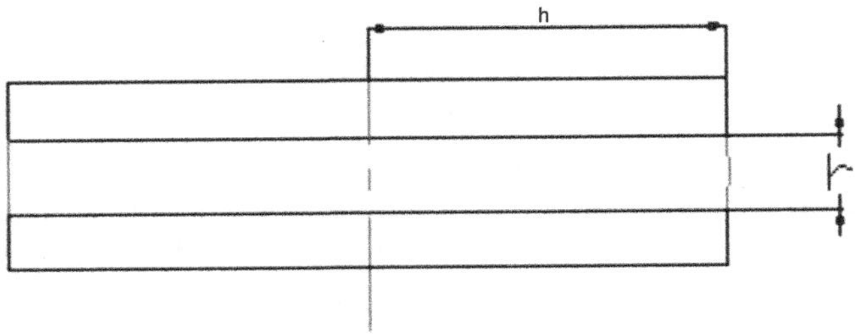

Figure 5.4 Schematic of a parallel plate rheometer

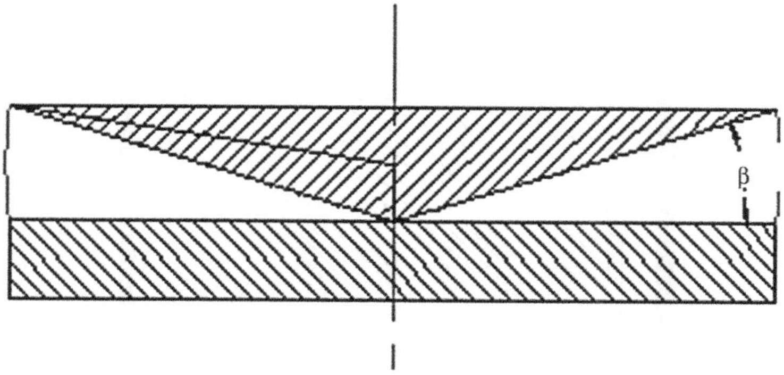

Figure 5.5 Schematic of a cone–plate rheometer

Then we have

$$\tau_{12} = \frac{3M}{2\pi R^3} \tag{5.23}$$

$$\dot{\gamma} = \frac{\Omega}{\beta} \tag{5.24}$$

$$\eta = \frac{\tau_{12}}{\dot{\gamma}} \tag{5.25}$$

$$\gamma = \frac{\phi}{\beta} \tag{5.26}$$

where M is the torque, Ω is the angular velocity of the rotational plate, γ is the shear strain, ϕ is the angular displacement, and R is the cone radius. The shear rate can be controlled by changing the angular velocity of the plate. The torque and normal force can be measured by the sensor installed on the cone.

Couette Cell

A schematic of a Couette (concentric cylinder) is shown in Figure 5.6.

Assume the following:

1. A steady isothermal laminar flow, $v_\theta = r\Omega$ and $v_r = v_z = 0$.
2. Gravity and end effect are neglected.
3. Symmetry in the θ direction, $\partial/\partial\theta = 0$.

Then we have

$$\tau_{12} = \frac{M_i}{2\pi R_i^2 L} \tag{5.27}$$

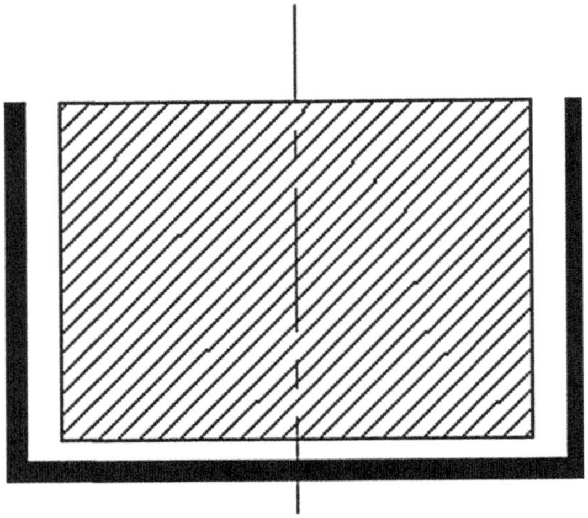

Figure 5.6 Schematic of a Couette (concentric cylinder)

$$\dot{\gamma} = \frac{2\Omega_i \overline{R}}{R_0 - R_i} \tag{5.28}$$

$$\gamma = \frac{\theta \overline{R}}{R_0 - R_i} \tag{5.29}$$

where subscript *o* represents the outer cylinder, *i* the inner cylinder, *L* the depth of the Couette flow, and $\overline{R} = (R_0 + R_i)/2$. It should be noted that it is very difficult to measure normal stress for the couette flow, so the coquette flow is not used for measuring N_1.

4. Double concentric cylinder.

   To measure small viscosity, one can increase the contact surface of a sample to improve the measurement resolution. The geometric shape of a double concentric cylinder can increase the contact surface area by one time, and can be used to measure small viscosity well. The schematic of a double concentric cylinder is shown in Figure 5.7.

   A steady sweep method is usually used for measuring shear viscosity. A constant shear rate is applied to the sample to get the shear viscosity at this shear rate, and then the shear rate is changed for the next measurement. The change of shear rate can be obtained by changing rotational speed (for a parallel plate, the gap can also affect the shear rate), as shown in Figure 5.8.

5.2.2.2 Measurement of Linear Viscoelasticity

The measurement tool of linear viscoelasticity is similar to that of shear viscosity. The difference is that the measurement of linear viscoelasticity uses a dynamic

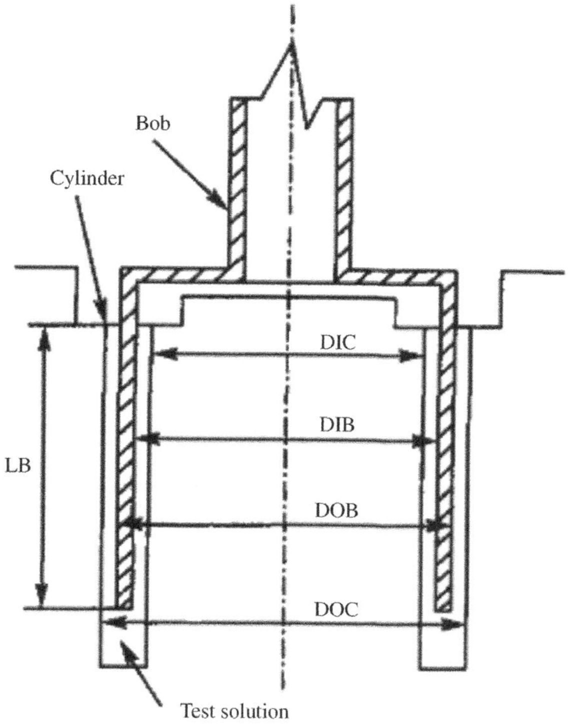

Figure 5.7 Schematic of a double concentric cylinder (Reprinted with permission from Wei, J.J., Kawaguchi, Y., Yu, B. and Feng, Z.P. , "Rheological characteristics and turbulent friction drag and heat transfer reduction of a very dilute cationic surfactant solution," *Journal of Heat Transfer* **128**, 977–983, © 2006 ASME International.)

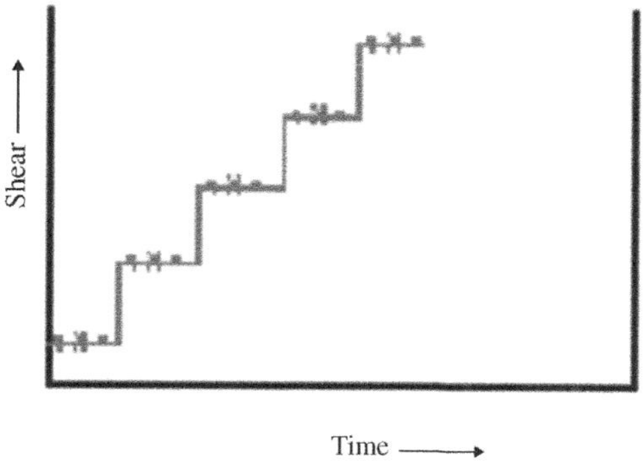

Figure 5.8 Shear rate sweep

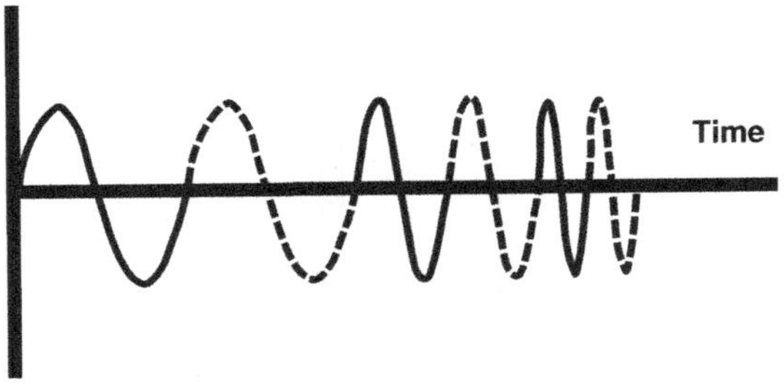

(a)

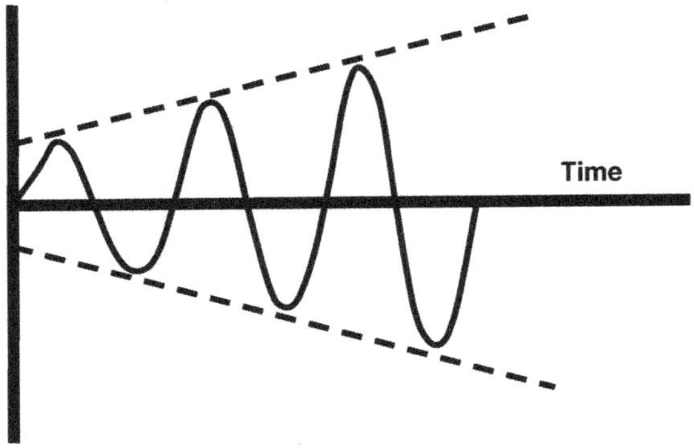

(b)

Figure 5.9 Dynamic sweep module. (a) Frequency sweep, and (b) strain sweep

sweep module. One is frequency sweep (measure the dynamic modulus at a constant strain rate and different frequencies, as shown in Figure 5.9a); another is strain sweep (measure the dynamic modulus at a constant frequency and different strain rates, as shown in Figure 5.9b). Strain sweep must precede frequency sweep to determine a suitable strain range to ensure that the material is in the linear viscoelasticity region, or the results will be meaningless.

5.2.2.3 Measurement of Normal Stress Difference

Usually the first normal stress difference N_1 is measured, and the parallel plate or cone–plate rheometer is selected for the measurement.

For a parallel plate:

$$(N_1 - N_2)|_{\dot{\gamma}_a} = \frac{2F}{\pi a^2}\left(1 + \frac{1}{2}\frac{\mathrm{d}\ln F}{\mathrm{d}\ln \dot{\gamma}_a}\right) \tag{5.30}$$

where F is the total normal force acting on the plate. N_1–N_2 can be used as N_1 since N_2 is very small.

For a cone–plate:

$$N_1 = \tau_{11} - \tau_{22} = \frac{2F}{\pi R^2} + 0.15\rho\Omega^2 R^2 \tag{5.31}$$

where F is the normal force measured on the cone, and $0.15\rho\Omega^2R^2$ is the inertia correction term.

5.2.2.4 Measurement of Elongational Viscoelasticity

The measurement of elongational viscosity is one of the most difficult problems in rheology, especially for low viscosity fluid. For high-viscosity polymer and surfactant solutions, there are several effective ways to measure elongational viscosity, such as homogeneous stretching, fiber spinning, contraction flow, and open-siphon method. However, these methods require that the sample have sufficiently high viscosity and thus they are not suitable for dilute low-viscosity surfactant solution. The best choice at present is a Rheometrics RFX instrument, which employs two opposing nozzle jets to generate elongational flow; it was first proposed by Fuller *et al.* [19].

RFX is a two opposing jets instrument. The solution is sucked into the nozzles (for uniaxial elongational flow) or is ejected from the nozzles (for biaxial elongational flow). The flow field shown in Figure 5.10 is formed. Numerical analysis shows that the flow along the center line between the two nozzles is pure elongational flow, and it is a mixed one of elongational flow and shear flow at other places. The apparent elongational viscosity can be calculated by measuring the torque action on the fixed arm of the instrument:

$$\sigma = \frac{4F_0}{\pi D^2} \tag{5.32}$$

$$\dot{\varepsilon} = \frac{8Q}{\pi G D^2} \tag{5.33}$$

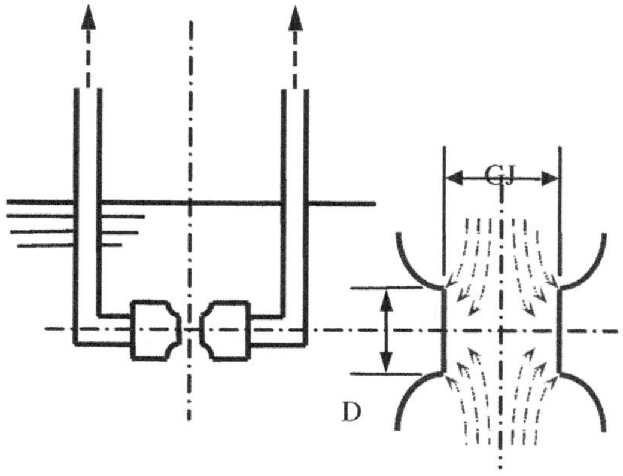

Figure 5.10 RFX instrument

The measured value by use of the two opposing jets instrument is not the actual elongational viscosity of the solution. It includes the effect of shear of nozzles and fluid inertia. Even if these effects are considered, a quantitative correlation cannot be established. Dontula *et al.* [20] suggested that the RFX can be used as an indexer rather than using measured elongational viscosities quantitatively for the shear viscosity of the solution less than 50 mPas.

5.2.3 Rheological Characteristics of Dilute Drag-reducing Surfactant Solution

Wei *et al.* measured the shear rheology of CTAC solutions [21]. Shear viscosity measurements were carried out in the stress-controlled mode by use of an Advanced Rheometric Expansion System (ARES) rheometer, which was equipped with a double-wall Couette cell and was manufactured by Rheometrics Scientific F.E. Ltd., Japan. The geometry of the ARES rheometer is shown in Figure 5.8 and the major dimensions are listed in Table 5.2. The solution temperature was set from 20 °C to 40 °C and was controlled by Julabo F25 Refrigerated Circulators, which were manufactured by Julabo Labortechnick GMBH, Germany. A pre-shear at $\dot{\gamma} = 10$ 1/s was applied for 20 minutes before the rate sweep test. However, 40 minutes of pre-shear was applied before the first measurement due to the large gap between the required temperature and the starting temperature. Measurements were performed for shear rates $\dot{\gamma}$ ranging from 10 up to 100 1/s. Both the delay time and measurement time were equal to 5 minutes.

The measurement of elongational viscosity is one of the most difficult problems in rheology, especially for low-viscosity fluid. For high-viscosity polymer and surfactant solutions, there are several effective ways to measure elongational viscosity, such as homogeneous stretching, fiber spinning, contraction flow, and the open-siphon method. However, these methods require that the sample have sufficiently high viscosity and thus they are not suitable for dilute low-viscosity surfactant solution. The best choice at present is a Rheometrics RFX instrument, which, as mentioned above, employs two opposing nozzle jets to generate elongational flow and was first proposed by Fuller *et al.* [19]. The schematic of the two opposing jets instrument is shown in Figure 5.11. Nozzles with jet diameters of 1, 2, and 3 mm were used to cover the elongation rates from 100 to 9000 1/s. The separation gap between two jets was set to be equal to the

Table 5.2 Rheometer geometry (Reprinted with permission from Wei, J.J., Kawaguchi, Y., Yu, B. and Feng, Z.P., "Rheological characteristics and turbulent friction drag and heat transfer reduction of a very dilute cationic surfactant solution," *Journal of Heat Transfer* **128**, 977–983, © 2006 ASME International.)

Rheometer	Dimensions
ARES (double-wall Couette)	DIC = 27.94 mm, DIB = 29.50 mm, DOB = 32.00 mm, DOC = 34.00 mm, and LB = 31.90 mm

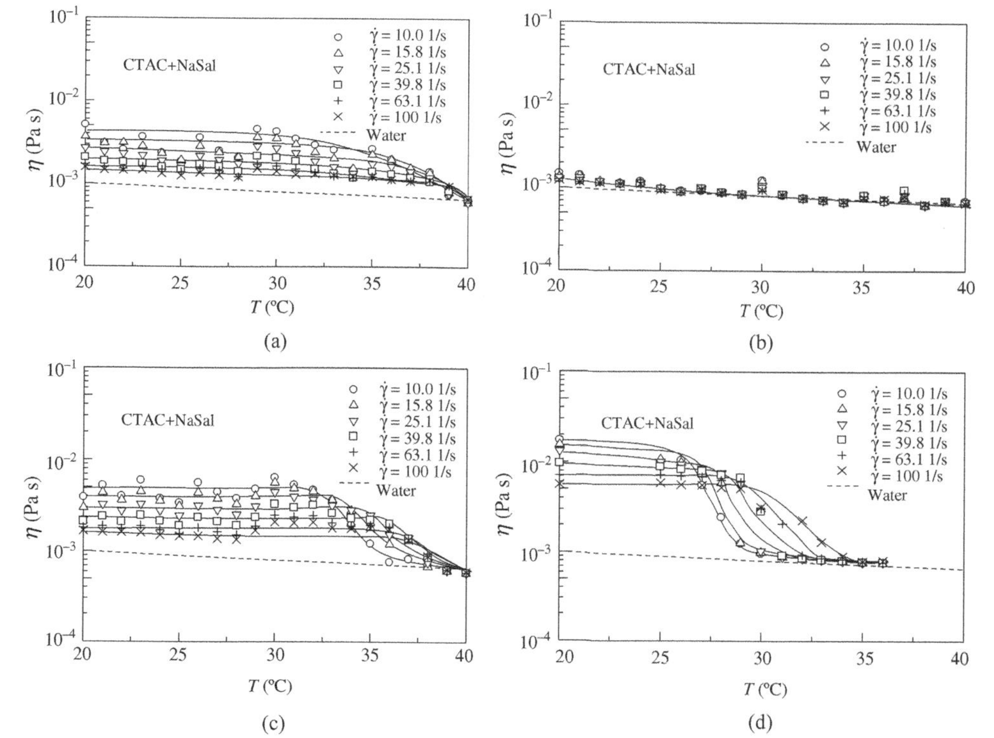

Figure 5.11 Shear viscosity versus temperature. (a) $C_m = 5 \times 10^{-5}$; (b) $C_m = 7.5 \times 10^{-5}$; (c) $C_m = 1 \times 10^{-4}$; and (d) $C_m = 2 \times 10^{-4}$ (Reprinted with permission from Wei, J.J., Kawaguchi, Y., Yu, B. and Feng, Z.P., "Rheological characteristics and turbulent friction drag and heat transfer reduction of a very dilute cationic surfactant solution," *Journal of Heat Transfer* **128**, 977–983, © 2006 ASME International.)

nozzle diameter. The RFX instrument was immersed into a beaker of the test surfactant solution. One of the jets was installed onto a movable arm. The other jet was installed onto a torque rebalance transducer (TRT). Surfactant solution was sucked into nozzles by a syringe pump with a known constant flow rate to generate elongational flow between two nozzles. The instrument computer measured the TRT force output and then calculated an apparent elongational viscosity. In order to understand the effect of fluid temperature, the temperature was controlled to 20 °C, 30 °C, and 40 °C by Julabo F25 Refrigerated Circulators. During the measurements, the waiting, delay, and measurement times were 2, 2, and 4 s, respectively. Experiments were always repeated three times, and the average value was used as the experimental data.

The tested cationic surfactant was cetyltrimethyl ammonium chloride (CTAC) (Toho Chemical Industry Co. Ltd, Japan). Sodium salicylate (NaSal) was used as the counterion for aiding the formation of rod-like micelles, whose formation is considered to be the reason for drag reduction. The surfactant solution was prepared by adding the same mass concentrations of surfactant and counterion to the desired amount of distilled water for the rheological measurements. The surfactant concentration was marked by the concentration of CTAC, and five kinds of mass concentrations (50, 75, 100, 150, and 200 ppm) of the solutions were made for the rheological measurements.

Figure 5.11a–d shows the measured shear viscosity η versus surfactant solution temperature T with shear rate $\dot{\gamma}$ as a parameter at surfactant mass concentration $C_m = 50$, 75, 100, and 200 ppm, respectively. For comparison, the shear viscosity of water indicated by a dash line is also shown in Figure 5.11a–d. The solid lines shown in Figure 5.11a–d represent the correlation lines of the experimental data for different shear rates. From Figure 5.11a, we can see that the shear viscosity η decreases gradually with increasing temperature T. The effect of shear rate $\dot{\gamma}$ on η is very small, and the measured shear viscosity is close to that of water. Thus the surfactant solution with a mass concentration of 50 ppm shows Newtonian characteristics of shear viscosity. However, for the higher concentrations shown in Figure 5.11b–d, all the measured shear viscosities are much higher than that of water. It can also be seen from Figure 5.11b–d that for a certain shear rate, there exists a critical temperature above which shear viscosities decrease rapidly to that of water. The effect of temperature on shear viscosities before the critical temperature is not obvious except for the 200 ppm case at lower shear rates where shear viscosities decrease gradually with increasing temperature. Generally speaking, if temperature does not exceed the critical temperature, the shear viscosities decrease with increasing shear rate, showing shear-thinning characteristics. The much higher viscosities of surfactant solution are probably due to the formation of a micelle network structure, and the critical temperature corresponds to the start of the network destruction. The critical temperature decreases with the increasing mass concentration of surfactant solution. Concretely, it is about 38 °C at $C_m = 75$ irrespective of shear rate, but it is in the range of 31–36 °C and 26–29 °C at $C_m = 100$ and 200 ppm, respectively, depending on the shear rate. This result is out of one's expectation since it is generally considered that the network strength increases with increasing surfactant concentration, resulting in an increased critical temperature to break down the network. But for this very dilute solution with surfactant concentration one or two orders of magnitude smaller than the other researchers' surfactant concentrations, this phenomenon is actually found in our measurements. For $C_m = 100$ and 200 ppm, we can see that the critical temperature increases with increasing shear rate. Beyond the critical temperatures, the solution temperature increases with increasing shear rate at the same shear viscosity. This is different from the Newtonian fluid, the temperature of which does not change with shear rate. In the very dilute surfactant solution, the network structure is weak or incomplete. The role of the shear stress in the formation of micelle structures is more important than that in the dense solution case. We may consider that the shear-induced structure probably occurs in the solution and this structure makes the internal energy of the surfactant solution change greatly with the addition of shear energy.

Drag-reducing surfactant solutions are very complex systems in terms of rheological properties. We should choose a complicated model, the Giesekus model with nonlinear stress terms, to describe the rheological characteristics of these surfactant solutions.

Giesekus [22] derived a constitutive equation as follows:

$$\tau = \tau_s + \tau_p = \tau_p + \eta_s \dot{\gamma} \tag{5.34}$$

$$\alpha \frac{\lambda}{\eta_p} \tau_p^2 + \tau_p + \lambda \dot{\tau}_p = \eta_p \dot{\gamma} \tag{5.35}$$

where subscripts s and p represent solvent and surfactant, respectively. To simplify the solution, the contribution from the solvent is set to be zero and we can get the steady shear viscosity solution as

$$\frac{\eta}{\eta_0} = \frac{4(1-\alpha)}{\sqrt{f+1}\left[\sqrt{f+1} + \sqrt{2}(1-2\alpha)\right]} \tag{5.36}$$

where

$$f = \sqrt{1 + 16\alpha(1-\alpha)\lambda^2\dot{\gamma}^2} \tag{5.37}$$

There are three parameters in Equations 5.36 and 5.37: relaxation time λ, zero shear viscosity η_0, and mobility factor α ($0 \leq \alpha \leq 1$). When the value of α is set to zero, the Giesekus model reduces to the Maxwell model.

Figure 5.12 shows the least-square fitting results of the measured shear viscosity with the Giesekus model by well-adjusting model parameter values for different

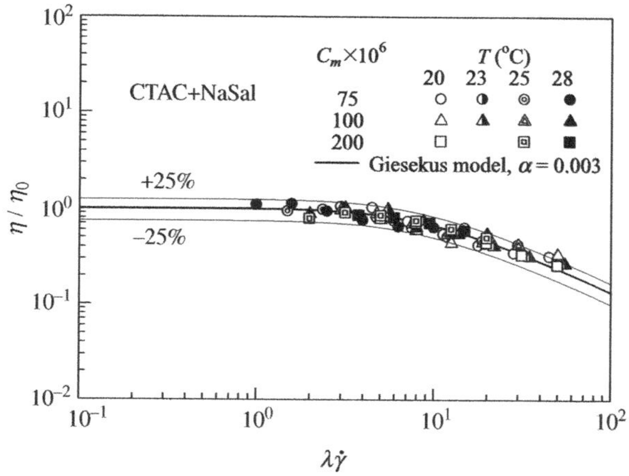

Figure 5.12 Comparison of the measured shear viscosity with the Giesekus model (Reprinted with permission from Wei, J.J., Kawaguchi, Y., Yu, B. and Feng, Z.P., "Rheological characteristics and turbulent friction drag and heat transfer reduction of a very dilute cationic surfactant solution," *Journal of Heat Transfer* **128**, 977–983, © 2006 ASME International.)

solution temperatures and concentrations. It can be seen that the prediction of the Giesekus model as plotted by a thick solid line can fit the experimental data well by selecting suitable model parameter values after trial and error. In the temperature range of 20–28 °C and concentration range of 75–200 ppm, the mobility factor α obtained by fitting is a very small constant of 0.003 and does not change with C_m and T, whereas both the zero shear viscosity η_0 and relaxation time λ are functions of C_m and T. Figure 5.13 shows η_0 versus inverse T with C_m as a parameter. The data of water indicated by a dash line are also shown for comparison. We can see that η_0 increases greatly with decreasing T as compared to the gradual change of water. η_0 increases monotonically with C_m. For the highest concentration of 200 ppm, η_0 decreases much more rapidly with increasing T than for 75 and 100 ppm cases, resulting in a smaller critical temperature as indicated in Figure 5.11b–d. Figure 5.14 shows λ versus inverse T with C_m as a parameter. λ increases greatly with decreasing T as η_0 does, but it increases with increasing C_m from 75 to 100 ppm and then decreases with further increasing C_m up to 200 ppm as compared with the monotonic change of η_0 with C_m. The relaxation time $\lambda = (\eta_0/G)$ can be used to describe the viscoelasticity of the surfactant solution, which is infinite for a Hookean elastic solid and zero for a Newtonian viscous liquid. In the present study, the value of λ is in the range of 0.1–1.0 s, and the elasticity increases with increasing λ. We can see from Figure 5.14 that λ increases with decreasing T. Therefore, the viscoelasticity will increase with decreasing temperature, resulting in the increase of turbulent drag reduction and the corresponding heat transfer reduction. DNS studies of the drag-reducing Giesekus fluid flow [23,24] also indicated that the turbulent drag reduction increases with the

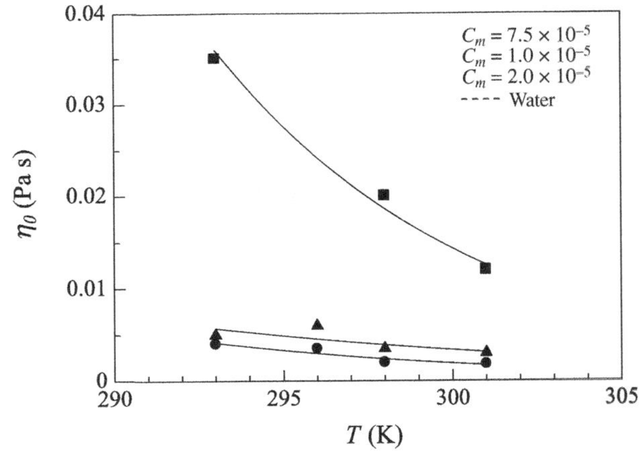

Figure 5.13 η_0 versus T and C_m (Reprinted with permission from Wei, J.J., Kawaguchi, Y., Yu, B. and Feng, Z.P., "Rheological characteristics and turbulent friction drag and heat transfer reduction of a very dilute cationic surfactant solution," *Journal of Heat Transfer* **128**, 977–983, © 2006 ASME International.)

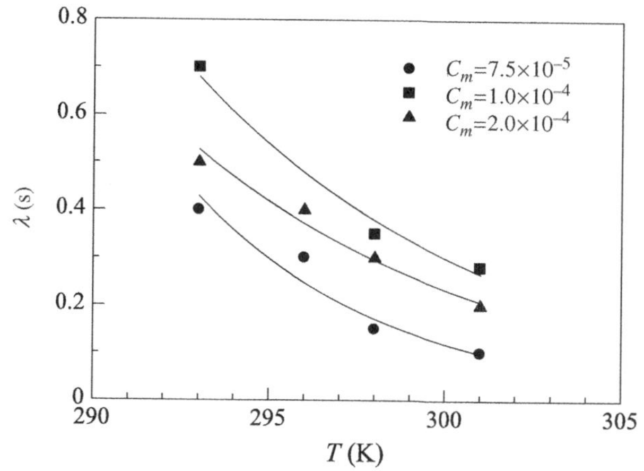

Figure 5.14 λ versus T and C_m (Reprinted with permission from Wei, J.J., Kawaguchi, Y., Yu, B. and Feng, Z.P., "Rheological characteristics and turbulent friction drag and heat transfer reduction of a very dilute cationic surfactant solution," *Journal of Heat Transfer* **128**, 977–983, © 2006 ASME International.)

relaxation time λ, which is consistent with the above analysis. The exponential correlations of η_0 and λ with T at $C_m = 75$, 100, and 200 ppm are listed in Table 5.3.

Figure 5.15 shows the comparison of experimental data of shear viscosity with the prediction of the Giesekus model. The values of η_0 and λ for the surfactant solutions with different concentrations and temperatures are obtained from the correlations listed in Table 5.3. For the concentration of 150 ppm, the correlations of η_0 and λ with T are not given, thus linear interpolation is used to get the values of η_0 and λ. Although the correlations are made in the temperature range from 20 °C to 28 °C, the experimental data of surfactant solutions at $C_m = 75$, 100, and 150 ppm for $T = 30$ °C are also plotted in Figure 5.15 for comparison. The values of η_0 and λ are obtained by extrapolation, which results in slightly larger deviation of the experimental data from the Giesekus model. Seventy percent of the experimental data agree with the prediction of the Giesekus model to within ±27.3%. Thus it seems that the Giesekus model can describe the shear viscosity characteristics of the CTAC–NaSal surfactant solutions at

Table 5.3 Correlations of η_0 with λ at different C_m (Reprinted with permission from Wei, J.J., Kawaguchi, Y., Yu, B. and Feng, Z.P., "Rheological characteristics and turbulent friction drag and heat transfer reduction of a very dilute cationic surfactant solution," *Journal of Heat Transfer* **128**, 977–983, © 2006 ASME International.)

C_m	η_0 (Pa.s)	λ (s)
7.5×10^{-5}	$4.62 \times 10^{11} \exp(-0.1104T)$	$9.55 \times 10^{22} \exp(-0.1835T)$
1×10^{-4}	$2.64 \times 10^{7} \exp(-0.07595T)$	$5.24 \times 10^{14} \exp(-0.1163T)$
2×10^{-4}	$1.99 \times 10^{-15} \exp(-0.1316T)$	$3.28 \times 10^{14} \exp(-0.1170T)$

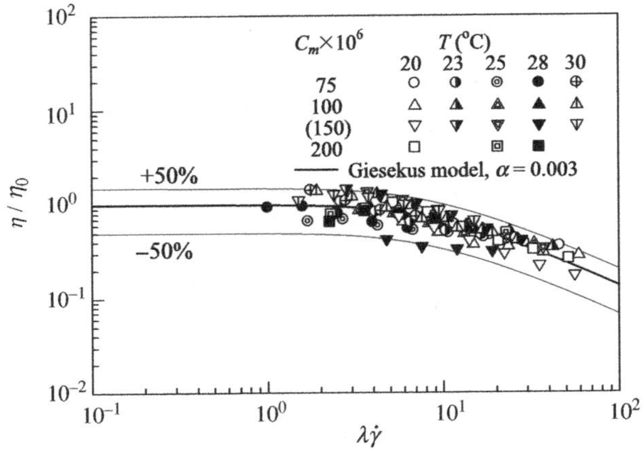

Figure 5.15 Prediction of the Giesekus model (Reprinted with permission from Wei, J.J., Kawaguchi, Y., Yu, B. and Feng, Z.P., "Rheological characteristics and turbulent friction drag and heat transfer reduction of a very dilute cationic surfactant solution," *Journal of Heat Transfer* **128**, 977–983, © 2006 ASME International.)

temperatures within 20–30 °C and concentrations within 75–200 ppm. The correlations of η_0 and λ can be used for the direct numerical simulation of turbulent drag-reducing CTAC–NaSal surfactant solution flow.

The elongational viscosity was calculated from the force exerted on the fixed-nozzle arm of the opposing jet rheometer RFX. The full-scale of the force transducer is 1.3 g, and it is reasonable to suppose that an error as large as 1/1000 of the full scale is contained in the detected force. Therefore, the data obtained from the force less than 0.01 g were considered to be comparable with errors and were eliminated from the original data during the data reduction process. Figure 5.16 shows the measured elongational viscosity η_E versus the elongation rate $\dot{\varepsilon}$ with temperature T as a parameter

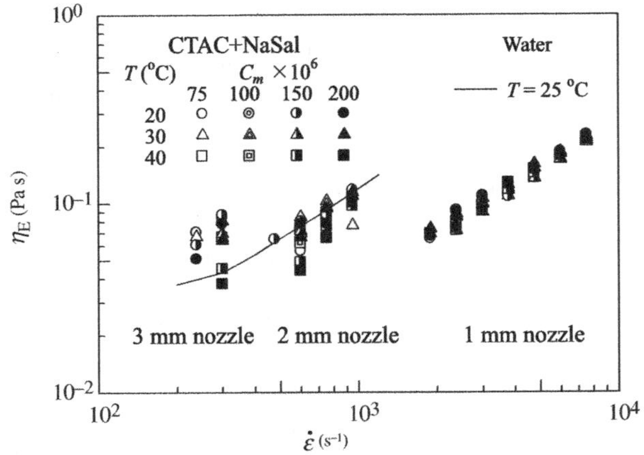

Figure 5.16 Elongational viscosity versus elongation rate

at $C_m = 75, 100, 150$, and 200 ppm. Nozzles with jet diameters of 1, 2, and 3 mm were used to cover elongation rates from 100 to 9000 1/s. The smaller the jet diameter, the larger the elongation rates can be measured. From this figure, it can be seen that the measured elongational viscosities are affected by the jet size to some extent. The dilute surfactant solution shows complicated rheological characteristics, and the measurement itself causes a modification of sample characteristics. This may be the reason that the measured values of elongational viscosity are dependent on the measuring geometry. The temperature T has no severe influence on elongational viscosity, especially at the large elongation rates. At the small elongation rates, despite a small effect of T, the elongational viscosity decreases with increasing T. The effect of surfactant concentration on the measured elongational viscosities is also small. The measured elongational viscosities of water by the RFX instrument are also shown in Figure 5.16 for comparison. We can see that there is no clear difference between the elongational viscosities of water and the dilute surfactant solutions in our measurement range. Even for the water case, we got an impractical elongation rate dependence of elongational viscosities. It is well known that the RFX is limited to measuring the elongation viscosities in a certain range of lower shear viscosities. Dontula *et al.* [20] suggested that the RFX can be used as an indexer rather than using the measured elongational viscosities quantitatively. How to measure quantitatively and accurately the elongational viscosity of fluid with very low shear viscosity in order to discuss these regarding the relation with drag reduction phenomena is still an unraveled problem we are facing.

5.3 Factors Affecting the Rheological Characteristics of Surfactant Solution

Microstructures formed in the surfactant solution have a direct influence on the rheology. For spherical micelles, the solution is isotropic and shows Newtonian fluid performance. The existence of rod-like or thread-like micelles changes the stress distribution of the solution, showing non-Newtonian rheological characteristics (some solutions do not show viscoelasticity even if they have network structures). Therefore, the effects of external conditions are mainly caused by the change of solution microstructures.

The rheological characteristics of dilute surfactant solutions are affected by many factors, and the main factors are surfactant concentration, temperature, surfactant type, and added counterion.

5.3.1 Surfactant Concentration

As mentioned above, different microstructures are formed at different surfactant concentrations. The shear viscosity does not simply increase with increasing concentration, and it shows a complex variation trend. For Ethoquad O12/NaSal solution, when the surfactant-to-counterion ratio $\xi = 1$, both the 5 mM/5 mM and 10 mM/10 mM solutions show highly shear-thinning characteristics. For the surfactant concentration above 5 mM, the solution shows Newtonian fluid characteristics with increasing shear rate up to 10 1/s, and then shows shear-thinning characteristics. The shear viscosity of the solution of $\xi = 2.5$ does not show any obvious change with the shear rate. For the solution of $\xi = 1$, zero-shear viscosity increases up to the maximum value (at 10 mM) with increasing surfactant concentration and then decreases; for the solution of $\xi = 2.5$, zero-shear viscosity decreases down to a minimum value and then increases. Before the surfactant concentration reaches 100 mM, zero-shear viscosity decreases with increasing surfactant-to-counterion ratio ξ [25]. There are three possible reasons for the decrease of zero-shear viscosity with surfactant concentration in the semidilute solution: (1) micellar branches are formed in the worm-like network, enhancing the solution flow; (2) micelles are connected to one chain and move independently at high concentration, reducing shear viscosity; and (3) micellar network structures reach a saturated state.

5.3.2 *Temperature*

Generally, increasing solution temperature can hinder the growth of micelles, and thus weaken the rheological phenomena of shear thinning and normal stress. When the temperature of surfactant solution is increased up to a critical value, the surfactant molecules will not form rod-like micelles, and the solution become Newtonian fluid.

5.3.3 *Type of Surfactant*

The structures of surfactant molecules determine the micelle formation and hence affect the rheology. All micelles are formed not by covalent bonding but by hydrophobic interaction which is very weak. The chemical and geometrical characteristics of surfactant molecules determine the micellar structure approximately. The main factors affecting the rheology are the alkyl chain length and parity of carbon atom number, saturation of alkyl chain, cis-trans structure in an unsaturated alkyl chain, and hydrophilic head group. These factors also determine the temperature range and critical shear stress for drag reduction.

The effective upper temperature limit increases with increasing alkyl chain length. For the surfactants with parity of carbon atom number, the parity effect can be observed at the Kraftt point, the critical precipitation temperature, or much lower temperatures.

With increasing hydrophobic alkyl chain length, more surfactant molecules tend to keep in the micelles to avoid them contacting water. The alkyl chain is hydrophobic, so the longer the alkyl chain length, the stronger the interaction of the hydrophobic group is, and the more obvious the effects on solvent are. When the carbon number in the alkyl chain is reduced to a certain value, the solution shows Newtonian fluid characteristics.

The effect of parity of the carbon number in the alkyl chain comes from the different aggregation arrangements in the micelle.

Cis-form OMAC solution shows non-Newtonian rheology, but it almost shows Newtonian fluid characteristics after it changes to trans-form by ultraviolet irradiation [26,27]. Therefore, the rheology of the solutions with cis-trans structure changes greatly.

When a double bond exists in the alkyl chain, the effective temperature for drag reduction extends to a lower temperature while keeping the upper temperature limit unchanged. For the unsaturated solution with double bond in the alkyl chain, the critical shear stress of the surfactant solution with cis-form structure is much larger than that with trans-form structure at $\xi = 1$ and $\xi = 1.5$.

In the quaternary ammonium cationic surfactant head group, if some or all of the methyl group are replaced by the hydroxyethyl group, the critical shear stress can be reduced and the effective temperature range can be extended for drag reduction. When the head group contains the hydroxyethyl group and dimethyl group, the viscoelasticity of the surfactant solution can reach a maximum value.

5.4 Characterization of Viscoelasticity of Drag-reducing Surfactant Solution by Using Free Surface Swirling Flow

It is generally considered that the DR phenomenon is caused by the viscoelasticity of the surfactant solution. Therefore, rheological measurements can be used for screening an effective drag-reducing surfactant within a measurable range. However, the rheological measurements showed that the elasticity of the CTAC solutions is so small that the rheometers could not provide reliable data for the first normal stress and storage modulus. Therefore, the challenge to be addressed is how to determine a surfactant solution with small viscoelasticity having DR ability. From an economic consideration, the amount of the testing surfactant should be as small as possible. Siginer and coworkers initiated an idea of using the free surface shape of a viscoelastic swirling flow to generate information of the constitutive equation for the viscoelastic liquid [28]. With the help of this idea, Wei *et al.* [29] solved this problem by using a swirling flow to observe and analyze the effect of the drag-reducing surfactant solution on the vortex inhibition and relate this to the screening of effective low-viscoelastic drag reducers.

The experimental set-up is shown in Figure 5.17. A glass cylinder having an inner radius of 70 mm was nearly submerged in a cubic water bath constructed from flat transparent perspex walls 130 mm in side length. The purpose of the external water bath was to reduce optical distortion caused by the curved surfaces of the cylinder. A rotating circular disc having a diameter of 65 mm and a thickness of 8.5 mm was placed at the bottom of the cylinder. The water bath was fixed on the surface of a digital hot plate/stirrer, and the temperature of the water bath was controlled by use of the

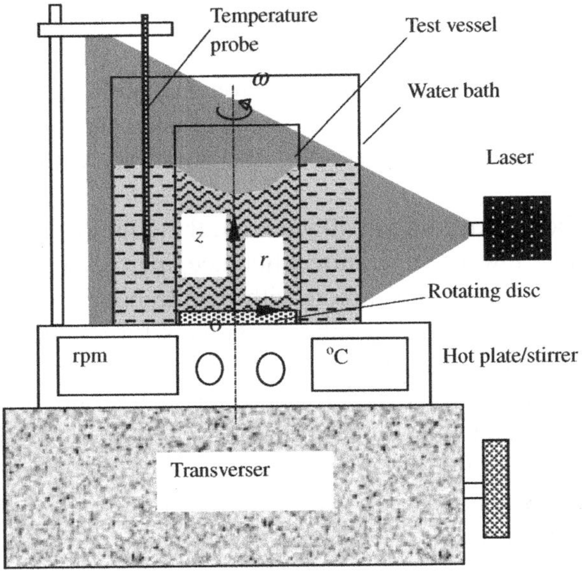

Figure 5.17 Schematic of the test facility (Reprinted with permission from Wei, J.J., Li, F.-C., Yu, B. and Kawaguchi, Y., "Swirling flow of a viscoelastic fluid with free surface, part I: Experimental analysis of vortex motion by PIV," *Journal of Heat Transfer* **128**, 69–76, © 2006 ASME International.)

heating function of the hot plate/stirrer. The temperature probe of the hot plate/stirrer was inserted into the water bath, and the test fluid temperature was measured with a digital thermometer before the PIV measurements. The temperature of the water in the water bath and that of the test fluid in the test vessel were both kept at a constant of 25 °C. A magnet bar (8 mm × 55 mm) was embedded into the rotating disc and the rotational speed was controlled by the magnetic force from the hot plate/stirrer acting on the magnet bar. The axis of the cylinder and the center of the disc were carefully adjusted to align with those of the hot plate/stirrer (magnetic center). The hot plate/stirrer was mounted on a flat stage that could be moved back and forth with an accuracy of 0.1 mm.

Water and surfactant solutions were used for the test. The surfactant was CTAC. NaSal was selected as the counterion. The surfactant solution was prepared by adding the same mass concentration of surfactant and counterion to the desired amount of distilled water. The surfactant concentration was marked by the concentration of CTAC, and three mass concentrations of 50, 200, and 1000 ppm were tested. The test fluid was filled into the glass cylinder up to a height H equal to the inner radius of the cylinder, R_w, and thus the aspect ratio of the fluid height to radius, H/R_w, was fixed to unity in the present experiments. The shear viscosities of the three kinds of surfactant solutions at 25 °C were also measured in the shear rate range of 10- 00 s^{-1} by use of an ARES rheometer equipped with a double-wall Couette cell.

The secondary flow pattern in the meridional plane and the tangential velocities were obtained by use of a PIV system consisting of a double-pulsed laser, laser sheet optics, a charge-coupled device (CCD) camera, a timing circuit (TSI Model 610032), an image-sampling computer, and image-processing software (TSI Insight, version 3.3). The double-pulsed laser (MiniLase-II/20 Hz, New Wave Research Co., Ltd., United States) is a combination of a pair of Nd-YAG lasers, each having an output of 25 mJ/pulse and a maximum repetition rate of 20 Hz. By changing the combination of cylindrical lenses, the laser sheet thickness can be modified from 0.14 to 0.6 mm and the beam spread angle from 4.3° to 13.3°. The timing circuit communicates with the CCD camera and computer and generates pulses to control the double-pulsed laser. The CCD camera used (PIVCAM, 10–30, TSI Model 630046) has a resolution of 1280×1024 pixels.

The meridional plane of the swirling flow was illuminated by the double-pulsed laser for measurement of the secondary flow pattern. The tangential velocities were also measured by illuminating the vertical planes parallel to the meridional plane at eight equally spaced radial locations as shown in Figure 5.18. The flat stage could be controlled to move back and forth to obtain the desired vertical cross-section for illumination and measurement. The interrogation area was set at 32×32 pixels (with 50% overlap in each direction) for velocity vector analyses. PIV images were acquired for 1000 dual frames (2000 double-exposed PIV photographs) for the meridional plane measurements and 100 dual frames for the other measurements. Due to the presence of the cubic water tank, the optical distortion caused by the cylindrical wall was negligible up to a radius of 0.9 R_w. The photograph acquisition rate was 8 Hz, indicating that the velocity field was recorded at 4 Hz. The tracer particles used to seed the flow were acrylic colloids 0.1–1 μm in diameter. The particle concentration was adjusted so that on average at least ten particle pairs were observed in an interrogation window for each case.

After a vector field was calculated by interrogating a dual frame, TSI Insight software was used to validate it and remove any erroneous velocity vectors that might have been incorrectly detected during interrogation due to random noise in the correlation function. Empty data cells (holes) that appeared after the removal of erroneous velocity vectors were filled using a Fortran program created by the interpolation of neighboring velocity vectors.

The Reynolds number and elasticity number (the ratio of elastic to viscous time scales) of the fluid in the present study are defined as

$$\mathrm{Re}_0 = \rho\omega R_d^2/\eta_0 \tag{5.38}$$

$$\mathrm{Re} = \rho\omega R_d^2/\eta \tag{5.39}$$

$$E = \lambda_G\eta/\rho R_d^2 \tag{5.40}$$

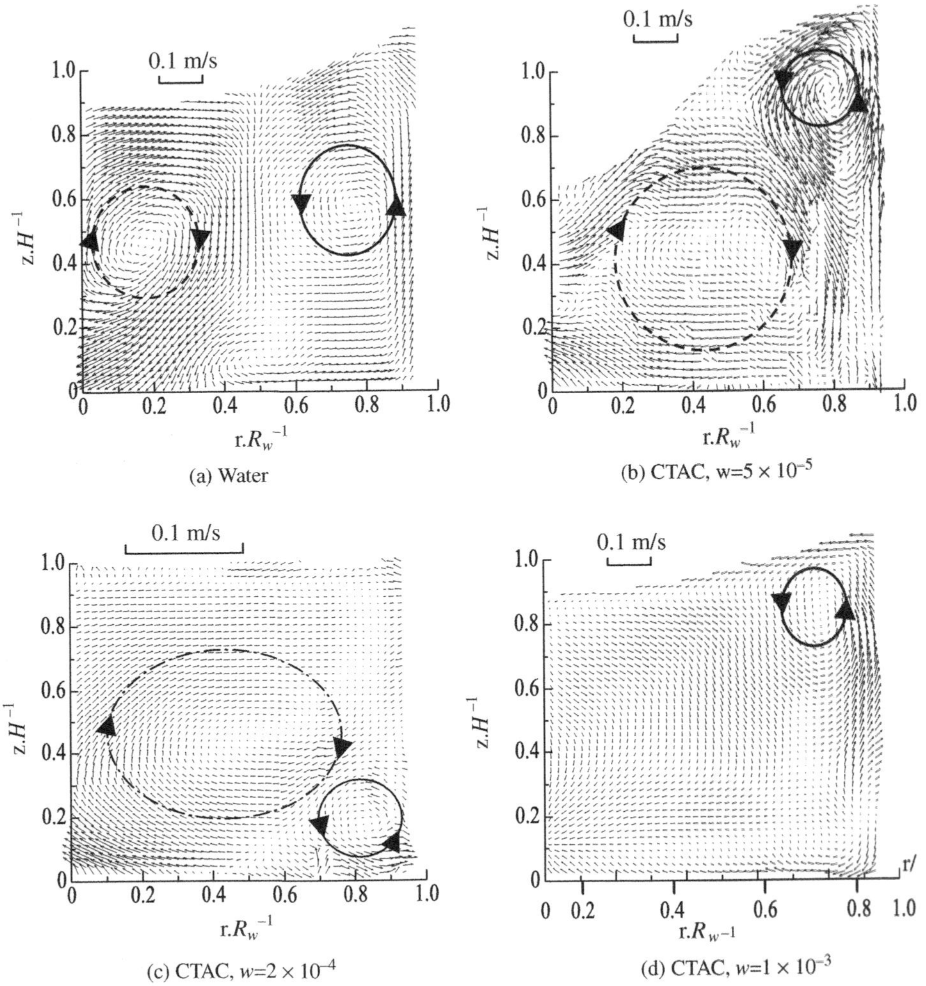

Figure 5.18 Secondary flow patterns in the meridional plane. (a) Water; (b) CTAC, $w = 5 \times 10^{-5}$; (c) CTAC, $w = 2 \times 10^{-4}$; and (d) CTAC $w = 1 \times 10^{-3}$ (Reprinted with permission from Wei, J.J., Li, F.-C., Yu, B. and Kawaguchi, Y., "Swirling flow of a viscoelastic fluid with free surface, part I: Experimental analysis of vortex motion by PIV," *Journal of Heat Transfer* **128**, 69–76, © 2006 ASME International.)

where η_0 is zero-shear viscosity and λ_G is the relaxation time in the Giesekus model. Shear viscosity η, which is a function of shear rate $\dot{\gamma}$, is evaluated by assuming shear rate $\dot{\gamma}$ equal to the angular velocity ω. The elasticity of the CTAC solutions in the present study is so small that the rheometers could not provide reliable data for the first normal stress and storage modulus. However, our ARES rheometer was able to provide accurate measurements of the shear viscosity data. For evaluating the elasticity of the three surfactant solutions, the relaxation times were extracted by fitting the measured

Table 5.4 Values of Re_0, Re, and E for water and three CTAC solutions

	Water	CTAC		
		$w = 5 \times 10^{-5}$	$w = 2 \times 10^{-4}$	$w = 1 \times 10^{-3}$
Re_0	4.3×10^4	3.9×10^4	3.2×10^3	8.6×10^2
Re	4.3×10^4	3.7×10^4	3.9×10^3	1.8×10^3
E	0	2.8×10^{-5}	2.0×10^{-3}	1.0×10^{-2}

shear viscosity curves to the Giesekus model [30] instead of the Maxwell relaxation time used by the other researchers. Table 5.4 lists the values of η_0, λ_G, Re_0, Re, and E for water and the three CTAC solutions. The elasticity number indicates the elasticity effect against the inertia effect.

The Froude number indicates the gravity effect against the inertia effect and is defined as

$$Fr = \omega^2 R_d^2 / gH \tag{5.41}$$

where g is the gravitational acceleration, and H is the liquid height. In the present study, the Froude number is 4.80, which is large for generating a sizeable deformation of the free surface against the gravity effect in Newtonian swirling flow.

Figure 5.18a–d respectively shows the secondary flow patterns in the meridional plane for water, CTAC 50 ppm, CTAC 200 ppm, and CTAC 1000 ppm. The vectors shown in Figure 5.18d are enlarged to three times the size of those in Figure 5.18a–c. Since the flow at the high Reynolds number in the present study is an unsteady one, the results shown are the average of the instantaneous velocities obtained from 1000 dual frames in the measurements. The average of 100 dual frames shows the same secondary flow pattern, indicating the statistically stationary nature of the swirling flow. It is well known that for Newtonian swirling flows at a low Reynolds number, the inertia-driven vortex plays a dominant role in the meridional plane. Centrifugal forces cause the fluid to flow radially outward along the rotating disc, and then the flow is deflected upward along the cylindrical wall. At the free surface, the flow bends toward the center and turns downward to finish its path of circulation. However, the flow pattern in the meridional plane for water at the present high Reynolds number differs significantly from that at low Reynolds numbers, and the inertia-driven vortex is pushed to the corner between the free surface and cylindrical wall by a large counter-rotating vortex. It is considered that the vortex breakdown phenomenon leads to the appearance of the counter-rotating vortex. According to the observations conducted by Spohn *et al.* [31] on vortex breakdown in an open cylindrical container with a rotating bottom, there is no upper Reynolds number limit at which breakdown disappears, which is very different from the observations of the closed cylindrical container flow in which the breakdown bubbles disappear beyond a certain Reynolds number.

The breakdown bubble increases in size with increasing Reynolds number. Therefore, it is supposed that the large counter-rotating vortex is the breakdown bubble. For the case of CTAC 50 ppm, as shown in Figure 5.18b, the size of the counter-rotating vortex becomes smaller and the size of the inertia-driven vortex increases but the intensity decreases compared to the water case. The size increase of the inertia-driven vortex is probably caused by the decrease of the free surface deformation. With further increasing surfactant concentration up to 200 ppm, as shown in Figure 5.18b, the counter-vortex disappears and the inertial-driven vortex decreases in both size and intensity. This is probably due to the large decrease of local Reynolds number and the competing interaction between elasticity and inertial force, which weakens both the inertia-driven and breakdown vortices. For the CTAC 1000 ppm surfactant solution flow, the inertia-driven vortex locates near the disc wall intersection, whereas an elasticity-driven reverse vortex governs the majority of the flow field. However, the vortex intensity is much smaller compared to that in the water flow case. This differs from a closed disc cylinder system, in which a large Re and small elasticity number produce Newtonian-like characteristics in the secondary flow in the meridional plane.

Figure 5.19 shows the free surface shapes for water and surfactant swirling flows. The free surface shapes were also averaged over 1000 frames. We can see that the height of the dip, h, at the center of the free surface (the maximum deflection of the free surface from the horizontal at the central axis) is much smaller for surfactant solutions than that for water and decreases slightly with increasing surfactant concentrations (elasticity numbers). In the present study, although the elasticity number is very small, the elasticity still plays an important role in the shape of the free surface and the secondary flow pattern. Since the local Reynolds number for CTAC 50 ppm is very close to that of Newtonian fluid flow, we do not consider the decrease of the dip to be

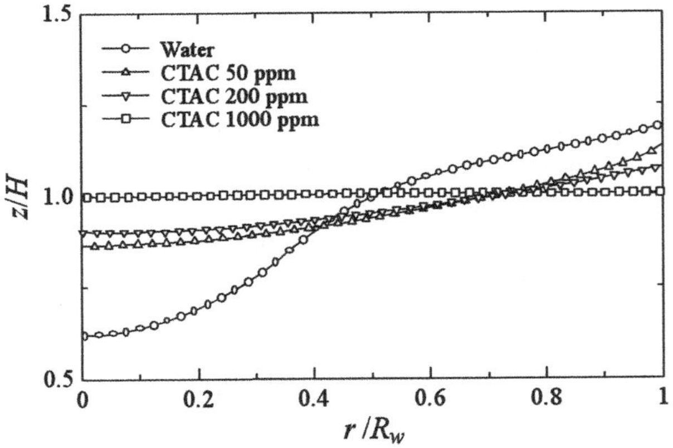

Figure 5.19 Free surface shapes (Reprinted with permission from Wei, J.J., Li, F.-C., Yu, B. and Kawaguchi, Y., "Swirling flow of a viscoelastic fluid with free surface, part I: Experimental analysis of vortex motion by PIV," *Journal of Heat Transfer* **128**, 69–76, © 2006 ASME International.)

due to the decrease of the local Reynolds number. The decrease of h/H with elasticity number is the so-called Quelleffekt. The sensitivity of the free surface deformation with small elasticity in the open swirling flow may be used as a simple method for the evaluation of DR ability instead of the rheology measurement. A critical value of h/H may exist for the judgment of the lowest viscoelasticity for the onset of DR to overcome the difficulty in screening effective drag reducers by use of a rheometer for the low-viscoelasticity case.

To testify this conjecture, we measured the free surface shapes of another three CTAC surfactant swirling flows with low mass concentrations of 25, 30, and 40 ppm, respectively. Figure 5.20 shows the relationship of h/H with mass concentration C_m. Our measured drag reduction data of CTAC–NaSal surfactant flow in a two-dimensional channel flow at a Reynolds number of 4.3×10^4 for different C_m are also shown in this figure. The values of h/H and DR are the same as that of water for $C_m \leq 30$ ppm. At a critical value of $C_m = 40$ ppm, the value of h/H decreases suddenly and the value of DR jumps from 0% to about 60%. This suggests that the viscoelasticity of a 40 ppm CTAC surfactant solution be a critical value for the onset of drag reduction at a Reynolds number of 4.3×10^4, and this critical viscoelasticity can be represented indirectly by the corresponding value of h/H in a free-surface swirling flow. Since the elasticity of the CTAC solutions is small, we did not observe the appearance of a bulge around the center of the free surface even for the 1000 ppm surfactant solution. However, when we inserted a rod into the center of the swirling flow of the 1000 ppm surfactant solution, we found the rod-climbing Weissenberg phenomenon.

Table 5.5 shows zero-shear viscosities, η_0, for surfactant solutions with different concentrations. We can see that η_0 increases with increasing concentration, but there is

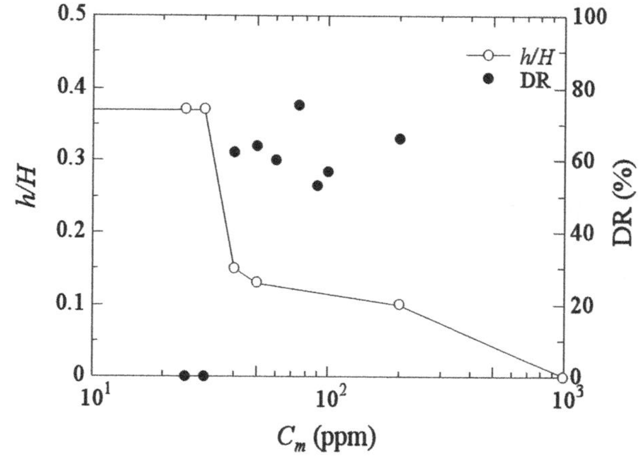

Figure 5.20 Relationship of h/H with DR and *w* (Reprinted with permission from Wei, J.J., Li, F.-C., Yu, B. and Kawaguchi, Y., "Swirling flow of a viscoelastic fluid with free surface, part I: Experimental analysis of vortex motion by PIV," *Journal of Heat Transfer* **128**, 69–76, © 2006 ASME International.)

Table 5.5 Values of η_0 for surfactant solutions with different concentrations

ppm	25	30	40	50	200	1000
(mPa.s)	0.83	0.89	0.96	1.05	12.1	45.0

no sudden change at 40 ppm. Therefore, it is considered that the sudden change of the free surface shape shown in Figure 5.20 is not caused by shear viscosity, but by elasticity of the solution.

5.5 Molecular and Brownian Dynamics Simulations of Surfactant Solution

The rheology of surfactant solution is closely related to the solution structures. The existence of micelles in the solution leads to the change of fluid stresses, showing distinctive rheological characteristics. For the effect of micelles on fluid, it is better to consider the forces and motions of all molecules in the selected domain, and thus to understand how the rod-like micelles are formed by surfactant molecules and how the network structures are formed by rod-like micelles. However, very small time and size scales in the simulation make the work too large to be finished by today's computers. Because the micelles are relatively larger than solvent molecules, the present simulation methods are to simulate the formation of micelles by use of molecular dynamics simulation, and to simulate the formation of network structure and its effect on rheology by use of Brownian dynamics simulation.

Different simulation methods use different time and size scales. Figure 5.21 shows the length–time plot for different simulation methods. The more accurate the simulation method is, the more accurate the time and size scale is required, and the smaller the computational domain is for the same computational ability.

5.5.1 Brief Introduction of Simulation Methods

At present, there are three main methods to describe microparticles: the Monte Carlo method, molecular dynamics simulation method, and Brownian dynamics simulation method. The Monte Carlo and Brownian dynamics methods are random simulation methods, while the molecular dynamics method is a deterministic method.

5.5.1.1 Molecular Dynamics Simulation (MDS)

Conception and Equations

Molecular dynamics simulation uses a computer to simulate the motion of atoms in a multipole system composed of a nucleus and electrons, and then calculates the structure and characteristics of the system. Each atom can be considered to move according to Newton's law by the potential force provided by other atoms. For an equilibrium system, molecular dynamics simulation can be used to calculate the

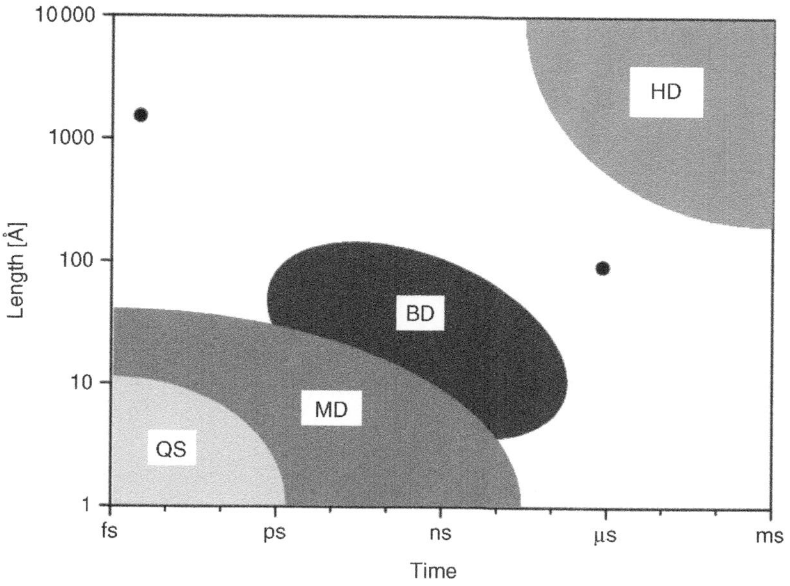

Figure 5.21 Time and size scales for different simulation methods

statistical average quantities of a physical variable by suitable time average. For a non-equilibrium system, the physical phenomena occurring during a molecular dynamics observation time can be simulated by MDS.

In MDS, a series of positions of atoms can be obtained by integrating the Newton motion equation. The differential equation of Newton's Second law is expressed by $\boldsymbol{a}_i = \frac{d^2\boldsymbol{r}_i}{dt^2} = \frac{\boldsymbol{F}_i + \boldsymbol{f}_i}{m_i}$, where $m_i, \boldsymbol{r}_i, \boldsymbol{F}_i, \boldsymbol{f}_i$ are the mass, the position vector, the potential force obtained from potential function U, $\boldsymbol{F}_i = -\nabla U_i$, and other forces, respectively.

Algorithms

To obtain the motion of atoms, many finite difference schemes can be used to solve the motion equations. There are several commonly used algorithms as follows:

1. Verlet algorithm

$$\boldsymbol{r}(t+\delta t) = 2\boldsymbol{r}(t) - \boldsymbol{r}(t-\delta t) + \delta t^2 \boldsymbol{a}(t) \tag{5.42}$$

2. Velocity–Verlet algorithm

$$\boldsymbol{r}(t+\delta t) = r(t) + \boldsymbol{v}(t)\delta t + \frac{1}{2}\delta t^2 \frac{\boldsymbol{f}(t)}{m} \tag{5.43}$$

$$\boldsymbol{v}(t+\delta t) = \boldsymbol{v}(t) + \frac{\boldsymbol{f}(t) + \boldsymbol{f}(t+\delta t)}{2m}\delta t \tag{5.44}$$

3. Leap-frog algorithm

$$\boldsymbol{r}(t+\delta t)=\boldsymbol{r}(t)+\boldsymbol{v}(t)+\boldsymbol{v}(t+\delta t/2)\delta t \tag{5.45}$$

$$\boldsymbol{v}(t+\delta t/2)=\boldsymbol{v}(t-\delta t/2)+\boldsymbol{a}(t)\delta t \tag{5.46}$$

Potential Function

There are many potential models for the interactions among molecules, such as the Lennard–Jones potential interaction.

$$\text{Lennard-Jones potential}\quad \phi_{ij}(r_{ij})=A_m\left(\frac{d}{r_{ij}}\right)^m-B_n\left(\frac{d}{r_{ij}}\right)^n \tag{5.47}$$

Ensemble of Equilibrium System

Equilibrium ensembles that are usually used are microcanonical, canonical, and isothermal–isobaric ensembles. A microcanonical ensemble is called NVE, and it is isolate and conservative. In the procedure of MDS, the atom number N, volume V, and energy E of the system are all kept unchanged. A canonical ensemble is also called NVT, in which the atom number N, volume V, and temperature T are all kept constant, and the total momentum is zero. An isothermal–isobaric ensemble is also called NPT, in which the atom number N, pressure P, and temperature T are all unchanged.

Initial Conditions of MDS

The initial configuration and velocities of the system can be determined by experimental data, a theoretical model, or a combination of both. If the simulated system has an initial density distribution of $n(\boldsymbol{r})$ and temperature distribution of $T(\boldsymbol{r})$, but has no fixed lattice structure, then the position of each atom can be obtained from $n(\boldsymbol{r})$ by using an acceptance–rejection method or metropolis. The initial velocity of each atom can be selected randomly from the Maxwell–Boltzmann distribution at the initial temperature distribution of $T(\boldsymbol{r})$. The Maxwell–Boltzmann distribution can be obtained from the random numbers evenly distributed in the region of 0~1 by using a simple transformation.

Periodic Boundary Conditions

A small amount of molecules can be used to study the macroscopic characteristics by implementing periodic boundary conditions [32]. The cubic box is replicated throughout space to form an infinite lattice. In the course of simulation, as a molecule moves in the original box, its periodic image in each of the neighboring boxes moves in exactly the same way. Thus, as a molecule leaves the central box, one of its images will enter through the opposite surface. Therefore, the number density in the central box (and hence in the entire system) is conserved.

Molecular Dynamics Simulation of Surfactant Micelle Formation [33,34]

When the surfactant concentration is higher than CMC, the monomers can aggregate to form micelles. In MDS, it is usually considered that a certain number of water and surfactant molecules interact with each other and reach an equilibrium state finally. In general, the number of surfactant monomers is larger than the minimum molecule number required for the formation of a micelle. For example, at least 60 sodium dodecyl sulfate (SDS) molecules are required for the formation of an SDS micelle.

In the simulation, atoms or radicals with a certain function can be regarded as basic particles, such as CH_x radicals ($-CH_3$ or $-CH_2-$) in a hydrophobic tail group, the central atom in a hydrophilic head group, or the atoms or radicals connected with the central atom. The position relationship among all particles in the simplified molecule is the same as that in a real molecule. The simulation time step is usually 2∼5 fs, and the simulation time is usually 1∼10 ns. NPT or NVT is selected for the ensemble. The particle mesh Ewald technique is employed.

Micelle Shape

The simulation shows that the micelle is not a regular sphere, but is like an ellipse. A spherical micelle can be formed at a lower number of monomers, and rod-like micelles can be formed as the number of surfactant monomers increases up to a certain value. This agrees very well with the experimental finding.

Micelle Size

Micelle size is one index to evaluate the micellar structure. The effective micellar radius can be defined as the distance between the central atom in the head group (such as the S atom in SDS) and the center of mass of the micelle, plus the distance between the central atom in the head group and the first maximum value of the radial distribution function of an oxygen atom in a water molecule, minus the radius of the water molecule. The value of the effective micellar radius of SDS is 2.20 nm.

Through MDS, one can testify the existence of spherical or rod-like micelles at surfactant concentrations higher than CMC, and can also calculate the structure, shape, and size of micelles. Some data which cannot be obtained or cannot be obtained easily from experiments can be obtained by MDS, and they can provide a basis for larger scale simulation such as the Brownian dynamics of micelles.

5.5.1.2 Brownian Dynamics Simulation (BDS)

Conception and Equations

The basic physical model of BDS is coarse-grained model, and the force balance equation is the Langevin equation, which is based on Newton's Second Law. In the equation, all forces acting on the microparticles (micelles much larger than solvent molecules in the surfactant solution) are considered, and the motions of the microparticles are analyzed. Generally, there are at least three forces acting on the particles: (1) a friction force between the particle and solvent; (2) a Brownian force, that is, a random force; and (3) a potential force among particles.

For a single particle in a one-dimensional case, Newton's Second Law can be expressed as

$$m\frac{d^2x}{dt^2} = \Sigma F \tag{5.48}$$

Potential force, friction force of the solvent, and random force arising from the surrounding solvent molecules are applied on the single particle, so we have

$$m\frac{d^2x}{dt^2} = \frac{\partial U}{\partial x} - \zeta\frac{dx}{dt} + R(t) \tag{5.49}$$

where U is the potential; ζ is the friction coefficient, for spherical particle, $\zeta = \frac{6\pi\eta a}{m}$; $R(t)$ is random force; and $<R(t)>=0$, $<R(t)\ R(t')>=2\xi k_B\delta(t-t')$.

The generalized Langevin equation of multiple particles in a three-dimensional Cartesian coordinate system is

$$m_i\frac{d\boldsymbol{v}_i}{dt} = -\frac{\partial V^{mf}}{\partial \boldsymbol{r}_i} - m_i\sum_j\int_0^t \gamma_{ij}(\tau)\boldsymbol{v}_j(t-\tau)d\tau + \boldsymbol{R}_i(t) \tag{5.50}$$

where $\langle R_i(0)R_j(t)\rangle = k_BT\zeta_{ji}(t)$

The macroscopic rheological phenomena are usually studied in pure shear flow or pure (uniaxial) elongational flow. To extend the study range and reduce computational task, periodic boundary conditions are usually used.

Surfactant Micelle Model

In the Brownian dynamics simulation of surfactant solution, a certain number of micelles (O (100)) are required. The model of micelles determines the scale and domain size of the simulation. The micelle can be assumed to be made up of a series of beads which connect linearly, and the bead diameter is equal to the micelle diameter. There are attractive and repulsive forces between different beads to ensure that the beads in a micelle are always tangent. This kind of bead–chain model is helpful to form micelles with different lengths and is easy to cope with regarding the collision and combination of beads, but it is not so easy to implement the model, and much more computer source and time will be consumed.

The contour lengths of a micelle in the actual solution are not the same, and break-ups and combinations of micelles exist. In the thread-like structure, there exists a persistence length which is rigid and cannot break up. So we can select the persistence length as the minimum length of micelles for investigation. In the dilute surfactant solution, many micelles are in a dispersed state, so the model is especially suitable for dilute solutions. The micelles with a persistence length can be considered as a linear arrangement of a model bead or a rigid rod [35,36]. If the basic unit is a bead or point particle, the rotation of the basic unit can be neglected; if it is a rod, the rotation of the basic unit should be considered [36].

5.5.2 *Brownian Dynamics Simulation by Using a WK Potential*

To simulate the network structures in the surfactant solution successfully, Wei *et al.* [36] employed a combined potential model to describe the interactions among rod-like micelles. They consider the surfactant solution as a system of N rigid rod-like micelles suspended in an incompressible continuum fluid medium of viscosity of η_s. The rod-like micelles interact through both interrod forces and hydrodynamic forces mediated via the continuum fluid. The micelles also receive fluctuating Brownian forces arising from the apparently random thermal bombardment by surrounding solvent molecules.

The rigid osculating multibead rod model is employed for the modeling of surfactant solutions of rod-like micelles. The micelle is assumed to be made up of n evenly spaced beads of diameter σ, which are linearly connected as shown in Figure 5.22. It represents a cylindrical rod of length $n\sigma$, diameter σ, and an aspect ratio of n. Different from a sphere case, the rod is multiscale and the BD simulation of it is a challenge. The positions of the beads are given in terms of their distances from the center of mass along the rod axis. The rods interact via the bead–bead potential interaction. Thus the interaction energy between two rigid linear rods is a sum of pairwise contributions from distinct bead i in rod k and bead j in rod l. Figure 5.22 shows the interrod interaction. As mentioned above, the micellar network structure is considered as the connection of rod-like micelles at the micelle ends. In order to simulate the formation of the network structure, for the end–end beads potential between two rods, the

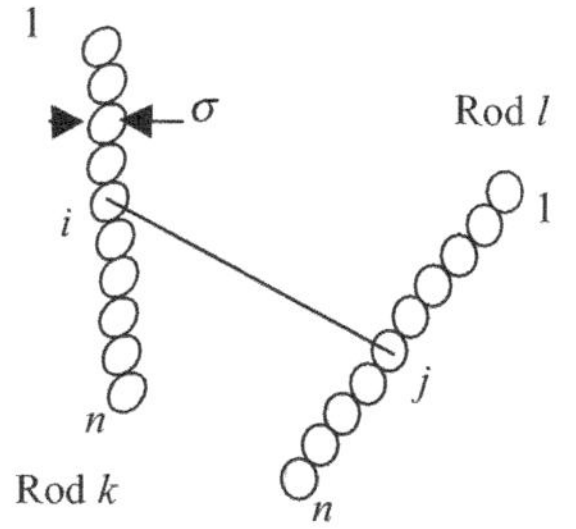

Figure 5.22 Schematic of rod model made up of n beads of diameter σ (With kind permission from Springer Science + Business Media: *Nonlinear Dynamics*, "Microstructures and rheology of micellar surfactant solution by Brownian dynamics simulation," **61**, © 2010, 503–515, Wei, J.J., Kawaguchi, Y., Yu, B., Li, F.-C., Zhang, C.W.)

Leonard–Jones potential is employed:

$$U_{klij}(r_{klij}) = 4\varepsilon[(\sigma/r_{klij})^{12} - (\sigma/r_{klij})^{6}] \tag{5.51}$$

where $\vec{r}_{klij}$ is the vector pointing from the center of bead i in rod k to bead j in rod l, and ε represents the strength of the interaction.

To prevent overlap among micelles, for interior–interior beads potential between two rods, a repulsive soft-sphere potential is assumed:

$$U_{klij}(r_{klij}) = 4\varepsilon(\sigma/r_{klij})^{6} \tag{5.52}$$

We call this kind of combined interrod potentials between micelles the "WK" (Wei–Kawaguchi) potential. This is different from the conventional BD simulation in which only one potential was used for interrod interaction. Here ε is the depth of potential. The selection of ε is a problem. If it is too large, the micellar structure will become very rigid; and if it is too small, the network structure cannot be formed. Both of the two cases are not consistent with the TEM image and rheological measurements. Here we selected $\varepsilon = 10\,KT$ for the simulations, where K is the Boltzmann constant.

The equations of the translational and rotational motions of a rod read

$$m_k \frac{d^2\vec{r}_{kc}}{dt^2} = \sum_{i=1}^{n}\left(\vec{F}_{ki}^{h} + \vec{F}_{ki}^{r} + \sum_{l=1;\, l\neq k}^{N}\sum_{j=1}^{n}\vec{F}_{klij}^{p}\right) \tag{5.53}$$

$$[\mathbf{I}]_k \frac{d^2\theta_k}{dt^2} = \sum_{i=1}^{n}\left[(\vec{r}_{ki} - \vec{r}_{kc}) \times \left(\vec{F}_{ki}^{h} + \vec{F}_{ki}^{r} + \sum_{l=1,\, l\neq k}^{N}\sum_{j=1}^{n}\vec{F}_{klij}^{p}\right)\right] \tag{5.54}$$

where m_k and $[\mathbf{I}]_k$ are, respectively, the mass and momentum of inertia of rod k; and

$$[\mathbf{I}]_k = \sum_{i} m_{ki}[(r_{ki} - r_{kc})^2[\mathbf{1}] - (\vec{r}_{ki} - \vec{r}_{kc})(\vec{r}_{ki} - \vec{r}_{kc})] \tag{5.55}$$

where $\vec{r}_{kc}$ and θ_k are, respectively, the position of center of mass and rotation angle of rod k. The definition of orientation angles of the rigid rod-like micelle in a 3D Cartesian coordinate system is shown in Figure 5.23. Here, $\vec{r}_{ki}$ is the position of bead i in rod k, and $[\mathbf{1}]$ is the unit tensor.

$\vec{F}_{ki}^{h}$ is the hydrodynamic force on bead i in rod k and is assumed to be proportional to the relative velocity of the sphere with respect to the macroscopic fluid flow.

$$\vec{F}_{ki}^{h} = -\xi(\vec{v}_{ki} - [\kappa]\vec{r}_{ki}) \tag{5.56}$$

where ξ is the Stokes friction constant, $\xi = 3\pi\eta_s\sigma$; and $[\kappa]$ is the velocity gradient tensor of the macroscopic fluid flow.

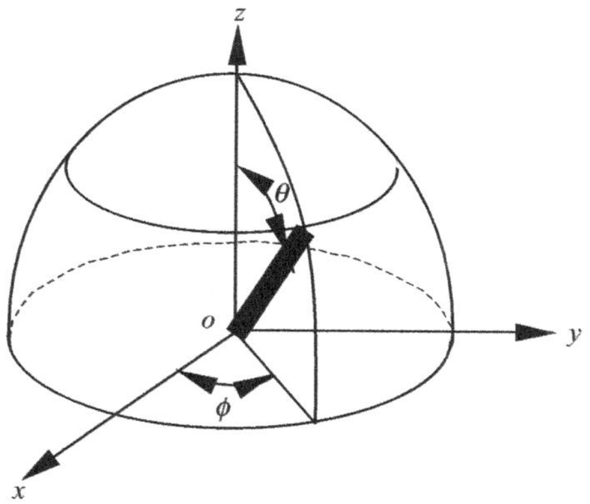

Figure 5.23 Orientation angles for a rigid rod-like micelle in a Cartesian coordinate system (With kind permission from Springer Science+Business Media: *Nonlinear Dynamics*, "Microstructures and rheology of micellar surfactant solution by Brownian dynamics simulation," **61**, © 2010, 503–515, Wei, J.J., Kawaguchi, Y., Yu, B., Li, F.-C., Zhang, C.W.)

$\vec{F}^{r}_{ki}$ is the random Brownian force on bead i in rod k due to the thermal motion of the solvent molecules. The short time steps needed to handle the fast motion of the solvent molecules, and the very long runs needed to allow the evolution of the slower mode of the rod-like micelles, make the simulation very time consuming. Here, we are not concerned with the motion of the solvent molecules, and the random force caused by the solvent molecules is assumed to be white noise, that is, it has a correlation time shorter than any process of interest. The amplitude of the random force is then given by the fluctuation–dissipation theorem:

$$< \vec{F}^{r}_{ki}(t)\vec{F}^{r}_{ki}(t') >= 2kT\xi[\mathbf{1}]\delta(t-t') \tag{5.57}$$

Here $< >$ denotes an average over the probability space on which $\vec{F}^{r}_{ki}$ is defined, and $\delta(t-t')$ is the delta function.

$\vec{F}^{p}_{klij}$ is the potential force on bead i in rod k arising from bead j in rod l.

$$\vec{F}^{p}_{klij} = \frac{dU_{klij}}{d\vec{r}_{ki}} = \frac{dU_{klij}}{dr_{klij}}\vec{n}_{klij} \tag{5.58}$$

where $\vec{n}_{klij}$ is the unit vector pointing from the center of bead i in rod k to bead j in rod l.

$$\vec{n}_{klij} = \frac{\vec{r}_{klij}}{r_{klij}} \tag{5.59}$$

The additive shear stress $\Delta\sigma_{\alpha\beta}$ due to the existence of rod-like micelles may be given by

$$\Delta\sigma_{\alpha\beta} = \frac{1}{2V}\left[\sum_{k=1}^{N}\left(\sum_{i=1}^{n}(F^h_{ki\alpha}(r_{ki\beta} - r_{kc\beta})) + \sum_{l=k+1}^{N}(F^p_{kl\alpha}r_{kl\beta})\right)\right] \tag{5.60}$$

where F^p_{kl} is the force on the center of mass of rod k arising from interactions with rod l, $\vec{r}_{kl}$ is the vector pointing from the center of mass of rod k to that of rod l, V is the volume of the computational cell, and α and β are the Cartesian components.

In this chapter, dimensionless quantities are used where the length (σ), energy (kT), and friction factor ξ of beads are set to unity. It follows that time is reduced by $\xi\sigma^2/kT$ and shear rate or elongation rate $kT/\xi\sigma^2$. The dimensionless shear rate or elongation rate is called the bead Peclet number, $Pe = \dot{\gamma}\xi\sigma^2/kT$ or $Pe = \dot{\varepsilon}\xi\sigma^2/kT$, which is the ratio of the time for a bead to freely diffuse a distance σ to the flow scale $1/\dot{\gamma}$ or $1/\dot{\varepsilon}$. At large Pe, the inverse shear rate or elongation rate is less than $\xi\sigma^2/kT$ and the structure rearrangement due to the shear or elongation will dominate. Conversely, at small Pe, the shear- or elongation-induced structure will be a perturbation of the Brownian structure.

For the steady shear flow, the Lee–Edwards sliding periodic image boundary conditions were used to maintain shear velocity continuity at the simulation cell boundaries. By periodic boundary conditions, we represent an infinite suspension as a spatially periodic array of identical cubic cells with a cell length of L. The replicas of the central cell move with the shear flow to ensure continuity in the fluid velocity across the boundaries of the unit cell. A bead leaves the central cell through a side $x = \pm L/2$, the nre-enters through the opposite side. A bead which leaves through an edge $y = \pm L/2$ or $z = \pm L/2$ will re-appear on the opposite edge but with a displacement in the x-direction according to the instantaneous position of the surrounding replicas.

In the simulation, the beads number in a rod was 10. The cutoff distance in the potential interaction was 5σ, above which the potential was set to zero. Three different micelle volume concentrations, 0.126%, 0.189%, and 0.453%, were simulated. They are very dilute surfactant solutions. The computational domain is a 3D cubic cell with a side length of 50σ.

At the start of a simulation, the rods were placed in the primary cell in an α-f.c.c. (face-centered-cubic) structure, which has been the starting configuration for many simulations. The ensemble of rods is allowed to equilibrate in the absence of shear flow for 10^7–10^8 time steps to make the interested quantities cease to show a systematic drift and start to oscillate about steady mean values. The shear flow field was then turned on at $t = 0$, and the statistics of the rod trajectories were taken at regular intervals. The nondimensional time step size is 10^{-4} in the simulation.

Figure 5.24 shows the micellar structures at the Peclet number (dimensionless shear rate) Pe of 0.01, 0.1, 1.0, and 10.0. It can be seen that the micellar network structure is formed at a low Pe of 0.01 and is destroyed completely at a high Pe of 10.0

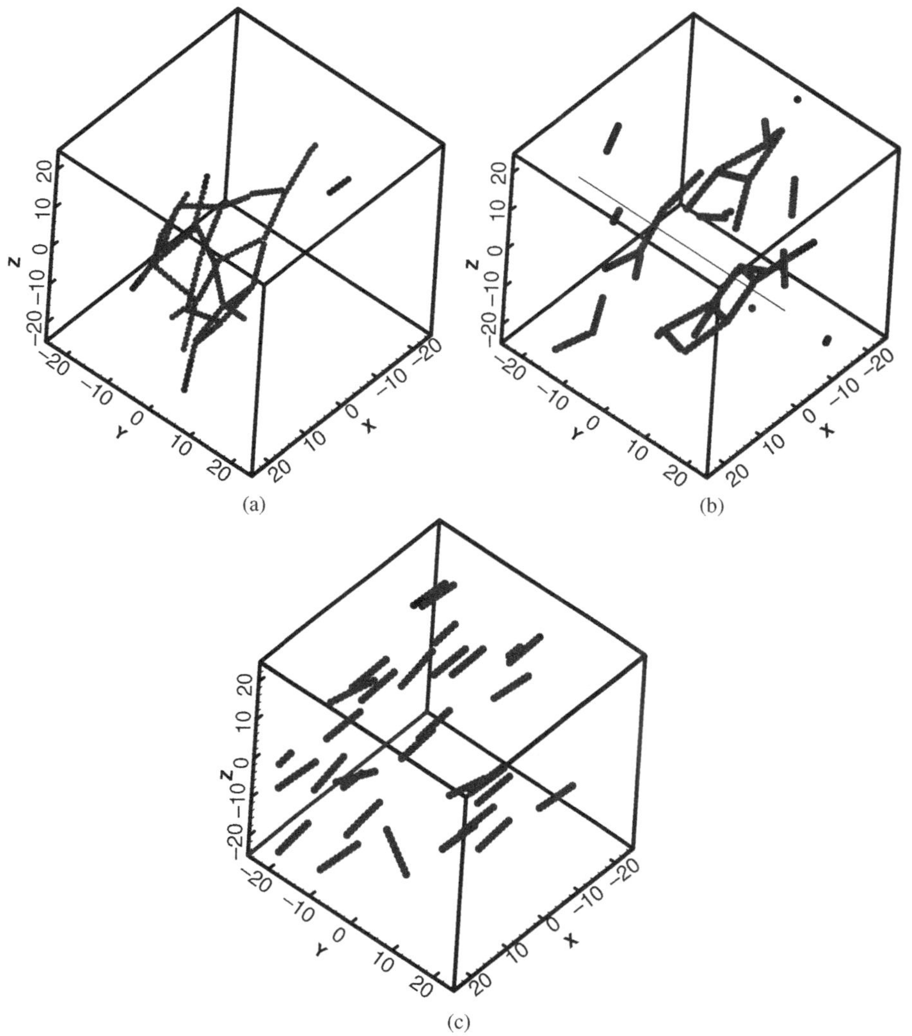

Figure 5.24 Micellar structures at different Pe. (a) Pe = 0.01; (b) Pe = 1.0; and (c) Pe = 10.0 (With kind permission from Springer Science+Business Media: *Nonlinear Dynamics*, "Microstructures and rheology of micellar surfactant solution by Brownian dynamics simulation," **61**, © 2010, 503–515, Wei, J.J., Kawaguchi, Y., Yu, B., Li, F.-C., Zhang, C.W.)

by shear stress. The rod-like micelles become more and more parallel to the shear flow direction with increasing Pe. This leads to a non-uniform orientation angular distribution of micelle directions, in other words, a preferred alignment angle.

Figure 5.25 shows the calculated dimensionless shear viscosity $(\eta - \eta_s)/(\eta_{0n} - \eta_s)$ versus dimensionless shear rate $n(n^2 - 1)/72$ Pe with surfactant volume fraction α_v as a parameter. η_{0n} is the zero shear viscosity in the absence of potential. It should be

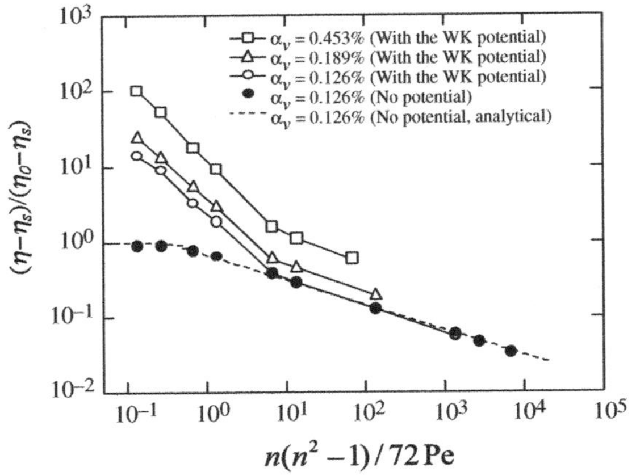

Figure 5.25 Shear viscosity (With kind permission from Springer Science+Business Media: *Nonlinear Dynamics*, "Microstructures and rheology of micellar surfactant solution by Brownian dynamics simulation," **61**, © 2010, 503–515, Wei, J.J., Kawaguchi, Y., Yu, B., Li, F.-C., Zhang, C.W.)

noted that it is difficult to probe the rheology at very small Peclet numbers due to the deteriorating signal response-to-noise ratio and the relatively slow re-orientational time scale of the rod. The difficulty in probing rheology at low strain rates has also been found in the NEMD simulations of molecular fluids and BD simulations of molecular colloid liquids. It can be seen that $(\eta - \eta_s)/(\eta_{0n} - \eta_s)$ increases with increasing surfactant concentration and decreases with increasing shear rate, showing shear-thinning characteristics. The shear-thinning characteristics are related to the shear aligning of rods shown in Figure 5.24. The analytical [37] and numerical results for the case of $\alpha_v = 0.126\%$ in the absence of potential are also shown. At shear rates less than 10 for the case of $\alpha_v = 0.126\%$, $(\eta - \eta_s)/(\eta_{0n} - \eta_s)$ shows a significant increase in the presence of WK potential interaction between rod-like micelles compared to in the absence of potential interaction. We can also see that there exists a critical shear rate above which the shear viscosity curves become parallel to the analytical curve for no potential case. This indicates that the micellar network structure caused by the WK potential interaction is completely destroyed and the contribution of WK potential to the rheology disappears. The drop from the viscosity curve to the parallel dotted line is the increase of viscosity due to the WK potential interaction.

Figure 5.26 shows the calculated dimensionless first normal stress coefficient ψ_1/ψ_{10n} versus the dimensionless shear rate $n(n^2 - 1)/72$ Pe with surfactant volume fraction α_v as a parameter. $\psi_{1,0n}$ is the first normal stress coefficient in the absence of potential. The first normal stress coefficient shows a shear-thinning characteristic and increases with increasing surfactant concentration. The analytical [37] and numerical results for the case of $\alpha_v = 0.126\%$ in the absence of potential are also shown. At shear

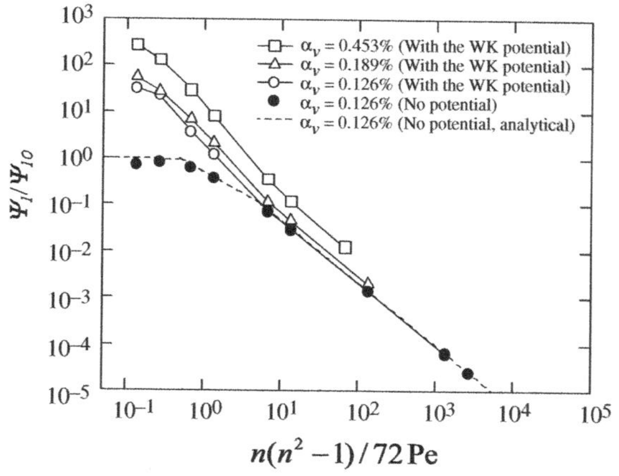

Figure 5.26 First normal stress coefficient (With kind permission from Springer Science +Business Media: *Nonlinear Dynamics*, "Microstructures and rheology of micellar surfactant solution by Brownian dynamics simulation," **61**, © 2010, 503–515, Wei, J.J., Kawaguchi, Y., Yu, B., Li, F.-C., Zhang, C.W.)

rates less than 10 for the case of $\alpha_v = 0.126\%$, ψ_1/ψ_{10n} with the WK potential interaction between rod-like micelles is much higher than without potential interaction.

For the steady elongational flow, periodic boundary conditions were also used. The fluid should extend along one dimension (x here) and contracts in the remaining orthogonal directions so as to maintain a constant density. In the most general treatment of this problem, one must follow the time evolution of a fluid element which in simulations is represented by a cubic box containing the micelles. All points on the boundaries of the box change at every time step according to the following equations during the simulation period.

$$\begin{aligned} r_x(t) &= r_x(t=0)\exp(\dot{\varepsilon}t) \\ r_y(t) &= r_y(t=0)\exp(-0.5\dot{\varepsilon}t) \\ r_z(t) &= r_z(t=0)\exp(-0.5\dot{\varepsilon}t) \end{aligned} \tag{5.61}$$

These equations are obtained by integrating $v_x = \dot{\varepsilon}r_x, v_y = -0.5\dot{\varepsilon}r_y, v_z = -0.5\dot{\varepsilon}r_z$. They ensure that the evolution of cell boundaries is compatible with the particle dynamics and that the system volume remains a constant of the motion.

Figure 5.27 shows the micellar structures evolution at Pe = 0.01. It can be seen that when steady elongational flow starts, the micellar network structure begins to elongate and orient itself along the axis of elongation. As the network structure orients itself, the rod-like micelles align with the elongation axis. The compressive nature of the flow in the directions transverse to the elongation direction appears to play an important role in the alignment process. This leads to a non-uniform orientation angular distribution of micelle directions, in other words, a preferred alignment angle.

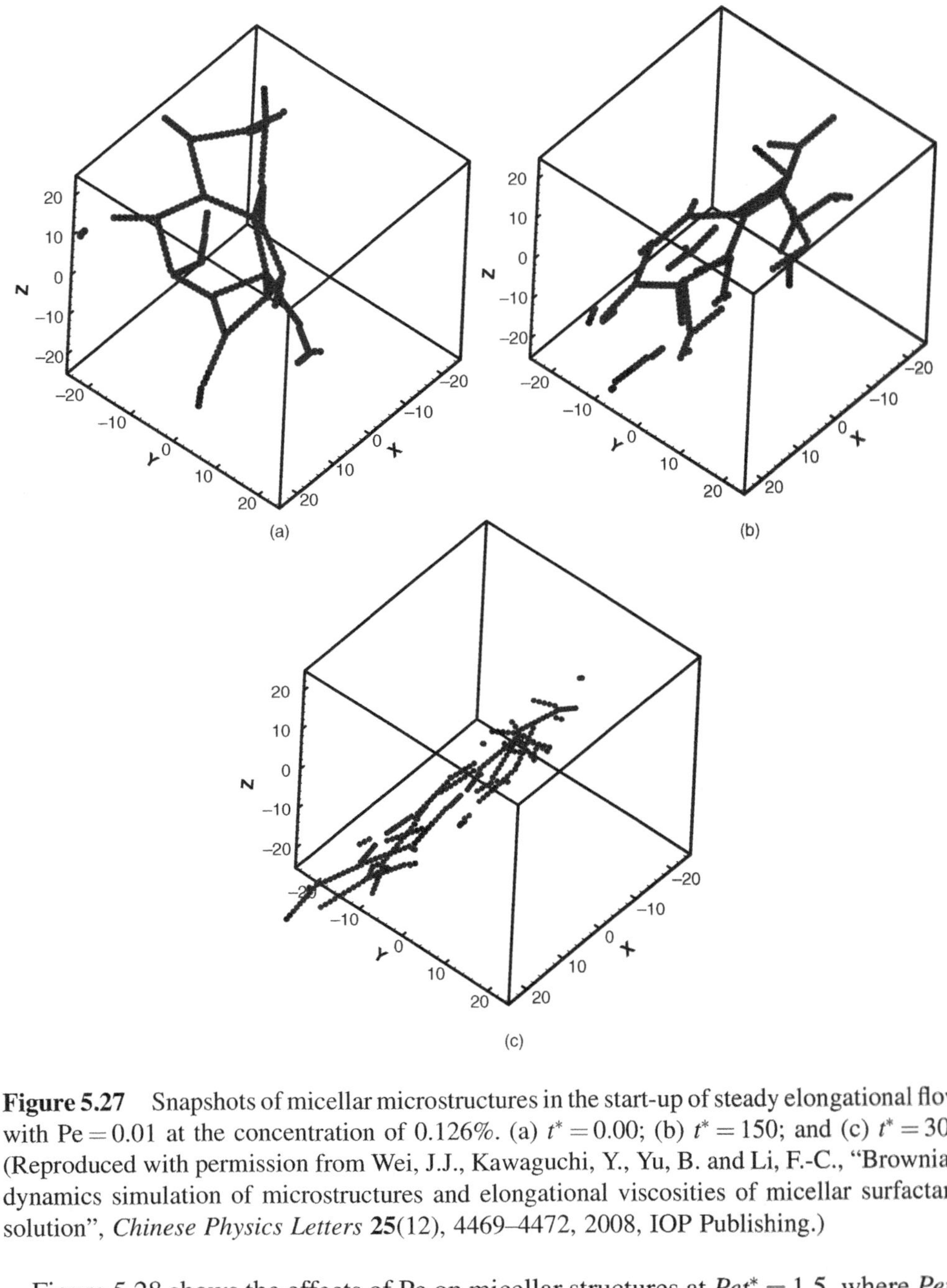

Figure 5.27 Snapshots of micellar microstructures in the start-up of steady elongational flow with Pe = 0.01 at the concentration of 0.126%. (a) $t^* = 0.00$; (b) $t^* = 150$; and (c) $t^* = 300$ (Reproduced with permission from Wei, J.J., Kawaguchi, Y., Yu, B. and Li, F.-C., "Brownian dynamics simulation of microstructures and elongational viscosities of micellar surfactant solution", *Chinese Physics Letters* **25**(12), 4469–4472, 2008, IOP Publishing.)

Figure 5.28 shows the effects of Pe on micellar structures at $Pet^* = 1.5$, where Pet^* is a dimensionless elongation strain. We can see that when the elongation rate exceeds a critical value, the micellar network structure starts to be destroyed and the rod-like micelles become more and more parallel to the elongation flow direction with increasing elongation rate.

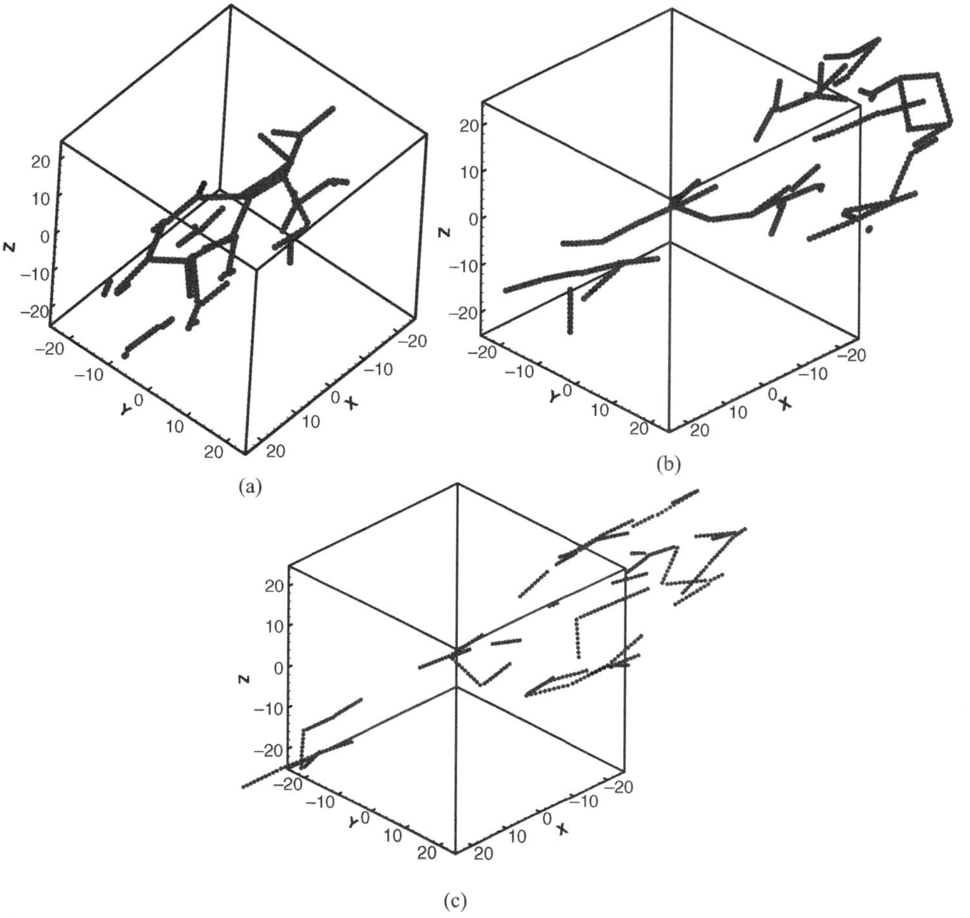

Figure 5.28 Snapshots of micellar microstructures in the start-up of steady elongational flow with Pet^* = 1.5 at the concentration of 0.126%. (a) Pe = 0.01; (b) Pe = 0.1; and (c) Pe = 1.0 (Reproduced with permission from Wei, J.J., Kawaguchi, Y., Yu, B. and Li, F.-C., "Brownian dynamics simulation of microstructures and elongational viscosities of micellar surfactant solution", *Chinese Physics Letters* **25**(12), 4469–4472, 2008, IOP Publishing.)

Figure 5.29 shows the calculated dimensionless elongational viscosity $(\eta_E - 3\eta_s)/3(\eta_{0n} - \eta_s)$ versus dimensionless elongation rate $n(n^2 - 1)/72\,\mathrm{Pe}$ with surfactant volume fraction α_v as a parameter. η_{0n} is the zero shear viscosity in the absence of potential. The analytical [37] and numerical results for the case of 0.126% in the absence of potential are also shown. It can be clearly seen that there exists a transition from a low-elongation viscosity plateau ($(\eta_E - 3\eta_s)/3(\eta_{0n} - \eta_s) = 1$) at low elongation rates to a high one ($(\eta_E - 3\eta_s)/3(\eta_{0n} - \eta_s) = 2$) at high elongation rates for no potential case. For the WK potential case, $(\eta_E - 3\eta_s)/3(\eta_{0n} - \eta_s)$ decreases with

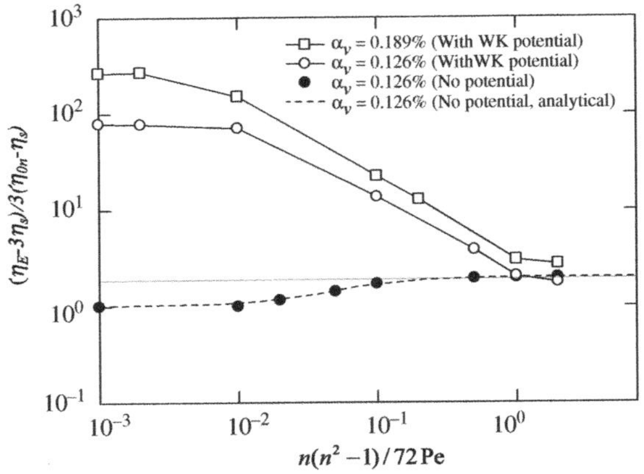

Figure 5.29 Elongational viscosities at different surfactant concentrations (Reproduced with permission from Wei, J.J., Kawaguchi, Y., Yu, B. and Li, F.-C., "Brownian dynamics simulation of microstructures and elongational viscosities of micellar surfactant solution", *Chinese Physics Letters* **25**(12), 4469–4472, 2008, IOP Publishing.)

elongation rates, leveling out at high elongation rates. The strain-thickening characteristics in the absence of potential are due to the increasing alignment of rods with elongation. The elongation viscosity may be evaluated using a viscous dissipation argument. The rate of viscous dissipation due to the presence of the rods increases as the alignment of rods increases, and takes a maximum value of $0.5N\xi(n-1)^2\sigma^2\dot{\varepsilon}^2/V$ when the rods completely align with flow direction. Here V is the volume of the computational box. The strain-thinning behavior in the presence of potential is related to the increasing destruction of the micellar network structure. The rods in solution interact with each other more during weak elongation than during strong elongation, causing the elongational viscosities to be higher at low elongation rates and to drop as the rods become increasingly aligned in the flow direction at higher elongation rates. This strain-thinning phenomenon has also been found by Cathey and Fuller [38] in their uni-axial and bi-axial elongational viscosity measurements of semidilute solutions of rigid rod polymers. At elongation rates less than 0.1 for the case of 0.126%, $(\eta_E - 3\eta_s)/3(\eta_{0n} - \eta_s)$ shows a significant increase in the presence of WK potential interaction between rod-like micelles compared to in the absence of potential interaction. We can also see that there exists a critical elongation rate above which the elongational viscosity curves become parallel to the analytical curve for no potential case. This indicates that the micellar network structure due to the WK potential is completely destroyed and the contribution of WK potential to the rheology disappears. The drop from the viscosity curve to the parallel dotted line is the increase of elongational viscosity due to the WK potential interaction. The elongational viscosities increase with increasing surfactant concentration. The increase of elongational viscosities by the formation of micellar network structure in the surfactant solution can suppress the vortex stretching, thus resulting in the reduction of turbulence energy production and friction drag in turbulent flow.

References

1. Lu, B. (1997) Characterization of drag reducing surfactant systems by rheology and flow birefringence measurements, Ph.D dissertation. The Ohio State University.
2. Santos, F.K.G., Neto, E.L.B., Moura, M.C.P.A. *et al.* (2009) Molecular behavior of ionic and nonionic surfactants in saline medium. *Colloids and Surfaces A*, **333**, 156–162.
3. Rosen, M.J. (1989) *Surfactant and Interfacial Phenomena*, 2nd edn, John Wiley & Sons, New York.
4. Lin, Z., Cai, J.J., Scriven, L.E. and Davis, H.T. (1994) Spherical-to-wormlike micelle transition in CTAB solutions. *Journal of Physical Chemistry*, **98**, 5984–5993.
5. Zhang, Y. (2005) Correlations among surfactant drag reduction, additive chemical structures, rheological properties and microstructures in water and water/co-solvent systems, Ph.D. dissertation, The Ohio State University.
6. Geer, R.D., Eylar, E.H. and Anacker, E.W. (1971) Dependence of micelle aggregation number on polar head structure. I: light scattering by aqueous solutions of decylammonium salts and related surfactants. *Journal of Physical Chemistry*, **75**, 369–374.
7. Zhang, Y., Qi, Y. and Zakin, J.L. (2005) Headgroup effect on drag reduction and rheological properties of micellar solutions of quaternary ammonium surfactants. *Rheologica Acta*, **45**, 42–58.
8. Rehage, H. and Hoffmann, H. (1991) Viscoelastic surfactant solutions: model for rheological research. *Molecular Physics*, **74**, 933–973.
9. Tanford, C. (1972) Micelle shape and size. *Journal of Physical Chemistry*, **76**, 3020–3024.
10. Elson, T.P. and Garside, J. (1983) Drag reduction in aqueous cationic soap solutions. *Journal of Non-Newtonian Fluid Mechanics*, **12**, 121–133.
11. Hoffmann, H., Rehage, H. and Reizlein, K. (1985) Viscoelastic detergent solutions from red-line micelles. Proceedings of the ACS Symposium on Macro- and Microemulsions, Washington DC (ed. D.O. Shaw,), pp. 41–66.
12. Hoffmann, H., Rehage, H., Schorr, W. *et al.* (1984) Viscoelastic detergent solutions, in *Surfactants in Solution*, vol. **1**, Plenum Press.
13. Zheng, Y., Lin, Z., Zakin, J.L. *et al.* (2000) Cryo-TEM imaging the flow-induced transition from vesicles to threadlike micelles. *Journal of Physical Chemistry B*, **104**, 5263–5271.
14. Lu, B., Zheng, Y., Davis, H.T. *et al.* (1998) Effect of variations in counterion to surfactant ratio on rheology and microstructures of drag reducing cationic surfactant systems. *Rheologica Acta*, **37**, 528–548.
15. Varade, D., Rodrguez-Abreu, C., Shrestha, L.K. *et al.* (2007) Wormlike micelles in mixed surfactant systems: effect of cosolvents. *Journal of Physical Chemistry B*, **111**, 10438–10447.
16. Schubert, B.A., Wagner, N.J. and Kaler, E.W. (2003) The microstructure and rheology of mixed cationic/anionic wormlike micelles. *Langmuir*, **40**, 4079–4089.
17. Qi, Y.Y. and Zakin, J.L. (2002) Chemical and rheological characterization of drag-reducing cationic surfactant systems. *Industrial and Engineering Chemistry Research*, **41**, 6326–6336.
18. Doi, M. and Edwards, S.F. (1986) *The Theory of Polymer Dynamics*, Oxford Science.
19. Fuller, G.G., Cathey, C.A., Hubbard, B. *et al.* (1987) Experiment viscosity measurements for low-viscosity fluids. *Journal of Rheology*, **31**, 235–249.
20. Dontula, P., Pasquali, M., Scriven, L.E. *et al.* (1997) Can extensional viscosity be measured with opposed nozzle devices? *Rheologica Acta*, **36**, 429–448.

21. Wei, J.J., Kawaguchi, Y., Yu, B. *et al.* (2006) Rheological characteristics and turbulent friction drag and heat transfer reduction of a very dilute cationic surfactant solution. *ASME Journal of Heat Transfer*, **128**, 977–983.
22. Giesekus, H. (1982) A simple constitutive equation for polymer fluids based on the concept ofdeformation dependent tessorial mobility. *Journal of Non-Newtonian Fluid Mchanics*, **11**, 69–109.
23. Yu, B. and Kawaguchi, Y. (2002) Effect of Weissenberg number on the flow structure: DNS study of the drag-reducing Giesekus fluid flow with MINMOD scheme. The 5th International Symposium on Engineering Turbulence Modelling and Experiments, Mallorca, Spain, pp. 617–626.
24. Yu, B., Kawaguchi, Y., Takagi, S. *et al.* (2002) Numerical investigation on turbulent structures in a drag reducing flow with surfactant additives in a 2D channel comparison of artificial diffusion scheme and MINMOD scheme. The 5th JSME-KSME Fluids Engineering Conference, Nagoya, Japan.
25. Lin, Z.Q., Lu, B., Zakin, J.L. *et al.* (2001) Influence of surfactant concentration and counterion to surfactant ratio on rheology of wormlike micelles. *Journal of Colloid and Interface Science*, **239**, 543–554.
26. Ketner, A.M., Kumar, R., Davies, T.S. *et al.* (2007) A simple class of photorheological fluids: surfactant solutions with viscosity tunable by light. *Journal of American Chemistry Society*, **129**, 1553–1559.
27. Ge, W., Kesselman, E., Talmon, Y. *et al.* (2008) Effects of chemical structures of para-halobenzoates on micelle nanastructure, drag reduction and rheological behaviors of dilute CTAC solutions. *Journal of Non-Newtonian Fluid Mechanics*, **154**, 1–12.
28. Siginer, A. (1984) General Weissenberg effect in free surface rheometry, part 1: analytical consideration. *Journal of Applied Mathematics and Physics*, **35**, 545–558.
29. Wei, J.J., Li, F.C., Yu, B. and Kawaguchi, Y. (2006) Swirling flow of a viscoelastic fluid with free surface, part I: experimental analysis of vortex motion by PIV. *ASME Journal of Fluids Engineering*, **128** (1), 69–76.
30. Kawaguchi, Y., Wei, J.J., Yu, B. *et al.* (2003) Rheological characterization of drag-reducing cationic surfactant solution: shear and elongational viscosities of dilute solution. Proceeding of the 4th ASME/JSME Joint Fluids Engineering Conference, Hawaii, July.
31. Spohn, A., Mory, M. and Hopfinger, E.J. (1993) Observation of vortex breakdown in an open cylindrical container with a rotating bottom. *Experiments in Fluids*, **14**, 70–77.
32. Allen, M.P. and Tildsley, D.J. (1987) *Computer Simulation of Liquids*, Oxford Science, Oxford.
33. Bruce, C.D., Berkowitz, M.L., Perera, L. *et al.* (2002) Molecular dynamics simulation of sodium dodecyl sulfate micelle in water: micellar structural characteristic and counterion distribution. *Journal of Physical Chemistry B*, **106**, 3788–3793.
34. Gao, J., Ge, W., Hu, G.H. and Li, J.H. (2005) From homogeneous dispersion to micelles: a molecular dynamics simulation on the compromise of the hydrophilic and hydrophobic effects of sodium dodecyl sulfate in aqueous solution. *Langmuir*, **21**, 5223–5229.
35. Padding, J.T., Boek, E.S. and Briels, W.J. (2008) Dynamics and rheology of wormlike micelles emerging from particulate computer simulations. *Journal of Chemical Physics*, **129**, 074903.
36. Wei, J.J., Kawaguchi, Y., Yu, B. *et al.* (2008) Brownian dynamics simulation of microstructures and elongational viscosities of micellar surfactant solution. *Chinese Physics Letters*, **25**, 4469–4472.
37. Bird, R.B., Curtiss, C.F., Armstrong, R.C. *et al.* (1987) Dynamics of polymeric liquids, in *Kinetic Theory*, vol. **2**, John Wiley & Sons.
38. Cathey, C.A. and Fuller, G.G. (1988) Uniaxial and biaxial extensional viscosity measurements of dilute and semi-dilute solutions of rigid rod polymers. *Journal of Non-Newtonian Fluid Mechanics*, **30**, 303–316.

6

Application Techniques for Drag Reduction by Surfactant Additives

6.1 Problems That Need to Be Solved in Engineering Applications

Surfactant drag reducers, which differ from the high polymer applied in long-distance petroleum transportation pipelines (once-through flow passage), are more appropriate for use in liquid circulation systems with a pump, due to the self-repairing ability of the microstructures responding to the turbulent drag-reducing effect in surfactant solutions. On the basis of successful applicable experiences, most often considered for surfactant drag reducers is their great potential application in district heating and cooling systems. When considering the real application of surfactant drag reducers in district heating and cooling systems, it is necessary to solve any relative problems at first, such as the compatibility of drag reducers with the environment, the corrosiveness of drag reducers to equipment of the flow system, the negative influences on heat-exchanging equipment together with drag reduction (DR), the scaling up of the turbulent drag-reducing effect from the laboratory scale to a real engineering application scale, the economy of the application of drag reducers, and so on. However, there are always some other unpredictable issues that happen in the real application system. This also needs to be considered when putting surfactant drag reducers into applications. For a district heating and cooling system, although it should be constructed according to its design at the beginning stage, there should be some uncontrolled factors in the construction of its details that cannot follow the design strictly for a large-scale system. Furthermore, for those systems that have been in use for a long time, there are the possibilities that the design papers no longer exist, and it is usually impossible to inspect the factors of how the system has corroded, how rusty the inside of the piping system becomes, fouling in heat exchangers, and so forth.

Turbulent Drag Reduction by Surfactant Additives, First Edition. Feng-Chen Li, Bo Yu, Jin-Jia Wei and Yasuo Kawaguchi.

Because of the above-mentioned many reasons, it is of particular importance to perform field tests in a real system at first, in order to estimate how much benefit or how much negative effect can be generated from the application of surfactant drag reducers to a thermal transportation system. In the following sections, several typical issues related to the application of surfactant drag reducers to a real system are briefly analyzed [1].

6.1.1 Influencing Factors of Drag-reducing Surfactant Additives on the Heat Transfer Performance of Heat Exchangers and Its Counter-measures

6.1.1.1 Issues about Heat Transfer Reduction

A heating system is taken here as an example. For a power plant running in a style of cogeneration of heat and power, after the generation of power, the low-pressure steam is used to heat water supplying a district heating system in winter. If the heat transfer ability of the heat exchangers at the hot water station is decreased (heat transfer reduction, or HTR), to obtain the same amount of heat supply, it needs to increase the steam temperature. This will inevitably decrease the efficiency of the power plant and influence its economics. The situation is also similar for a cooling system. The decrease of the heat transfer rate of heat exchangers will cause the chiller to work in a lower temperature mode in order to maintain the same amount of cooling supply, and finally will cause a great increase of running costs. If one is considering the application of surfactant drag reducers in currently running systems, the problems associated with HTR in heat exchangers will be surely pointed out by the system owner, which calls for full discussions and demonstrations from a variety of aspects.

A large amount of experimental studies have demonstrated that the application of turbulent drag reducers will greatly decrease the heat transfer ability of tube–tube heat exchangers due to the strong suppression effect of surfactant drag reducers on turbulence. For other types of heat exchangers with relatively simple flow passages such as tube–shell, helical pipe, and so on, since the flow in it is similar to that in a straight pipe, obvious DR and HTR can also coexist. But for plate-type heat exchangers, the heat transfer ability does not decrease remarkably due to the complicated changes of flow direction within the plates.

However, for a central cooling system with the application of a surfactant drag reducer in Japan, there were no reports of any problems associated with significant HTR in heat exchangers in their chillers. This conflicts with the above-mentioned experimental results. On the other hand, in a district heating system that has been in operation for many years in the Czech Republic, when applying a surfactant drag reducer, deterioration of the heat transfer ability in the heat exchangers did not happen on the whole. Post-test analysis was then performed to clarify this abnormal phenomenon. The possible reason is that the original heat exchangers have relatively serious fouling (precipitation of calcium carbonate), resulting in a relatively large heat transfer resistance; therefore, after adding a drag reducer to the system, the HTR phenomenon does not have much influence on the heat transfer functions of the whole system. In general, some factors such as the complexities of the structure of heat exchangers, the properties of heat transfer media, and so on will make estimation of the heat transfer performance of heat exchangers difficult. Hence, it is of particular importance to test how much it will influence the real heat exchanger performance after using surfactant drag reducers.

6.1.1.2 Counter-measures for Heat Transfer Reduction

Due to the occurrence of HTR in turbulent drag-reducing surfactant solution flows, there is a need for the design of new exchangers with complexly shaped flow passages, or those with enough redundancy in the beginning design of heat exchangers, in order to escape from the influence of HTR on the overall performance of the whole thermal flow system. But if one is applying turbulent drag reducers to a used system, it is relatively difficult to choose this way. In Denmark, a heat transfer enhancement method was tested using helix insertion into the heat transfer pipes of a heat exchanger. Since the insertion shrinks the effective area of flow passage, local pressure loss is greatly increased. Therefore, when choosing the heat transfer enhancement method, it is necessary to consider what influences of the possible great increase of local pressure loss will be exerted on the performance economics of the whole thermal flow system. Several typical heat transfer enhancement approaches for turbulent drag-reducing flow by additives have been described in Chapter 2.

6.1.2 Influences of Drag-reducing Surfactant Additives on the Environment

For a newly constructed district heating and cooling system, it is required to be closed. Thus the circulating water does not need to be exchanged in such a system and can be used everlastingly. To exchange water frequently, on one hand, the system's owner needs to pay costs for the large amount of water and some supplementary equipment; on the other hand, the relatively high oxygen content in fresh water may accelerate the corrosion of the piping system. However, at the seal location of the circulation pump of a real liquid circulating system, a small amount of water leakage always occurs, resulting in a slow decrease of the water inventory in the system. It needs to supply additional water to the system periodically to compensate for the leakage. There are often such cases that the circulating water in the system is exchanged yearly. For the district heating and cooling system with an application of surfactant drag reducer, the discharge of circulating water might cause pollution to rivers or lakes. Moreover, to discharge the drag reducer solution to the sewer may also lead to negative influences on microorganisms in sewage treatment plants. Such factors must be thought over within the real applications of surfactant drag reducers. In reality, for those above-mentioned possible negative influences stemming from the discharge of surfactant drag reducer solution, an adsorption method by means of activated carbon, bentonite powder, or the like, or a membrane isolation method, can be used to deal with it.

Normally, it is considered that circulating water in a district heating system will begin to leak to the soil after operating 20~30 years. Therefore, leakage of the ingredients of a drag reducer will happen immediately after the injection of surfactant drag reducer into a very old system. Although the normally used surfactant drag reducers belong to the fourth class of ammonium salt, are also the main ingredient of daily-use chemical products, and have very weak toxicity, if one is considering the possibility of a large amount of leakage to the soil, it is necessary to test the compatibility between surfactant drag reducers and the natural environment.

The fourth-class ammonium salt is very stable in surfactant molecules. In case surfactant drag reducers flow into the soil or river, one has to face the problem that surfactant molecules are very difficult to degrade. The so-called SPE drag reducer produced by Akzo-Nobel of Sweden is a synthesized additive with consideration of the above-mentioned issues. In Denmark, all drinking water comes from underground, and so issues associated with the pollution of underground water are very sensitive. However, after the field test of the SPE drag reducer in a district heating system in Herning, Denmark, the application of surfactant drag reducer in the district heating system was approved by the Danish government at the end of 1999. This can be looked at as having overcome a big obstacle related to compatibility with the natural environment in the real application of surfactant drag reducers.

From the viewpoint of environmental suitability, betaine-type surfactant drag reducers, such as SPE, have been highly expected. However, together with the advantage of its organic degradation nature, this kind of surfactant is apt to become rotten. This is the big shortcoming of such surfactants. Some efforts for increasing the stability of SPE drag reducers in laboratory studies have been reported, such as the addition of a chelating agent, bactericide, and so forth. But if one adds too much bactericide into surfactant drag reducers, it will again elevate the sustainability of the natural environment. Therefore, one needs proper limits when supplementing other additives for increasing the stability of drag reducers, in order to reach a balance with consideration of a variety of factors.

6.1.3 Scale-up Problem

Simple Newtonian fluids, such as water and air, satisfy the similarity law. Flow frictional resistance can be easily estimated based on the similarity law when the flow passage diameter and flow velocity are changed. But for complex non-Newtonian fluids, such as surfactant drag reducer solutions, it does not satisfy the similarity law anymore. Therefore, big difficulties will be encountered in determining how to estimate flow characteristics in different flow passages and under different flow conditions when introducing a drag reducer solution into an already running liquid circulation system or designing a thermal flow system with an application of drag-reducing fluid.

There have been many publications talking about direct numerical simulations of turbulent drag-reducing flows based on momentum equations for drag reducer solutions by means of computer simulation techniques, for example references [2,3]. Nevertheless, limited by the insufficient capacity of today's computers and by the imperfection of the constitutive equations for describing the very complex rheological properties of surfactant solutions, it is still impossible to estimate the turbulent DR characteristics at a Reynolds number as high as that in a real application system only through numerical simulations. Only through experimental tests for individual pipes with different diameters and by obtaining the features of flow frictional resistance at different DR rates in different pipes can the characteristics of overall flow resistance of a real application system be anticipated based on those experimentally tested data. However, surfactant drag reducers have a variety of categories, and the turbulent DR rate also changes at different solution concentrations and temperatures. On the other hand, a different real application system may be composed of piping at different diameters and a variety of other fittings. For all different real systems, it is unrealistic to perform laboratory tests on the flow characteristics of surfactant drag reducer every time, let alone run up costs for the construction of a large-scale facility for experimentally testing turbulent drag-reducing flow features in pipes with large diameters.

In summary, for real applications, it is preferable to simplify all the problems. Based on the measured data for the characteristics of turbulent DR in some small-sized pipe flows, such data should be able to estimate turbulent DR in a large pipe flow and in different pipe fittings. Therefore, it is of great importance to realize such a kind of scaling-up method and continuously accumulate some relatively simple empirical correlations as the design database for spreading out the application of surfactant drag reducers in real application systems.

6.2 Separation Techniques for Surfactant Solution [4]

Even though the surfactant and counterion concentrations used in drag reduction are less than 5000 ppm, they may be high enough to kill fish if directly discharged into a river or lake due to their ability to reduce surface tension and affect fish gills' ability to obtain oxygen. On the other hand, large amounts of surfactants may seriously affect wastewater-processing systems. For practical applications, such as district heating and cooling systems, the absolute amount of surfactants is usually very large due to the large volume of the system. In the case of system cleaning, system breakdown, and other emergency situations, surfactants and counterions have to be separated from the solution before it can be discharged into the sewer system.

Table 6.1 Adsorbents used in the separation

Adsorbents	Maker
Mineral colloid BP	E.C.C America, Inc., Southern Clay Products Subsidiary, Gonzales, TX
Bentonite	Sigma Chemical Company, St. Louis. MO
Darco G-60 Activated Carbon (100 mesh)	Aldrich Chemical Company, Milwaukee, WI
Activated Carbon Type 207A (8–30 mesh)	Barnebey and Sutcliffe Co., Columbus, OH
Activated Carbon, Untreated Powder (100–400 mesh)	Sigma Chemical Company, St. Louis, MO
Activated Carbon, G-60 (8–16 mesh)	Nucon International, Inc., Columbus, OH

There are three separation methods to separate surfactants and counterions from the solution: adsorption, ultrafiltration, and reverse osmosis. In the following sections, the three separation methods will be introduced briefly.

6.2.1 Adsorption

The adsorption method uses solid adsorbent to adsorb surfactants and counterions to make them separate from solution, and the remaining surfactant and counterion concentration of the filtered liquid can be measured and analyzed by the spectrophotometric method. The adsorbents usually used are listed in Table 6.1.

The test showed that adsorption is a feasible method for separating the cationic surfactant from the solution. The contact time is short, the required devices are just a mixer and a simple filtration system to separate the adsorbents at the end of adsorption, and the operation is quite simple. It is also able to treat large quantities of solution. If activated carbon is used, the quaternary surfactant and counterion NaSal can be adsorbed simultaneously but the NaSal adsorption is incomplete. Other separation methods for the counterion following adsorption of the surfactant need to be studied, such as ultrafiltration and reverse osmosis. A drawback to adsorption is that the surfactant cannot be recoverable from the adsorbent.

6.2.2 Ultrafiltration

Ultrafiltration (UF) is the use of an ultrafiltration membrane to separate large surfactant micelles from the solution. The solution is forced through the membrane by compressed air. The filtrate is collected, and the concentration of cationic surfactant in the filtrate is analyzed.

A "dead-end" method can be used to determine which ultrafiltration molecular weight cut-off (MWCO) membrane will be most suitable for the solution. The test showed that the UF membrane can effectively separate the surfactant with the smaller sized membrane. But even for the large pore size membrane, the flow rate of the permeation is too low to treat large amounts of the solution, and this membrane does not separate excess NaSal effectively. The general advantages of using ultrafiltration are that the operation can be continuous, and the surfactant can be recovered.

6.2.3 Reverse Osmosis

Ultrafiltration can separate particles from solution in the size range of 0.005–1 μm, whereas reverse osmosis can separate particles in the size range of 0.0005–0.01 μm. So ultrafiltration can separate surfactant micelles, while reverse osmosis may be effective to separate excess NaSal. The process for reverse osmosis is similar to that for ultrafiltration, except that the cartridge used here is a hollow fiber reverse osmosis cartridge. Also the operating pressure can be as high as 200 psi, so the pump used here is a high-pressure pump.

Only NaSal was tested in the reverse osmosis system to check its ability to separate this counterion. The concentration of NaSal is 900 ppm. It was found that the NaSal concentration in the permeation ranged from 20 ppm to 200 ppm. A spectrophotometer was used to analyze the NaSal concentration. However, the resulting solution was red in color in the reverse osmosis experiments. Because the analysis method was calibrated using clear solution, the analysis results may not be valid for solutions which are different in color from the calibration solution. It was generally concluded that reverse osmosis may be able to separate the sodium salicylate, but further study is needed.

6.3 Drag Reduction Stability of Surfactant Solutions [4]

A continuous long-term drag reduction test was performed in an iron pipe, and it was found that drag reduction decreased after a period and then finally dropped to near zero. In other different flow systems without iron pipes, it had also been observed that drag reduction of the same solution was lost after 3–40 days depending on the set-up of the system.

To clarify if this loss of drag reduction is due to mechanical effects or chemical effects, experimental analysis is needed.

6.3.1 Effect of Adsorption

It is considered that there are some chemical interactions between the fluid and the flow system hardware which affect the formation of rod-like micelles and the micellar networks needed for drag reduction. These interactions may either remove surfactant to decrease the surfactant concentration below the second critical micellar concentration, or increase the critical micellar concentration. First, the effects of possible adsorption of surfactant molecules on the walls were investigated. It was found that surfactant molecules are able to attach to the rust particles and keep them suspended in the solution. After rust was removed from the carbon steel walls of the flow system, the wall was still able to absorb surfactant molecules.

If all materials lost are assumed to be absorbed by the wall, the adsorption densities of surfactant and NaSal on the wall can be calculated. It was found that the adsorption densities here are three orders of magnitude higher than the results obtained by other researchers. The discrepancy must come from the surface area, that is, in addition to the internal wall surface, there are other surfaces adsorbing surfactant molecules as well. One must conclude that there is a certain amount of surfactant adsorbed by rust particles and they are not moving with the fluid and have settled down somewhere in the system. Drag reduction was observed soon after the first dosing, though that was short-lived due to low effective concentration. At higher concentrations, drag reduction was stable despite the loss. Thus, it can be concluded that the adsorption of surfactants on the wall should have no effect on drag reduction if the total concentration of surfactants is high enough to saturate the wall and rust in the system.

6.3.2 Effects of $Fe(OH)_3$

Because the wall of the flow system can remove only a limited amount of the surfactant from the solution, but drag reduction is lost after some time period, there must be some materials in the flow which interact chemically or physically with the surfactant. In the drag-reducing flow system, most of the material is iron pipes and fittings. Also, copper tubing, PVC pipes, and some rubber O-rings are present in the system. PVC is generally inert to this dilute surfactant solution. Rubber O-rings have very little surface area, and they should not have a large negative impact on drag reduction. The remaining two materials are iron and copper. Obviously, the large quantity of rust from iron pipes plays an important role. Rust is not likely to affect the solubility of the solution, so it must be removing or destroying surfactant and/or NaSal molecules. There are probably two different mechanisms for rust's effects. First, it adsorbs surfactant molecules; and, second, free ferric ions in solution may react with them.

Experiments showed that when the molar ratio of rust to surfactant reaches 3:1, 79% of surfactant in a 5 mM (about 2300 ppm) solution is adsorbed on the rust particles. However, the amount of surfactant can hardly be quantitatively determined as it depends on the total surface area of the particles rather than on the amount of the materials.

There may be a chemical reaction between the rust and the surfactant or salicylate. The main component of rust is ferric hydroxide. However, ferric hydroxide can be dissolved to a very limited extent in water based on its very small solubility product constant. From a chemical equilibrium calculation, at the conditions in the system, reaction between the free ferric ion and salicylate is insufficient to move the equilibrium between the ferric hydroxide and the ferric ions to any significant extent, that is, the rust is not being dissolved by salicylate, and it has no effect on NaSal. Also, a significant amount of surfactant is not likely to be removed due to the very small solubility product constant of $Fe(OH)_3$. Therefore, the negative effect of rust on drag reduction is from its adsorption of surfactant, not from chemical reactions by free Fe^{3+} ions.

6.3.3 *Effects of $Cu(OH)_2$*

In addition to the observation of the loss of drag reduction in the iron pipe flow system, it has also been repeatedly observed that drag reduction was lost in flow systems without any iron part. The solution after losing drag reduction is usually greenish, which is the characteristic color of copper compounds. Thus copper compounds may play an important role in the loss of drag reduction.

The product of copper corrosion under wet conditions is mostly copper hydroxide. In the iron pipe system, $Cu(OH)_2$ was not observed as its light greenish-blue color could be masked by the brown color of the rust. Experiments show that the presence of $Cu(OH)_2$ is fatal to drag reduction. The reaction between copper hydroxide and salicylate forms insoluble crystals and reduces the effective concentration available to form micelles. In turn, this solid sludge-like material, along with copper hydroxide particles, may adsorb surfactant molecules to a great extent. In addition, a reaction between surfactant and copper hydroxide cannot be excluded. The recovery of drag reduction can be obtained by adding ethylenediaminetetraacetic acid (EDTA).

6.3.4 *Recovery of Drag Reduction*

As the reasons for the loss of drag reduction have been clarified, some methods to recover drag reduction can be proposed, which include replenishment, chelating metal ions, and removing metal hydroxides.

6.3.4.1 Effects of Replenishment on Drag Reduction

For a surfactant solution which had lost drag reduction, its drag reduction ability may recover by adding either a surfactant or NaSal. Experiments show that drag reduction could always be regained with the addition of surfactant. The addition of NaSal could recover drag reduction when rust is present in relatively small amounts. The drag reduction recovery lasted a shorter time than that by the addition of surfactant. The addition of NaSal has no effect when the amount of rust is high. Drag reduction regained by replenishment is always somewhat lower than its previous levels.

6.3.4.2 Chelating Process to Remove Metal Loss

Since free metal ions are fatal to drag reduction due to their reactivity with surfactants and counterions, a method should be sought to remove these metal ions and to recover drag reduction.

Ethylenediaminetetraacetic acid is well known for its strong chelating ability with free metal ions. Even though a large amount of metal hydroxide, not free metal ions, exists in the solution, the addition of EDTA could remove these hydroids by shifting the chemical equilibrium between the hydroxides and the free ions. EDTA is very effective to remove $Cu(OH)_2$ due to the large solubility product constant of 10^{-18} and the large equilibrium constant ($10^{17.5}$) of the reaction between EDTA and Cu^{2+}. Even though the equilibrium constant of the reaction between EDTA and the ferric ions is as high as 10^{25}, the solubility product constant of ferric hydroxide is only 10^{-40}, and $Fe(OH)_3$ cannot be removed by EDTA.

6.4 Applications of Surfactant Drag Reduction

Surfactant drag reduction has been used in petroleum transport, fire control, district heating and cooling systems, and so on. Several examples of applying surfactant to hydronic heating and air-conditioning systems and the selection of good drag reducers are introduced in the following sections.

6.4.1 Application of Surfactant to Hydronic Heating and Air-Conditioning Systems

6.4.1.1 A Hydronic Heating System in a Czech City [5]

Description of the System

The workshop is a one-floor detached house with a total space coverage of 537 m^3. The heating system consists of a gas furnace with a water heater and blow-off tank; the tubes in two main circuits have diameters from 0.375 to 1 in. The system is provided with seven working cast iron radiators (the others were closed) with a total of 150 elements; each radiator element has a volume of 1.1 l. Water is forced into the heater by a pump, and warmed-up water from the heater flows through a thermostatic valve into the system. The furnace is operated by a thermal fuse which guarantees the desired and fixed temperature. There is also installed a space automatic regulation by which the furnace burner is on–off controlled. However, the regulation was fixed before measurement and was not changed during measurement. The total water content in the tubes, tanks, and radiators of the system is 0.25 m^3. The surfactant was solubilized and pressed into the system by a hand pump through a swing-check valve. Reading of the data was done just behind the furnace (designated by "M" in Figure 6.1). Temperatures at the inlet and outlet from the gas furnace (GF), the flowrate, and the pressure drop were recorded.

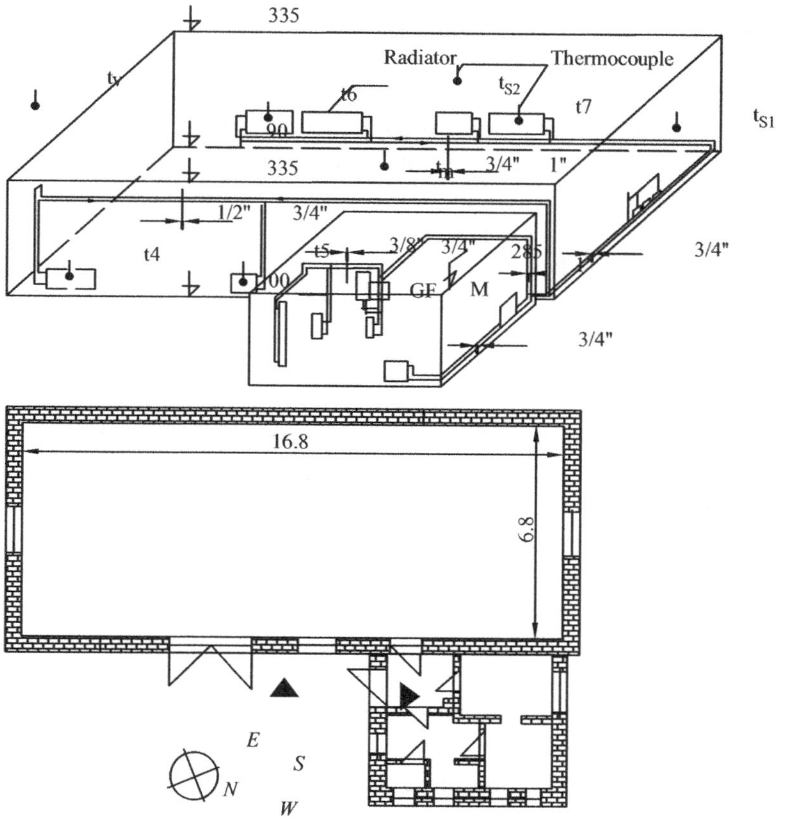

Figure 6.1 Ground floor of the workshop and the axonometric diagram of the heating system with tube dimensions and the position of radiators. GF: gas furnace; M: measuring instruments; t_s: thermocouples on the walls; t_4–t_7: thermocouples on the radiators; and t_v, t_m: thermometers

The discharge changed very much when the fires started, but the change was small when the burner was lighted by internal regulation of the furnace only. Thermocouples were installed on heating bodies, t_4–t_7, and on two inside walls, t_{s1} and t_{s2}; the outside t_v, and inside t_m, space temperatures were measured by thermometers. Surface temperature characterizes the influence of atmospheric conditions and thermal inertia of brick walls. The surface temperature on radiators was taken in the upper part of the middle of the body; it will characterize the heat transfer inside the radiator with water or with the solution. This quantity decreases with the distance from the furnace, and it is therefore necessary to compare the temperature on the same spots.

Handling of Chemicals

The temperature of water in the hydronic heating system varies at medium temperatures approximately between 35°C and 55°C. The zwitterionic surfactant named

SPE98330 is a product of Akzo-Nobel Surface Chemistry AB of Sweden; its basic (and, for drag reduction, decisive) components are betaines with the remarkable property of easy and quick biodegradability. SPE98330 worked well at low temperatures and up to 70°C, and the maximum drag reduction can reach 90%. It was stable as a drag reducer for several weeks of intermittent usage, but there was a decrease of drag reduction after some time of circulation. At the time of the present experiment, the concentrated surfactant was a little over 3 years old.

Untreated tap water has been always used in this heating system and therefore, before pouring the surfactant into the tubing, 0.5 g/l of Trilon-A was dosed in order to decrease the hardness of water, and 10 mg/l of Sagen was dosed as a disinfecting agent against undesired microbial growth. Experiments showed that both agents were effective. Some solution was lost from the system during the running experiment by manipulation such as withdrawal of samples or de-aeration of the tubing. This little loss of water was replaced by fresh tap water and by the addition of new doses of the two named chemicals.

It is necessary to check the concentration of the surfactant during a test of this type. It is well known that there is considerable adsorption of the surfactant on tube walls. It was already noticed earlier, during a successful field test which was an experiment in a large district heating system, that the weight concentration of the surfactant decreased almost by one third or even by one half of the added amount during the first 2 or 3 days after the addition was executed. This behavior was most expressive at the beginning of that test and lasted until the surfaces were saturated. Finally, the effective surfactant concentration in the heating liquid during the field test steadily increased. However, Figure 6.2 shows no concentration increase in the present experiment in spite of new additions of surfactant.

Viscosity measurements were used to determine the concentration of SPE98330. It is noted that the sensitivity of this method is lower at small concentrations due to very low viscosity. Nevertheless, later this was not found to be a great deficiency.

The gauge curve as a standard for the relation of kinematic viscosity to concentration was constructed by measuring the viscosity of calibrating solutions in two capillaries of different diameters (Figure 6.3). The same values of viscosity in both capillaries were obtained in calibrating solutions up to a concentration of 0.6 g/l; at higher concentrations, the gauge curves in both capillaries markedly differ. It means that below this limiting concentration, the solution behavior is purely viscous. At higher concentrations, however, the non-Newtonian properties are already distinctly manifested.

Hydrodynamic Results

Figure 6.4 shows the measurements of pressure loss of water and surfactant solution over the complete heating system in dependence on the overall rate of flow. The results in water can be characterized by a great scatter. In the maximum tube diameter of this system, the velocity is 0.34 m/s and the Reynolds number is probably somewhere

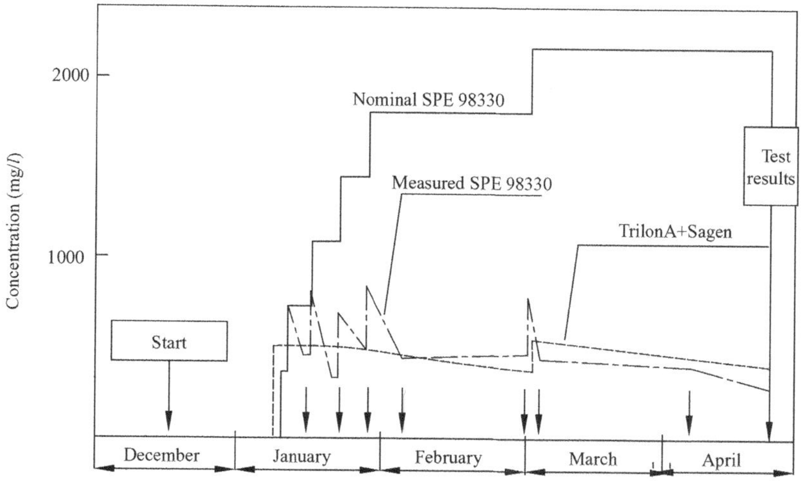

Figure 6.2 Time schedule of surfactant doses and samplings. Arrows show the date of sampling. Full line: addition of the surfactant; dashed line: concentration of treating agents; and dash-and-dot line: measured bulk (effective) concentration of the surfactant

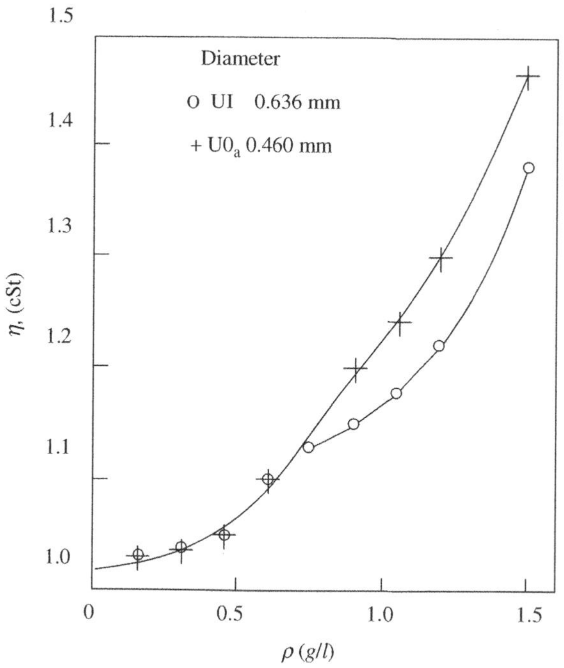

Figure 6.3 Calibration curves in two capillaries; the dependence of apparent viscosity on surfactant concentration

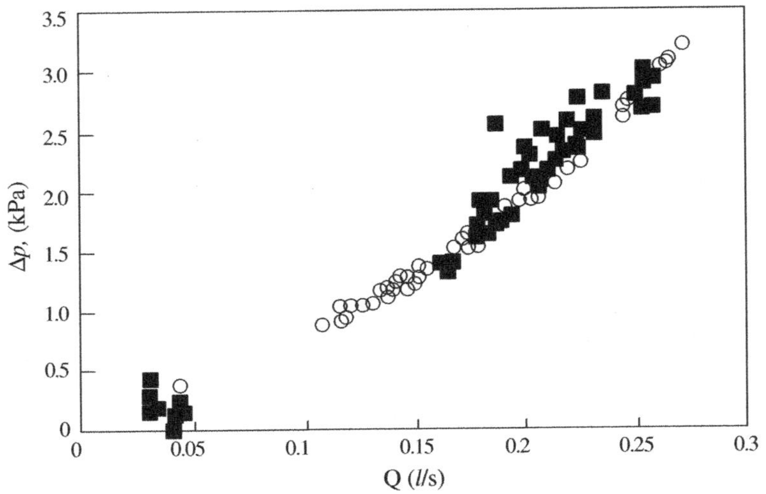

Figure 6.4 Pressure loss dependence on the discharge of surfactant solution from February to November. Solid square: measurement values of water; and open circles: measurement values of the surfactant solution

between 10 000 and 15 000 (depending on the temperature). At 0.2 l/s, the scatter is limited by pressures of 2.0 and 2.4 kPa in water, and by 1.75–2.05 kPa in surfactant. The data at a very low rate of flow, when the flow rate is approximately 0.04 l/s, are not taken into account. The flow of the solution with surfactant does not display such a scatter like that in water, but just the opposite: The curve is smooth and well centered, which is well shown in Figure 6.4.

The overall evaluation of results does not look bright. The low effectiveness can be explained by consideration of the following reasons:

1. *Low-level turbulence in the system*: The negligible effect of the surfactant may be caused by low turbulence intensity in the system. Maximum velocity achieved in larger pipes was under 0.5 m/s. From Figure 6.4 at 0.23 l/s, we see that the average loss of head in water is approximately 2.65 kPa and in the surfactant solution 2.4 kPa. Approximate drag reduction is only about 5%.
2. *Extremely large adhesion and low effective surfactant*: A large influence on impairing effectiveness is certainly the low concentration of the surfactant in the bulk of the heating liquid. Even though the concentrated agent several times added up to the total of 2.16 g/l, the measured concentration of samples always stayed low. After the heating season was over in April, the tubing was emptied and there was found a great amount of dirt in the circulating liquid. Adhesion on the tube walls was expected, but obviously the largest surface for adhesion offered tiny particles of dust and rust. Evidently, the enormous adhesion and therefore low bulk concentration comprised one of the main reasons for no drag reduction in this experiment.

3. *Biological degradation*: The possibility of biological degradation was also considered. Because the behavior of the surfactant was the same from the beginning to the end of the experiment, this alternative was therefore not suspected. At the beginning, we were persuaded that there was enough disinfecting material in the system because the used disinfection agent had proved earlier to be good enough and reliable over 2–3 months of intermittent experimenting with the same concentration in the laboratory facility. However, to be sure of preventing biodegradation, an overdose of the agent should probably be applied.
4. *Age of the surfactant*: Finally, the age of the surfactant was also considered and was suspected to be a little bit high. The surfactant was therefore tested in a small "diagnostic" facility with a 10.5 mm tube diameter.

In addition, heat transfer on the radiators is not decreased by the presence of the surfactant and the thermal comfort of the heated space is not affected either way.

6.4.1.2 A Hydronic Heating System in Qingdao, China [6]

In December 2007, we carried out the experiment of surfactant drag reduction in an actual hydronic heating system in Qingdao, China. The surfactant was CTAC with an equal mass concentration of NaSal as the counterion. The heat of the system was supplied from the primary network of a heat supply factory, and was transferred to water in the secondary network through a plate heat exchanger, and then the hot water was supplied to residential users. The total volume of the circulating hot water was about 70 m^3. The total heat supply area was about 115 000 m^2. The buildings had 6∼7 floors with a floor height of 3 m. The total heat load was about 5140 kW, and the average heat load was 44.7 W per unit area. The system was designed by Qingdao Enhui Energy-Saving Technology Corporation Ltd, it was an energy-saving type of hydronic heating system with electricity consumption of 290 kWh per day, and the electricity consumption per unit area was only one-fourth that of the same size heating system in China. The annual electricity is about 42 000 kWh. The surfactant addition and water circulation are shown in Figure 6.5. The surfactant concentration was 500 ppm, and obvious drag reduction was obtained: The flowrate of hot water was increased by 11.7% at a fixed rotational speed of pump, and the details can be seen in Figure 6.6. In Figure 6.6, the abscissa is the duration time of the whole test, and the time of just beginning to add surfactant was set at zero. We can see clearly that the drag reduction effect is obvious after the surfactant was added to the system, showing a rapid increase of flowrate. However, the duration of the drag reduction effect is short after stopping the addition of surfactant, and the drag reduction disappeared after 12∼24 h with the flowrate dropping to the value that exists without adding surfactant. If we assume that the flowrate stays unchanged by lowering the rotational speed of the pump, the maximum power of the pump (electricity consumption) can

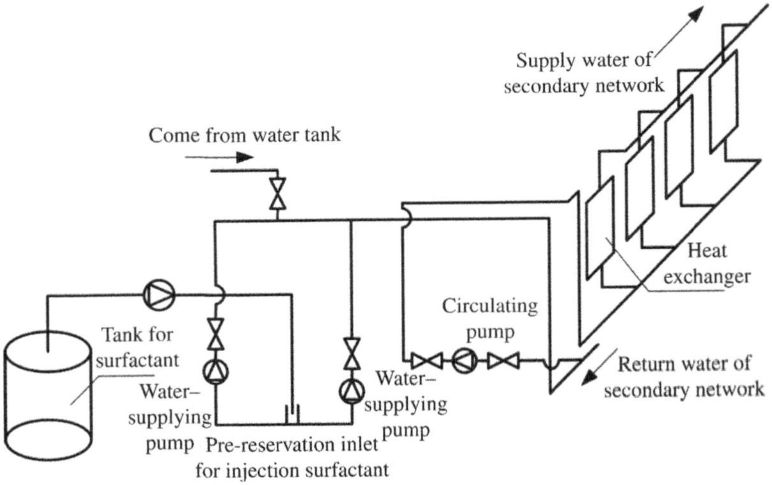

Figure 6.5 Schematics of surfactant injection and water circulation

be reduced by 28.4%. Detailed analyses are in reference [6]. Figure 6.7 shows the variation of the pump characteristic curves by changing the rotational speed to keep the flowrate unchanged.

The problem that effective drag reduction time was short is probably caused by the fact that the temperature range in the system (45–55°C) exceeds the effective drag reduction temperature range of the CTAC/NaSal surfactant solution.

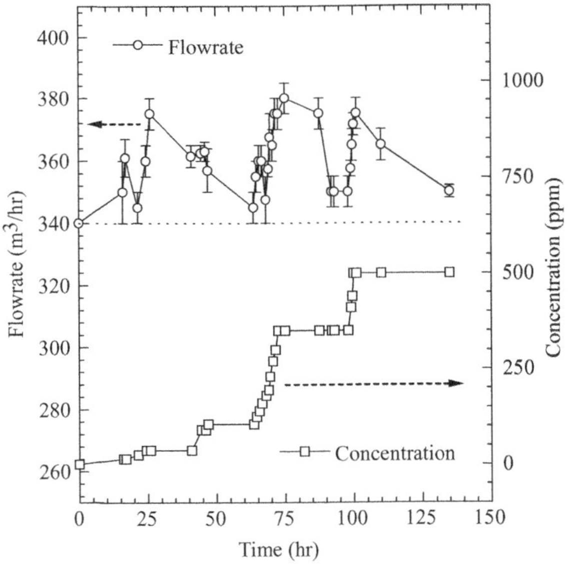

Figure 6.6 Changes of surfactant concentration and flowrate in the test procedure

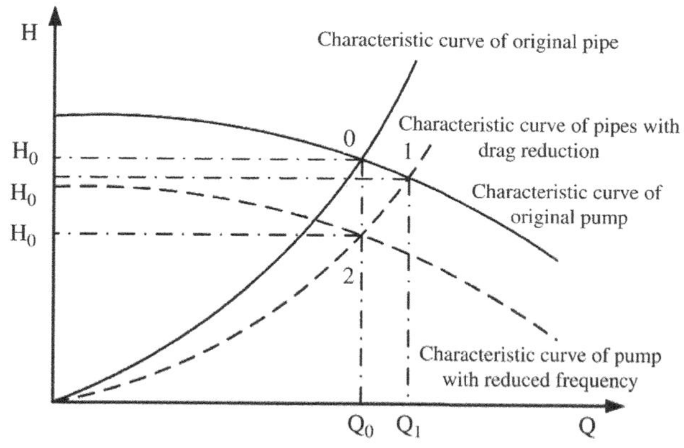

Figure 6.7 Variation of pump characteristic curves at different operating conditions

6.4.1.3 Hydronic Heating System in Herning, Denmark [7]

A surfactant drag reduction technique was successfully applied in a midsized hydronic heating system in Herning, Denmark. The system supplies heat to the Lind zone from a combined heat and power factory through a 5.6 km water-transporting pipe. The mixture of zwitterionic and nonionic surfactants made by the Akzo-Nobel Corporation was injected into the system pipe. The merits of the surfactant mixture are that its toxicity is very low and it has high biodegradability. The system has been working safely for more than 20 years. Power consumption of 4200 MWh per year can be saved after adding surfactant, and the emission of CO_2 can be decreased by 3000 t per year, having high economic and social benefits.

6.4.1.4 Central Air-Conditioning System in Kobe, Japan [8]

Considering that the cationic surfactant EO12 solution has very good drag reduction performance in cool water of 5–10°C, Professor Usui at Kobe University of Japan tested the drag reduction performance of the cationic surfactant EO12 solution with NaSal as a counterion in an air-conditioning system. Figure 6.8 shows the drag reduction effect. The pressure drop in the pipe decreases greatly at the same flowrate, and the maximum drag reduction can reach about 80%. The large decrease of pump power by adding surfactant means that the flowrate increases greatly at the same rotational speed of the pump. In the central air-conditioning system, the flowrate of refrigerant is always designed as a constant value, so the flowrate can be fixed by adjusting the pump frequency to lower the pump power. LSP Cooperation Company used EO12 as a matrix and added some rust inhibitors to make a drag-reducer product called LSP-01. A 55 kW pump was used in the system. The electricity consumption was 2.32 million Japanese yen per year before injecting the surfactant,

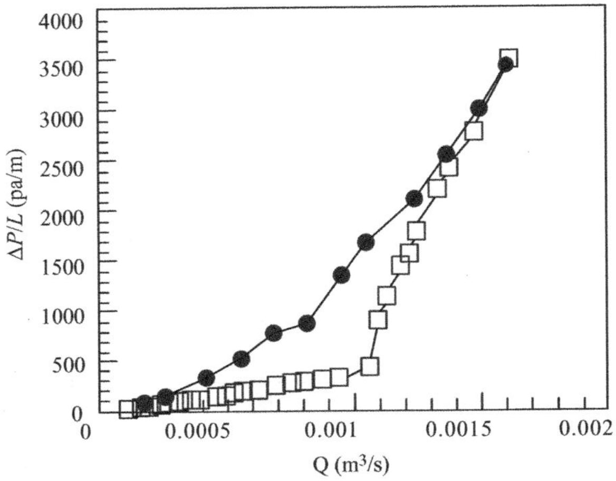

Figure 6.8 Measurement of the drag reduction effect

and it decreased by 54% after adding the surfactant, saving 1.26 million Japanese yen per year. There will be a very high energy-saving effect in the highly utilized air-conditioning systems of hotels. The problems of corrosion and its inhibition should be given attention for the system which has never used surfactant drag reducers. From the present test results, LSP-01 has a very good rust inhibition effect. In addition, for long-time running cases, it is very important for the surfactant to have a continuous drag reduction effect. Figure 6.9 shows one case of measurement data of the variation of power consumption over time after adding surfactant, and there is no obvious change during the test period. It was also found in other countries that the surfactant with a certain concentration or above can keep its drag reduction performance even after one year of operation.

6.4.1.5 Heating System of the Sapporo City Hall, Japan [9]

Professor Yasuo Kawaguchi of Tokyo University of Science and his colleagues have succeeded in reducing by 65% the energy consumption of a circulating pump in the heating system of the Sapporo City Hall in Japan. This building has 19 floors and two basement floors (Figure 6.10). This reduction in energy consumption was achieved by reducing the flow resistance of the circulating heating water by injecting a surfactant into it. The system was a hot-water circulating system that includes a pump with a 37 kW rated output, about 32 tons of water, and a water tank. Figure 6.11 shows the layout of the hot-water circulating pump and surfactant injection pump. The surfactant is LSP-01A with a concentration of 0.5%. It should be possible to save 58 000 kWh of electricity and to reduce expenditures by ¥630 000 annually, assuming that the system operates for 10 hours per day, 240 days per year, at a power cost of 11 JPY/kWh.

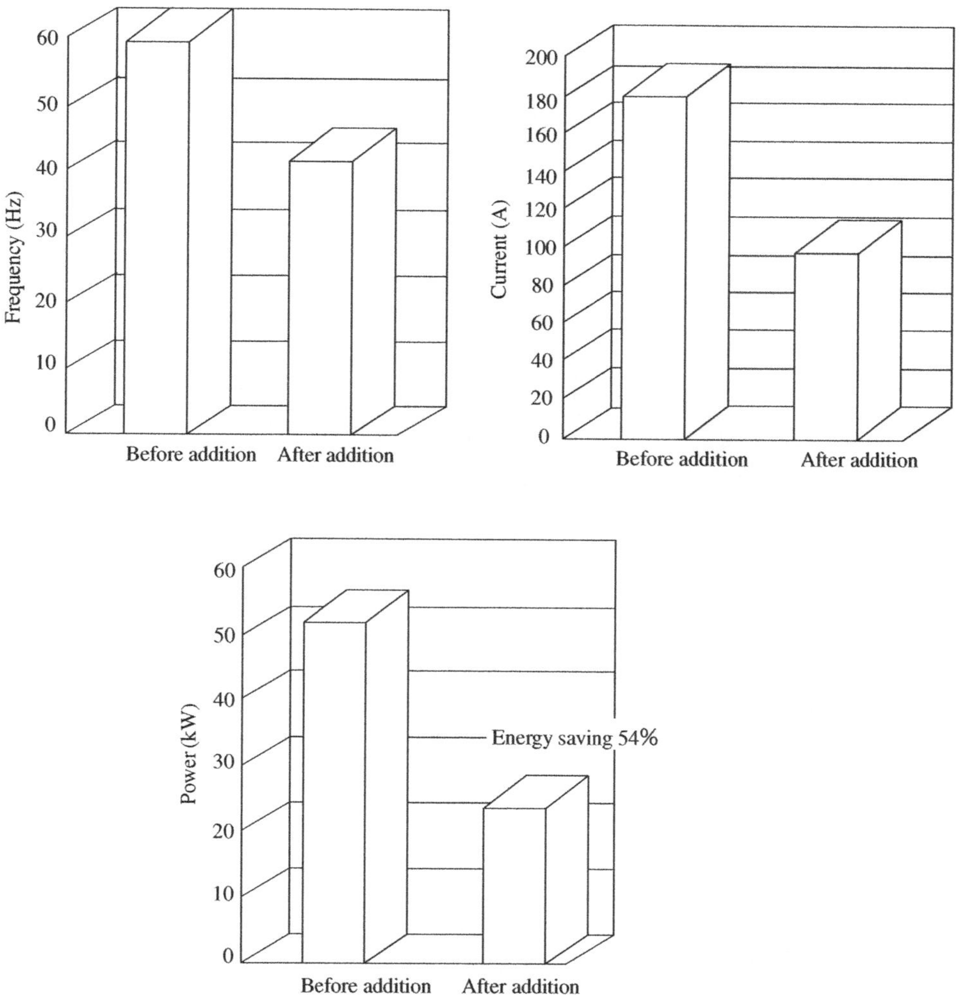

Figure 6.9 The energy-saving effect after adding surfactant

In addition, 32 tons/year of carbon dioxide emissions can be eliminated, taking the emission factor for carbon dioxide as 0.555 kg/kWh.

6.4.2 Surfactant Selection in Actual Applications [10]

In the above applications, the combination of cationic quaternary ammonium surfactants with the counterion of NaSal has a very good drag reduction effect. Therefore, they are widely researched in the countries of northern Europe where there is a large amount of hot water transportation systems. However, these kinds of surfactants have a very strong germicidal effect, and if they are diffused on the ground surface or

Figure 6.10 Sapporo City Hall, Japan, where the surfactant drag reduction test is conducted (Reproduced with permission from Advanced Industrial Science and Technology (AIST) (2007), "Successful 65% Reduction in the Energy Consumption of a Circulating Pump by Using a Surfactant," http://www.aist.go.jp/aist_e/latest_research/2007/20070618/20070618.html accessed 9 November 2011.)

Figure 6.11 Layout of a hot water circulating pump and a surfactant injection pump (Reproduced with permission from Advanced Industrial Science and Technology (AIST) (2007), "Successful 65% Reduction in the Energy Consumption of a Circulating Pump by Using a Surfactant," http://www.aist.go.jp/aist_e/latest_research/2007/20070618/20070618.html accessed 9 November 2011.)

permeate under the ground, they will have adverse effects on small animals and microorganisms; they will also have adverse effects on the human body in utilizing hot water in the secondary system. In addition, in the waste disposal stage of the water added with surfactant, there is a possibility that the surfactant may kill bacteria in the biological wastewater treatment system. Due to many unknown points of the cationic quaternary ammonium surfactants' effect on the environment, many countries don't permit them to be used for drag reduction.

6.4.2.1 Indexes for Selecting Surfactants

Compared with quaternary ammonium surfactants (the LD50 value of CTAC surfactant is 250 mg/kg, and thus possesses moderate toxicity; the smaller the LD50, the greater the toxicity; e.g., the LD50 of seawater is about 3750 mg/kg), we should select surfactants with an LD50 above 5000 (i.e., in the region of no toxicity), with corrosion inhibition, and with no Cl and Br elements.

Nonionic aminoxide surfactants (with a carbon atom number of 12–18) can meet the above conditions. Good drag reducers can be obtained by changing the carbon atom number in the hydrophobic head or additional molar number of aminoxide. Other conditions should also be considered, such as the surfactants suitable for 5–15°C cool water transportation and 60–80°C hot water transportation, and the surfactants suitable for external filtration membrane and that have the possibility of biogradation.

6.4.2.2 Effective Surfactants

According to the above selection conditions, the effective surfactant suitable for cooling water (5–15°C) is a mixture of oleyldimethylaminoxide (ODMAO) or oleyldihydroxyethylaminoxide (ODEAO) and zwitterionic surfactant for increasing solubility (alkylbetaine, etc.): the effective surfactant suitable for hot water (60–80°C) is a mixture of behenyldimethylaminoxide (BDMAO) and zwitterionic surfactant for increasing solubility. The LD50 of all these surfactants is above 5000, belonging to the classification of no toxicity. Quaternary ammonium surfactants have long rod-like or thread-like micelles, and can be separated from water by using a filtration membrane in a short time. According to experiments, 70% of the surfactant with the concentration above 1500 ppm can be recovered, meeting the organic emission reference requirement that biochemical oxygen demand (BOD) should be less than 300 ppm.

References

1. Kawaguchi, Y. (2003) *Reporter of "Smooth Water" Application Research Meeting*, Inner Report of Institute of Advanced Industrial Science and Technology, pp. 37–42.
2. Yu, B., Li, F-C. and Kawaguchi, Y. (2004) Numerical and experimental investigation of turbulent characteristics in a drag-reducing flow with surfactant additives. *International Journal of Heat and Fluid Flow*, **25**, 961–974.

3. Yu, B. and Kawaguchi, Y. (2004) Direct numerical simulation of the drag-reducing viscoelastic flow: finite-difference high-resolution method. *Journal of Non-Newtonian Fluid Mechanics*, **116**, 431–466.
4. Lu, B. (1997) Characterization of drag reducing surfactant systems by rheology and flow birefringence measurements, Ph.D. Thesis, The Ohio State University.
5. Myska, J. and Mik, V. (2003) Application of a drag reducing surfactant in the heating circuit. *Energy and Buildings*, **35**, 813–819.
6. Jiao, L.F., Li, F.C., Su, W.T. *et al.* (2008) Experimental study on surfactant drag-reducer applying to district heating system. *Energy Conservation Technology*, **26** (3), 195–201.
7. Kawaguchi, Y. (1999) Turbulence suppression phenomenon caused by adding surfactant and its application. *The Energy Conservation*, **51**, 18–24.
8. Usui, H. (1999) Application of surfactant drag reduction to central air-conditioning system in buildings. *The Energy Conservation*, **51**, 30–33.
9. Advanced Industrial Science and Technology (AIST) (2007) *Successful 65% Reduction in the Energy Consumption of a Circulating Pump by Using a Surfactant*, http://www.aist.go.jp/aist_e/latest_research/2007/20070618/20070618.html accessed 9 November 2011.
10. Inaba, H. (1999) Large scale system application of surfactant drag reduction in future cities. *The Energy Conservation*, **51**, 38–42.

Index

Turbulent Drag Reduction by Surfactant Additives, First Edition. Feng-Chen Li, Bo Yu, Jin-Jia Wei and Yasuo Kawaguchi.

Printed and bound by CPI Group (UK) Ltd, Croydon, CR0 4YY

30/09/2026

14973386-0001